Mundane Genomics

Violeta Argudo-Portal
Vincenzo Pavone • Mauro Turrini
Ayo Wahlberg
Editors

Mundane Genomics

DNA after the Hype

palgrave
macmillan

Editors
Violeta Argudo-Portal
Department of Social Anthropology
University of Barcelona
Barcelona, Spain

Mauro Turrini
Instituto De Políticas y Bienes
Públicos (IPP)
Consejo Superior de Investigaciones
Científicas (CSIC)
Madrid, Spain

Vincenzo Pavone
Instituto De Políticas y Bienes
Públicos (IPP)
Consejo Superior de Investigaciones
Científicas (CSIC)
Madrid, Spain

Ayo Wahlberg
Department of Anthropology
University of Copenhagen
Copenhagen, Denmark

ISBN 978-981-95-4549-0 ISBN 978-981-95-4550-6 (eBook)
https://doi.org/10.1007/978-981-95-4550-6

Cover Illustration: Joe Dunckley. In the cover image, a cyclist stands in front of a kindergarten area of a park filled with references to the genomic imaginary. One of the play structures is shaped like a double helix, the characteristic form of the DNA molecule, while some of the tiles on the ground display the four colors associated with the four nucleic bases of a gene.

This Palgrave Macmillan imprint is published by the registered company Springer Nature Singapore Pte Ltd.
The registered company address is: 152 Beach Road, #21-01/04 Gateway East, Singapore 189721, Singapore

To all the patients, participants, and individuals whose everyday lives are shaped subtly or profoundly by genomics. Your experiences ground this field in lived reality, rendering this scholarly inquiry both possible and necessary. To the scholars who, over the past two decades, have shaped the field of social studies of genomics. Your critical insights, enduring commitment, and intellectual companionship have paved the way for this volume.

Acknowledgements

Nourished by a wealth of inspiring paper presentations and talks held at the "Reproductive Futures" conference in Tampere in June 2022, we spent one of the coffee breaks reflecting on the various ways sequencing technologies and genetic testing were currently impacting the reproductive bioeconomy. These reflections morphed into an idea to think more generally about how widespread genetic testing had become throughout the world, a quarter of a century after the announcement of the first complete human genome sequence in June of 2000—in reproduction, clinical genetics, pharmaceutical medicine, forensics, genealogy, and more. We promptly prepared a call for an open panel on "Genetic testing: Materiality and temporality" to be convened at the November 2023 Society for the Social Studies of Science (4S) conference held in Honolulu. We could not have had a better response and following three highly generative hybrid sessions featuring no less than seventeen presentations, we proposed to keep the conversation alive. All talks had contributed to generating a broader picture of the multiple ways in which genetics is changing lives, in the ways we reproduce, connect, think of ourselves and others, as well as in the ways we make and imagine communities.

The majority of contributors to *Mundane Genomics* participated in our Honolulu panel sessions, while a few more joined as the book project took form. As always, collective volumes are substantial undertakings that involve the invaluable support of many people. We want to thank the organisers, presenters, and participants who joined the "Genetic testing: Materiality and temporality" panel sessions, which formed the foundation for this book. It was no coincidence that these sessions took place at the Society for Social Studies of Science conference, a global community of scholars that we continue to value and learn from. As will become clear in both the introduction to the volume and the ensuing chapters, our thinking has been inspired and informed by three decades of social studies of genomics and genetics conducted in Asia, Africa, the Americas, Europe, and more.

For their adept support during the production process, our warm thanks to Marie-Louise Karttunen for sharp copy-editing and to our assistant Nicholas Noah Wollesen for keeping track of and facilitating the steady flow of revised and edited chapters. We would also like to thank Marion Duval, our editor at Palgrave Macmillan, for guiding us through the publication preparations.

Finally, we acknowledge that part of this work was supported by the Ifgene Project, "Projecting the future of genetic prediction" (MCIN/AEI/10.13039/501100011033 - ID PID2020-115899GB-I00, PI Mauro Turrini and Ruben Blanco).

Violeta Argudo-Portal, Barcelona
Vincenzo Pavone, Madrid
Mauro Turrini, Madrid
Ayo Wahlberg, Copenhagen

Contents

Notes on Contributors

Flor Arias Aparicio is a lecturer in the Department of Public Law at the University of Extremadura. Her research work is carried out in the field of public law studies in different fields such as territorial cohesion and cross-border cooperation, environment politics, and judicial aspects of ARTs; she has participated in a number of different projects on the latter subject, supported by the National R+D+i Plan on assisted reproduction and regenerative medicine. She has also published various studies related to genetic tests, the most recent in Indret, *Journal for Analysis of Law*, titled "Legal aspects of obtaining, using and circulating human gametes" (with P. Nicolas & S. Romeu, 2018).

Violeta Argudo-Portal is a Serra Húnter Lecturer in the Department of Social Anthropology at the University of Barcelona. Her research intersects social anthropology, science and technology studies, feminist thought, and public policy. She examines how biomedical knowledge, products, services, and technologies are organized, produced, and distributed.

Rubén Blanco is Tenured Professor of Sociology at the Department of Sociology: Methods and Theory, Complutense University of Madrid. His research initially focused on the sociology of scientific knowledge and, more recently, on issues related to information and computer technologies.

Eliza Brown is Assistant Professor in Sociology at the University of California, Berkeley, specializing in reproduction. Her interests lie in temporality and anticipation, the role of money in medical interactions, and how medical providers and patients communicate about chance and risk. She uses qualitative methods, including ethnography, depth interviews, and discourse analysis. Her work has been published in *American Sociological Review*, *Journal of Health and Social Behavior*, *Social Science & Medicine*, and *Sociological Forum*.

Ricardo Urquizas Campello is a Professor at the Institute of Social Medicine of the State University of Rio de Janeiro (IMS/UERJ) and a fellow at the Center for the Study of Violence at the University of São Paulo (NEV-USP). His research focuses on new security and punishment technologies, prisons, surveillance, police violence, electronic monitoring of offenders, and forensic genetics. Author of the book *Short Circuit: electronic monitoring and the crisis of the Brazilian prison system* (2023).

Paula Bosch is a Junior Researcher at the Spanish Institute of Public Goods and Policies of the National Research Council (IPP-CSIC). She is an STS researcher working on reproductive bioeconomies, the provision of ARTs in public and private care and the role of add-ons on the commercialization and privatization of reproductive medicine—focusing specifically on genetic medicine and algorithms.

Vivette García-Deister is a Professor at the Faculty of Sciences, National Autonomous University of Mexico (UNAM). Her work explores the epistemologies of biomedical and forensic genetics, as informed by ethnographic methods, transnational history, and the philosophy of science in practice. She is committed to collaborative and interdisciplinary research, with a focus on the public role of STS and the impact of scientific infrastructures on racism, health, and justice. She currently serves as Editor-in-Chief of *Tapuya: Latin American Science, Technology and Society*.

Anne-Sophie Giraud is a CNRS Research Fellow at the LISST–Centre d'anthropologie sociale (UMR 5193). Her research examines reproduction, pregnancy, and the ontological and ethical status of the embryo and fetus in France, building on her book L'être anténatal (2024). Since 2018, she has focused on selective reproductive technologies, including preimplantation and prenatal diagnosis, and currently coordinates the ANR-funded NORPRO project (2023–2026) on contemporary reproductive norms and ethical negotiation in procreation.

Laura Louise Heinsen is an Assistant Professor in the Department of People and Technology, Roskilde University. Working at the intersection of anthropology and science and technology studies (STS), her research explores how biomedical and surveillance technologies shape clinical, legal, and care practices in the Danish welfare state. She has conducted ethnographic work on the routinization of reproductive and genetic technologies through studies of prenatal screening, selective abortion, predictive cancer genetics, and preventive medicine, and currently works on AI-based surveillance in dementia care.

Eva Hilberg is a Research Fellow at the iHuman Institute (School of Sociological Studies, Politics and International Relations) at the University of Sheffield. Her work analyses the use of new technologies and intellectual property rights in the definition of global health priorities, and has been published widely. Currently she is working on the roles of patients' narratives in the field of rare disease, as part of a Wellcome Trust-funded project on Orphan Drugs.

Lisa M. Hoffman is Professor in the School of Urban Studies at University of Washington Tacoma. Her PhD is in cultural anthropology (Berkeley) and she is also trained in China studies (MA) and philosophy (BA). Her scholarship has focused on questions of power, governing, and social change, with a particular interest in subjectivity and its intersections with spatiality. She has authored *Patriotic Professionalism in Urban China* (Temple, 2010), coauthored *Becoming Nisei* (UW, 2020), and coedited *Spaces of Danger: Culture and Power in the Everyday* (Georgia, 2015), and *Handbook on Urbanization in China* (Sage, 2026). Her current research project engages science and technology studies (STS) and considers how cancer risk, genetics, and precision health are shaping subjectivity and contemporary practices of living.

Amarpreet Kaur is an Assistant Professor in Health Technologies and Governance in the School of Social Policy and Society at the University of Birmingham. Her primary background is in the field of sociology. Amarpreet's research largely focusses on the intersection of reproductive technologies and genomic innovations for the prevention of genetic disease and reproductive choices in the United Kingdom. She is also a co-editor for the forthcoming *Handbook on the Sociology of Reproduction in England.*

Tineke Kleinhout-Vliek is a science, technology, and innovation studies scholar at University College Dublin who specialises in alternative modes of organising in the pharmaceutical innovation field. She researches and publishes on the nexus of pharmaceutical policy and access to medicines. Dr Kleinhout-Vliek is a co-founder of the Global Pharmaceuticals & Society Studies Network and an active member of the Public Pharma for Europe coalition.

Stephanie Lloyd research sits at the intersection of medical anthropology and science and technology studies. Her two primary research axes focus on (1) suicide as a lived experience and as modelled in neuropsychiatric research and (2) experiences with neurotechnologies, particularly deaf people's use of cochlear implants.

Anaïs Martin, PhD is a social anthropologist. Her work looks at kinship in contemporary Euro-American societies, with a particular focus on third-party reproduction. Her research lies at the intersection of new kinship studies, the anthropology of the body, and the anthropology of knowledge.

Paul Martin is Professor of the Sociology of Science and Technology at the University of Sheffield and Co-Director of the flagship multidisciplinary research centre – the Institute for the Study of the Human (iHuman). His work sits at the interface of STS and medical sociology, and he is currently working on a major five-year project on high priced "orphan" drugs for rare diseases funded by the Wellcome Trust. He has a long-standing interest in the expectations surrounding the development of gene therapy.

Vincenzo Pavone is a permanent research fellow at the Spanish Institute of Public Goods and Policies of the National Research Council (IPP-CSIC), and currently its Director. Since 2006, he has focused on critical STS, examining the social and political implications of emerging technologies, particularly assisted reproductive technologies (ARTs). His work explores the interplay between science, technology, and politics, using sociology, STS, and political economy to study bioeconomies and reproductive markets. He has led multiple national and EU-funded projects on ARTs, regenerative medicine, public engagement in science, bio-objects, and egg donation practices.

Helle Vendel Petersen is Nursing Research Manager in the Medical Department of Zealand University Hospital, Nykøbing F, and Associate Professor in the Department of People and Technology, Roskilde University. Her research has focused on cancer genetics—particularly hereditary colorectal cancer—and the psychosocial dimensions of living with predisposing genetic mutations. She has contributed extensively to national studies on genetic testing in Lynch syndrome families, patient perspectives on risk management, and ethical considerations surrounding individuals' rights to know or not know their genetic risks.

Roberta Raffaetá is Associate Professor of Socio-Cultural Anthropology at the Department of Philosophy and Cultural Heritage, Ca' Foscari University of Venice, and Deputy Director of NICHE (The New Institute: Centre for Environmental Humanities). She studies the multiple ways people assign value to life in a world where humans and non-humans are interdependent, with a focus on how technoscience is transforming in response to eco-social challenges and technological revolutions. Her work lies at the intersection of anthropology, political ecology, and science and technology studies. Since obtaining a PhD at the University of Lausanne, she has worked at various universities in Italy and abroad (UCLA, UCSD, Monash Melbourne, Lausanne). Her research has been funded by the European Commission (Marie Curie and ERC), Fulbright, The Italian Ministry of Research, Wenner Gren, and others.

François Romijn is a Marie Skłodowska-Curie Fellow (2024–2027) hosted at Deakin University (Melbourne) and the Max-Planck-Institute for Evolutionary Anthropology (Leipzig), where he carries out an ethnographic study of archaeogeneticists' practices ("The social life of ancient DNA"). He is also associate researcher at the Laboratory of Anthropology of Contemporary Worlds at the Université libre de Bruxelles (Belgium). Trained as a sociologist and cultural anthropologist and working within the field of social studies of science, his research focuses on social engagement with DNA knowledge. Specifically, he explores how newly accessible biological information is used—by scientists or lay-users—to say something about social relationships or individuals' sense of self, in terms of interpretations of ancestry, kinship, origins, but also in terms of constructions of personal and collective pasts.

Pablo Santoro is Associate Professor of Sociology at the Department of Sociology: Methods and Theory, Complutense University of Madrid. His research has focused on the social implications of contemporary biotechnologies.

Aleksandra Stelmach is a Research Associate at the Department of Global Health and Social Medicine at King's College London; and at the Department of Management, University of Exeter Business School. Her research combines sociological and STS perspectives on emerging fields in health and medicine, focusing on the language used to communicate and engage with novel technologies. In her current work, she explores social and ethical issues surrounding stem cell technologies, and her past research examined the fields of gene drive, synthetic biology, and epigenetics.

Mankei Tam graduated from the Department of Anthropology at the Chinese University of Hong Kong in 2018. He is a postdoctoral fellow for the FARE SHK_HealthXCross Project, studying the relationships between global health and microbiome technoscience evolving in Hong Kong and other key hubs that situate China globally. Before joining HealthXCross, his project concerned citizens' self-empowering practices in critically assessing radiation risks and their collaborations with scientists to explore new forms of agriculture after the Fukushima nuclear disaster.

Josh Tenenberg's interdisciplinary research is in the borderlands of the human and the technical, spanning education, human-computer interaction, design research, workplace studies, semiotics, and technical communication. He has a PhD in Computer Science, undertook advanced study at the Ostrom Workshop on Political Theory and Policy Analysis, Indiana University, and was visiting professor at the University of Victoria (Canada) and Indiana University. He served for six years as the founding Co-Editor-in-Chief of *ACM Transactions on Computing Education*, has published several dozen papers in the human sciences, and is the author of *Narratives of Qualitative Research: Making Praxis Visible*, published by Routledge (2024).

Mauro Turrini is a tenured researcher at the Spanish Institute of Public Goods and Policies of the National Research Council (IPP-CSIC). He works at the intersection of the socioanthropology of medicine and STS. His current lines of research concern genetics, assisted reproduction, and the critical appraisal of shared decision-making.

Ayo Wahlberg is a Professor in the Department of Anthropology, University of Copenhagen. Working in the field of social studies of (bio) medicine, his research has focused on the modernization of traditional herbal medicine (in Vietnam and the United Kingdom), the routinization of reproductive and genetic technologies (in China and Denmark), health metrics (in clinical trials and global health), and chronic living (through a multi-country study entitled "The Vitality of Disease—Quality of Life in the Making").

1

Introduction. Mundane Genomics: An Analytical Reorientation

Mauro Turrini, Violeta Argudo-Portal, Ayo Wahlberg, and Vincenzo Pavone

M. Turrini (✉) • V. Pavone
Instituto De Políticas y Bienes Públicos (IPP), Consejo Superior de Investigaciones Científicas (CSIC), Madrid, Spain
e-mail: mauro.turrini@csic.es; vincenzo.pavone@csic.es

V. Argudo-Portal
Department of Social Anthropology, University of Barcelona, Barcelona, Spain
e-mail: vargudoportal@ub.edu

A. Wahlberg
Department of Anthropology, University of Copenhagen, Copenhagen, Denmark
e-mail: ayo.wahlberg@anthro.ku.dk

V. Argudo-Portal et al. (eds.), *Mundane Genomics*,
https://doi.org/10.1007/978-981-95-4550-6_1

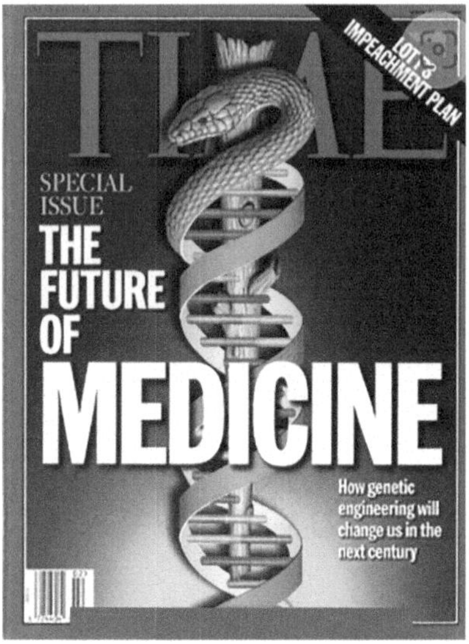

11 January 1999

20 May 2024

As the completion of the first human genome sequence was approaching, *Time* magazine published a special issue called "The Future of Medicine: How Genetic Engineering Will Change Us in the Next Century" in January 1999. In it Walter Isaacson speculated:

> But 20th century medicine did little to increase the natural life-span of healthy humans. The next medical revolution will change that, because genetic engineering has the potential to conquer cancer, grow new blood vessels in the heart, block the growth of blood vessels in tumors, create new organs from stem cells and perhaps even reset the primeval genetic coding that causes cells to age.
>
> Our children may be able (I hope, I fear) to choose their kids' traits: to select their gender and eye color; perhaps to tinker with their IQs, personalities and athletic abilities. They could clone themselves, or one of their kids, or a celebrity they admire, or maybe even us after we've died. (Isascon, 1999, p. 44)

In May 2024, *Bloomberg Businessweek* ran a cover story headlined, "The DNA Delusion: Mass-Market Genetic Testing Has Little to Offer Most People", highlighting that, while companies like 23andMe solicit the saliva of customers with the claim that they can "help you live a healthier life by predicting what diseases you are at risk of developing", "all of that spit isn't quite living up to the hype … When researchers unveiled a sequence of the human genome, we thought that decoding DNA would lead to the end of all sorts of diseases. It turned out to be way more complicated" (Bloomberg Businessweek, 2024). Twenty-five years have passed since Bill Clinton, accompanied by Francis Collins and Craig Venter, announced the completion of the Human Genome Project in June 2000 at a historic White House event.

A quarter into the "Biotech century" (Rifkin, 1998), the anticipated revolutionary shift in medicine driven by genomics may not have fully materialized. Still, genomics has undeniably become an integral part of everyday life for many, albeit in ever-stratified ways. The reductionist and essentialist view of genetics that dominated the 1990s is certainly no longer tenable (if it ever was), and the numerous failed promises associated with genomics have dampened expectations. Our understanding of DNA has also changed multiple times: it went from being the "code of life" to a more modest toolbox that cells mundanely use to perform their daily functions (Rheinberger & Mueller-Wille, 2018). At the same time, genomics continues to expand throughout society by interacting with national cultures, regulations, healthcare systems, patients, families, and more, making genomics much more complex, more "multiple".

The shift from the singular genome mapped by the Human Genome Project to the proliferation of genomic data in contemporary biomedicine, forensics, and genealogy marks a significant epistemic and institutional transformation. No longer confined to the rhetoric of revolution, genomics increasingly operates within the contours of the everyday. Today, genetic testing is being performed in numerous medical and non-medical settings, often as a routine procedure, generating a huge amount of data, some of which is classified and confidential, while some is relatively more accessible. All such data are, however, vulnerable to being hacked and/or commercialized to varying extents. As we write this text,

23andMe's bankruptcy, announced in March 2025, has spurred numerous concerns about what will happen with its genomic data if it is finally bought by Regeneron Pharmaceuticals for $256 million (Levenson, 2025; Jamali, 2025), reviving numerous concerns about customer privacy and genomic data as a biotech product. New medical treatments are being conceived or proposed, taking into account information proceeding from genetics tests (Pollock, 2012); prospective parents debate the pros and cons of having their embryos or fetuses genetically tested (Franklin & Roberts, 2006; Zhu & Dong, 2018); and family conversations are sparked by the most recent outcomes of ancestry genetic testing (Nelson, 2016), at times involving children way before the onset of a suspected hereditary illness. In a laboratory nearby, just as we write, researchers and medical staff may be discussing the next steps in a research project focusing on the role of epigenetics when it comes to human health (Saldaña-Tejeda, 2018).

The term "mundane" in this book's title points to the ambivalent and often contested process whereby genomic technologies and information have been routinized, normalized, and made personally relevant to an increasing number of people, in and out of the medical realm. The concept of the mundane redirects our gaze away from the exceptional, focusing instead on what happens after the hype fades—on the descent of DNA into the banal routines, infrastructures, and relations of ordinary life. It is a movement from ascendance to immanence, from the symbolic force of the "century of the gene" (Fox Keller, 2000) to the dispersed realities of biodigital entanglements, from one of the most sacred scientific icons to its incorporation into popular culture, the economy, and society.

Here, we play on the dual meaning of "mundane". On the one hand, DNA has transitioned from being a symbol of "the next medical revolution" at the core of a modern scientific mystique to one of many everyday elements spanning various domains, including healthcare, forensics, genealogy, and research. Second, thanks to remarkable technological advancements, genomic information has permeated societies extensively through the contingencies of local cultures, regulations, health care systems, and commercial and financial markets. As a result, more and more people encounter (are interpellated by) genomic data in their daily lives, shaping their self-perception, relationships, and lives in highly stratified ways. We employ the concept of "mundane genomics" to capture precisely this normalization and multiplicity of DNA as a central aspect of the wider dissemination of genomic cultures and practices into contemporary society. Without any intention of being exhaustive, our book aims to explore five distinct yet interconnected domains that constitute particularly generative landscapes of genomics: assisted and selective reproduction; "clinically actionable" genetic predisposition and the "genetically-at-risk" chronicities that ensue; direct-to-consumer genetic testing; forensics; and "frontier" omics research.

Attending to the Mundane

The analytical reorientation toward mundane genomics within and through these domains, we propose, requires recognizing multiple social science legacies that are companions in the crafting of this volume. Feminist scholarship, as Cynthia Enloe (2011, 2013) has argued, has been crucial in taking mundane matters seriously as an analytical entry point into social order. Rather than treating the mundane as peripheral or pre-political, feminist scholars have analyzed the everyday as a powerful lens that reveals power relations, explores invisibilized realms, and more richly contextualizes practices. Similarly, social anthropology has long emphasized the epistemic value of the mundane—of attending to what belongs to the world, to the materialization of practices and beliefs. As noted in the anthropological tradition, importance is placed on constituting a series of objects and privileging the most ordinary or banal. Marcel Mauss famously taught the "tin-can rule" according to which "a tin can characterizes our societies better than the most sumptuous jewel or the rarest stamp" (Musée d'ethnographie, 1931, p. 8).

Our reorientation also aligns with a broader STS insight into the power of seemingly minor techniques and infrastructures in scripting the social order. Bruno Latour's classic essay "Where Are the Missing Masses?" (1992) demonstrates how mundane artefacts are not merely underpinnings for social norms but also for their material enactments. These devices embed expectations and constraints directly into everyday practice. Mike Michael has been particularly influential in tracing the sedimentation of technoscience into daily life. In *Technoscience and Everyday Life: The Complex Simplicity of the Mundane* (2006), he argues that technoscience does not emerge solely through moments of rupture or discovery, but is constituted through bodies, routines, and infrastructures. He emphasizes the temporal interplay between novelty and routine: the spectacular and the banal are not opposites but mutually constitutive. The mundane stabilizes and gives texture to the new while shaping it (Michael, 2003).

This creative sedimentation process is particularly evident in the formation of infrastructures—networks, standards, supply chains—that mediate access, shape use, and stabilize expectations. Such infrastructures function as mundane technologies that strengthen or weaken specific socio-technical assemblages, embedding innovation into durable routines. They concretize the diffusion of novelty into technical arrangements that make science livable, durable, and governable. For instance, in his ethnographic study of post-transplantation tissue banking, Hoeyer (2010) shows that this infrastructural turn is especially visible in how biomedical innovation transitions from the dramatic realm of bioethical debate and media headlines into quieter "mundane practices". His analysis focuses on the layered, unglamorous effort to maintain a "safe and stable supply" of biological materials—a foundational, yet largely invisible labor that lacks the spotlight of "the new" and that requires our attention.

In the study of science and biomedicine, attention to mundane matters enables the analysis of hidden or invisible labor in contemporary biomedicine (Froger-Lefebvre et al., 2023; Pinel et al., 2025), maintenance work and fragility recognition of technoscientific artefacts and infrastructures (Denis & Pontille, 2015; Argudo-Portal & Domènech, 2022; Domínguez Rubio et al., 2025), care entanglements (Puig de la Bellacasa, 2017; Buch, 2015), everyday therapeutics processes (Skovgaard et al., 2024; Geelen et al., 2015; Heinsen et al., 2021), the perils of overplaying the role of emerging technologies without attending to mundane resistant technologies (Hollin, 2019), or the mundane layers of biopolitical governance (Lemke, 2013; Krupar & Ehlers, 2020; Deejay & Henne, 2024). This body of work shows how crucial it is to pay attention to everyday life to gain a better understanding of contemporary medical and scientific practices.

We turn to such mundane practices in our volume to reorient the analytical gaze on genomics and thereby expand our understanding of it. Drawing inspiration from Foucauldian genealogy, this volume approaches the mundane as a political analytic: a way to trace how power is embedded in the normalization, materialization, and routinization of innovation. The generative domestication of innovation is not politically neutral. As Rose (1999) argues, liberal governance increasingly functions through mundane techniques—administrative forms, calendars, checklists—that operate beneath the threshold of spectacle. In genomics, this might include consent templates, risk rubrics, billing codes, bureaucratic routines, or healthcare reorderings. These minor practices—or "microphysics"—do not merely support power, they embed it in the textures of meaning, identity, and care of everyday life (Rapp, 1999; Wahlberg, 2018).

Pickersgill et al. (2011) describe the brain as an object of "mundane significance": a background figure in ordinary conversation, medical routines, and media representations, no longer surrounded by mystique. Soulier et al. (2016) show that lay publics readily interpret genomic information within their everyday moral frameworks, connecting abstract ethical issues to practical concerns such as fairness, responsibility, and trust. Genomics, rather than being perceived as exotic or sacrosanct, is approached as just one among many tools to be evaluated and domesticated. What emerges here is not epistemic deference but mundane literacy: the capacity to interpret, negotiate, and even resist genomics through relational and practical reasoning. DNA, too, is being desacralized. Once held as a sacred icon of life's essence (Nelkin & Lindee, 1995), it now circulates in mundane formats: saliva test kits bought online, markers in ancestry reports, and probabilistic risk scores in clinical encounters and documents (O'Riordan, 2011).

These data, once generated through bodily sampling and algorithmic processing, begin to inhabit biodigital environments where corporeality and information converge. Genomic data are no longer abstract representations; they are incorporated into market practices, economic exchanges within the biotech sector, bodily routines, and affective landscapes (Fortun, 2023). What makes DNA specific is not only its biological distinctiveness but also its capacity to constitute a grammar of selfhood through heterogeneous, technologically mediated processes that are at once "very mundane and very high tech" (O'Riordan, 2010, p. 45). Harris et al. (2016) illuminate this convergence by analyzing how users of DTC genetic tests engage in forms of "autobiology", "biosociality", and/or personal data donation to biomedical progress: narrative and sharing practices in which people articulate personal and collective identities across digital media. These biodigital expressions show how DNA functions not just as information but as a material and affective component of everyday life, entangled with habits, infrastructures, and relations (Kelly et al., 2018). While the direct-to-consumer model has often been portrayed as a disruptive force in biomedicine, its normalization into mundane routines, everyday tracking apps, and personal wellness discourse reflects its ambivalent positioning—simultaneously banal and controversial (Turrini, 2018).

To further conceptualize this entanglement, Pink et al. (2017) refer to "mundane data" as the affective, material, and leaky dimensions of personal information. Applied to genomics, this approach reveals how DNA data are not simply stored or interpreted, but also handled, circulated, touched, and forgotten—folded into clinical, domestic, and digital routines. The data flow through infrastructures, cling to protocols, and shape daily decisions, with power that lies in their ambiguity and ubiquity: banal yet consequential, incorporated into bodies and the market, and abstracted into digital information. Genomic data are sometimes embedded through hype, sometimes absorbed quietly into the ordinary rhythms of contemporary life.

Our analytical shift to the study of genomics is inspired by other works, such as Pollock et al.'s (2021) special issue on *Race and Biomedicine Beyond the Lab: 21st Century Mobilisations of Genetics*, which decenters the focus from lab stories to explore how biomedical narratives are mobilized and operate outside the lab. We were also inspired by Hyde and Denyer Willis' (2020) editorial on storytelling and anthropology and the need to be aware of stories that do not only attend to jarring events; there are many more textures, such as the mundane, to be taken into account in ethnographic stories. Our analytical reorientation does not emphasize site displacement but delves into what different researchers consider has to do with mundane genomics in various countries, situations, and sites. This volume unpacks mundane genomics with research sites that include reproduction clinics, public hospitals, households, prisons, state-led infrastructures, direct-to-consumer tests, and patients' stories.

Thinking of genomics as mundane activates an analytical sensibility that reorients our attention and analysis, creating new categories, reworking some while leaving others. In the medical domain, for example, "genetic exceptionalism" designates a view that genetic information is qualitatively different from other forms of medical data, deserving special ethical and regulatory treatment. The notion of genetic exceptionalism has long shaped ethical, legal, and social debates (Rothstein, 2007), reinforcing the idea that DNA holds a privileged epistemic and moral status. Yet, in practice, genomics often unfolds within broader biomedical collectives that remain epistemically diverse and pragmatically plural. Echoing Cheung et al.'s (2025) ethical reasoning on differentiating between the distinctive and the exceptional, we might then refer to genomics as a distinctive realm, rather than belonging to the exceptional. Such a turn enables us to position genomics in relation to other biological and biomedical arenas—much like other types of medical information—and avoid decontextualized hype and clear takes on genomics as categorically different or isolated. Genomic tests might reveal sensitive information about a family member, but so might other types of medical tests. Of course, genomics is distinctive in many ways, as shown throughout this volume; however, despite early fears of "geneticization", DNA operates less as a sovereign force destined to colonize and overwhelm other techniques and knowledges, and more as a powerful, although contested, grammar of selfhood—identity, body, health—open to negotiation, resistance, and situated reinterpretation (Weiner et al., 2017).

By reframing genomics through the lens of the mundane, we see DNA not as a master code or a radical break, but as a material and semiotic actor embedded in everyday arrangements. Genomics becomes intelligible not only in moments of innovation but also through the quiet labor, infrastructural friction, bureaucratic conventions, and ethical shortcuts that enable it to function and govern. Each of the book's sections engages with empirical cases and theoretical insights to show how genomics operates beyond moments of controversy or celebration, shaping and being shaped by the textures of everyday life. These facets offer a nuanced framework for understanding how genomics persists, transforms, and governs the everyday. Drawing on the research of this volume, and inspired by scholarly conceptualizations of mundanity rather than treating the mundane as a singular quality, we explore its multiple manifestations through five categories: as *ordinary*, embedded in families and groups; as *routinized*, normalized, banalized, and ethically invisible; as *modest*, situated in alliance with others and linked to selective ignorance; as *institutional*, reshaping individual and collective identities, industries, and biopolitics; and as *leaky*, *multiple*, and *de-centered*. Far from being mutually exclusive, the goal of these categories is to delve into some of the most important aspects of the mundanity/mundaneness of genomic techniques and information.

Mundane as Ordinary and Enculturated

The first dimension of mundane genomics refers to the integration of genetic information—such as disease-linked variants and ancestry markers—into ordinary social and cultural life. While genetic disease often carries connotations of rarity and biomedical exceptionality, it frequently becomes routinized and familiar within families and communities. The concept of *biosociality*, introduced by Paul Rabinow (1996), has been foundational in setting the terms for understanding how genetic variants are not only medical facts but also social materials—interpreted, negotiated, and lived. This idea paved the way for a rich body of work exploring how people domesticate and socialize genomic knowledge, particularly when such knowledge is woven into kinship, care, and identity practices (Gibbon & Novas, 2008). Within this field, a significant subset of studies has emphasized how even severe or rare conditions are normalized—not through denial, but through mundane, affective, and pragmatic accommodations. Karen-Sue Taussig (2009), in *Ordinary Genomes*, shows how Dutch families come to accept and incorporate diagnoses of serious, rare genetic syndromes. "Normalization" is not a form of negation, but a cultural process of recognition and belonging, echoing the long history of tolerance and religious and cultural plurality in the Netherlands. Duana

Fullwiley (2011), in *The Enculturated Gene*, traces how sickle cell anemia in Senegal is shaped through biosocial entanglements: biomedical knowledge, kinship networks, economic precarity, and local moral economies co-produce the condition as "mild" and livable, rather than deviant or catastrophic. Importantly, she shows that genetic markers—like the "Senegalese haplotype"—are interpreted not only as biological data but as cultural signs. In this context, genomic information becomes a vehicle for affirming collective resilience, not just individual burden. Gendered dimensions of this process are brought to the fore in Sarah Gibbon's (2007) ethnography on BRCA1/2 in the UK, wherein she highlights how women incorporate genetic risk into decisions about reproduction, prevention, and family responsibility, showing that genomics is deeply entangled with cultural norms around motherhood, care, and bodily autonomy. In a related theoretical vein, Margaret Lock's (2013) concept of *local biologies* further disrupts universalist assumptions, proposing that biological processes such as gene expression or disease manifestation are shaped "from the inside out" by historical, cultural, and environmental contexts.

The chapter by **Wahlberg, Heinsen, and Vendel Petersen** shows how genomic information does not simply enter the family—it grows within it, branching through everyday conversations, silences, hesitations, and the careful choreography of care. Their research traces how families grappling with Lynch syndrome—a rare inherited predisposition to early-onset colorectal cancer—navigate the ethics of disclosure, not in abstract terms of autonomy or bioethics but in the thick of kinship. Here, the "right to know" becomes a matter of the right moment: when a child is "old enough", when a parent feels "ready", when family rhythms allow. Genomics unfolds not through formal scripts but in the texture of parenting and the everyday negotiations of childhood, protection, and trust. In this account, risk is not a fixed probability but a lived relation—entangled with the moral weight of care, the emotional labor of preparing a child, and the timing of intergenerational responsibility.

"Mundane" as ordinary and enculturated is central to **Lloyd**'s critical analysis of neuroepigenetics, particularly in suicide research. In this chapter, she uses speculative fiction (notably Ann Leckie's *Ancillary Justice* series) to question what is lost when emotional experiences are reduced to biomarkers. She defines "mundane neuroepigenetics" as the everyday integration of genomic and epigenetic knowledge into understandings of health, risk, and identity. This normalization is described as an ambivalent process through which such knowledge becomes personally relevant, both within and beyond medicine. Neuroepigenetics becomes "mundane" when its explanations of suicide are embedded in personal narratives and public discourse—when individuals use genetic reasoning to explain patterns of suicide in their families. This shift carries consequences: suicides are reframed as biologically determined outcomes rather than complex, social acts. Human experiences risk being reduced to variables like early life adversity (ELA), erasing context and nuance. Lloyd warns that normalizing neuroepigenetic models may marginalize non-biological explanations of suicide, such as socio-political injustice, emotional trauma, and personal agency.

This biosocial domestication also extends to ancestry genomics, where markers of genetic origin—often framed as "personalized history"—are absorbed into collective identity narratives and cultural imaginaries of belonging. A range of studies (Nelson, 2016; Wailoo et al., 2012; Nash, 2015) has shown how ancestry testing is appropriated by individuals and communities not merely as a recreational or genealogical tool, but as a socially embedded practice through which historical trauma is narrated, ethnic and national identities are negotiated, and claims to recognition and belonging are articulated. In these contexts, genetic data are not treated as objective or external but become emotionally and morally charged, anchored in everyday forms of storytelling, political positioning, and intergenerational memory. In this way, ancestry genomics becomes mundane—not because it is insignificant, but because it is made familiar, intimate, and actionable in ordinary life. **Romijn**'s chapter illustrates this process by examining a white American community of Belgian descent, showing how genomic ancestry knowledge becomes entangled with narratives of collective memory, migration, and identity. Here, DNA is not about health but about ethnic heritage and political discourse—in debates about immigration or public health, for instance. Romijn's notion of "shallow time" captures how genomic ancestry is woven into present-day concerns through emotionally and morally resonant stories that bridge historical distance. He shows that even non-medical forms of genetic knowledge become embedded in the everyday as ordinary tools for imagining self, community, and belonging.

Kaur's chapter uses the concept of *mundane genomics* to examine how the growing normalization of genomic technologies intersects with societal views on equity, access, and ableism. He considers "mundane" the routine use of genomic data in everyday healthcare, especially in human and reproductive health. Focusing on genetic screening and testing—now common in prenatal and neonatal care—he argues that such normalization creates expectations that genomics should be used to prevent or manage conditions before symptoms appear, reinforcing ableism by promoting certain physical or cognitive traits as preferable. However, it may also resist ableism by empowering reproductive decisions for those with genetic conditions. Empirical data show how public views on Heritable Human Genome Editing (HHGE) vary with the perceived severity of conditions. Some participants accepted genetic ailments as a normal part of life, indicating that certain conditions themselves have become mundane. While widespread screening may imply devaluation of lives with genetic differences, many—especially those affected—view genomics as a means to improve quality of life, not deny identity. The chapter concludes that mundane genomics reflects and shapes societal norms, highlighting the need for global governance to ensure equitable access and avoid reinforcing structural ableism. It is not merely a technical, but a sociocultural phenomenon that reshapes how we understand health, disability, and human worth.

Mundane as Routine and Ethically Invisible

The previous section explored the intimate dimension of mundane genomics—how genetic knowledge becomes embedded in the personal, familial, and communal lives of individuals carrying specific variants. Here, we shift focus to the broader diffusion of genomic practices among populations not directly linked to inherited conditions. This spread is particularly evident in health optimization, preventive care, and reproductive planning. In such contexts, genomics becomes mundane not through familiarity with illness, but through routinization, standardization, and what we term "ethical invisibility".

To understand this transition, we can distinguish between two modes of normalization. The first involves incorporating genetic conditions into everyday life—what we might call *normalization through incorporation and enculturation*. The second, more recent, involves *normalization through diffusion*: genomics as a generalized technology for managing health risks. Within this frame, genetics becomes just one tool amongst many others in the governance of health, bound up with responsibilization, atomization, and the moralization of choice.

Foucauldian scholars have long traced the molecularization of life as a new form of governance (Rabinow & Rose, 2006; Rose, 2007). Genomic risk exemplifies this logic: it multiplies thresholds for intervention and mobilizes individuals—often asymptomatic—toward practices of self-surveillance, health optimization, and reproductive planning. Clarke et al. (2010) frame this within "biomedicalization", where medicine increasingly focuses on controlling, managing, reshaping, and enhancing life itself, rather than solely on caring for or curing illness. Yet, as genomics continues to circulate as a powerful symbolic and scientific force—evoked in promises of personalized medicine, risk anticipation, and biomedical innovation—it is increasingly marked by a paradox: it is everywhere, yet ethically invisible. This invisibility is not related to a decline in media visibility or public interest. On the contrary, genetic knowledge continues to be instrumentalized across domains. What has become invisible, rather, are the ethical stakes of its application and the processes of subjectivation genomics might once have enabled.

The rise of what could be termed mundane genomics—low-friction, routinized, and infrastructural forms of genetic testing—has rendered genomic knowledge at once banal and ethically non-productive. The capacity of such technologies to constitute "biological citizens" or "somatic selves" (Rose, 2007) is no longer guaranteed. This tendency is particularly salient in the domain of selective reproductive technologies (Wahlberg & Gammeltoft, 2018). Two central trends exemplify this mundane diffusion: the proliferation of polygenic risk scores, within or outside clinics, as part of DTC genetic testing, and the integration of genomic technologies into reproductive care—notably in selective reproductive technologies involving fetuses, embryos, or gamete donors.

While bioethical perspectives emphasize informed consent and autonomy, a mundane lens foregrounds the infrastructural, cultural, and temporal processes by which technologies become "normal". Here, routinization precedes understanding, and consensus is enacted through logistics rather than deliberation. The mundane, then, refers not to insignificance, but to the banalization and ethical distancing of complex technologies. For example, prenatal diagnosis has long been shaped by regulatory frameworks, institutional arrangements, cultural meanings, and tacit norms around motherhood, disability, and reproductive responsibility (Rapp, 1999; Löwy, 2017). These attitudes consolidate in shared imaginaries—expectations that legitimize and naturalize selective reproductive techniques. Metzler (2020) describes these imaginaries as moral infrastructures, facilitating the rapid uptake of tools like non-invasive prenatal testing (NIPT). NIPT's appeal lies in its routinization and affective neutrality: it is "just another blood test". Its non-invasiveness is not simply technical; it reconfigures the epistemic and ethical status of fetal genetic knowledge by obscuring its decision-making consequences. As a result, the pregnant subject is not constituted as a moral agent weighing the gravity of reproductive risk, but rather as a consumer of routine screening. This "screening logic", now largely detached from the dilemmas of selective reproductive technologies (Wahlberg & Gammeltoft, 2018), narrows the space for deliberation and attenuates the ethical weight of what is, at its core, a selective act. As Thomas et al. (2019) argue, this normalizing of NIPT is achieved not just by technological ease but through discursive framings that emphasize reassurance, neutrality, and emotional comfort, transforming a complex ethical choice into an almost ambient element of prenatal care.

Spanish IVF clinics similarly present preimplantation genetic testing for aneuploidies (PGT-A) as both routine and cutting-edge: an uncontroversial investment in reproductive success (Turrini & Otín-Gavín, 2025). **Pavone, Bosch, and Turrini** show how PGT-A has become routine in Spanish fertility clinics, with patients often accepting the test as part of a "complete" package, unaware of its limitations. The routinization of genetic screening of embryos tends to limit women's decisional costs, framing this choice as a good clinical and economic investment to speed up fertility treatments. The promise of certainty—selecting a "healthy" embryo—renders the technology nearly invisible. This process, however, comes with consequences, enrolling patients in complex reproductive journeys, full of hurdles, characterized by an almost obsessive quest for information and by an increasing pressure that makes PGT-A a "compulsory" option that is increasingly impossible to reject. In the end, the technique becomes embedded with a powerful set of hopes and desires that pushes patients to try repeated IVF cycles in an increasingly expensive slippery slope which is very difficult to stop.

Similarly, expanded carrier screening (ECS)—especially in third-party gamete donation—illustrates the ethical invisibilization of selection. ECS operates at a distance from reproduction itself: it anticipates risk before conception. Donor-recipient "matching" becomes a form of reproductive hygiene, stripped of narrative, encounter, or deliberation, particularly in systems where gamete donation is anonymous. The production and implementation of norms, protocols, and rules (including the lack or fragmentation of them) is a well known and established way to make techniques, practices, and data routine and banal, and, occasionally, invisible. **Arias'** chapter deepens the analysis of the regulatory framework surrounding this practice in Spain. In this context, the absence of clear legislation has enabled ECS and genetic matching to become normalized clinical practices. These genetic tests have evolved through fragmented regulatory mechanisms—including court decisions, professional guidelines, and oversight by regulatory bodies—rather than through coherent legislation or public debate.

A similar logic shapes the rise of predictive and preventive testing. **Hoffman and Tenenberg**'s chapter examines polygenic risk scores, which, despite uncertain clinical value, circulate widely in consumer health platforms and institutional prevention programs. These tools do not diagnose but, instead, foster anticipatory behaviors, speculative futures, and individualized responsibility. Users fold test results into everyday health routines—self-tracking, lifestyle adjustments, or speculative reflection. Far from producing epistemic rupture, these practices reinforce wellness logics. In doing so, they exemplify the increasing *mundane-ness* of genomic technologies—how they quietly settle into everyday routines and responsibilities, shaping health imaginaries without drama or disruption.

Together, these examples point to a new regime of visibility: while genomics remains hyper-visible as innovation and premium service, it becomes ethically invisible through routinization and infrastructural absorption. Genomics becomes mundane not just by repetition, but by its ethical reframing as uncontroversial common sense. Technologies are marketed as tools of risk reduction, embedded in personal decision-making, and shielded from critical scrutiny. What disappears in this process is not risk itself—but controversy. As genomic medicine becomes standard, its political, epistemic, and ethical stakes are quietly effaced by routine, reassurance, and imagined inevitability. In this sense, we witness a mutation of biopower, one in which genomic governance is enacted not through explicit normalization or public pedagogy, but through its purposeful banalization. Ethical reasoning is not abolished, but dispersed and pre-empted—outsourced to infrastructures, rendered implicit in workflows, and thus stripped of deliberative depth. As mundane genomics grows in reach, its ethical invisibility becomes not a void, but a mechanism of power.

Mundane as Institutional and Infrastructural

Various scholars have addressed genomics to unpack institutionalization processes. Some foundational works on the institutionalization of genomics have shown how genomics reshapes citizenship by encouraging individuals to understand themselves and act in terms of their biological risk and genetic identities, leading to new forms of self-governance and responsibilization (Rose & Novas, 2005). Similarly, others have explored how genomics became institutionalized through regulatory bodies, biotech industries, and public health frameworks, leading to the emergence of "biosociality"—new social groupings formed around shared genetic traits (Rabinow, 1996). Addressing what reproductive technologies reproduce beyond babies, Sarah Franklin (2013) has powerfully shown how genomic knowledge has become embedded in institutions, reshaping concepts of kinship, inheritance, and family norms. Studies on the institutionalization of genomics, however, have not only addressed social and cultural domains, but economics and financial ones too. In a pioneer work, Sunder Rajan (2006), for instance, traced how genomics became institutionalized through global capitalism, showing how scientific innovation was and remains deeply entangled with market forces, shaping research agendas and access to genomic medicine. More recently, Van de Wiel (2020) has published a compelling analysis of how the institutionalization of IVF and reprogenomics has undergone a speculative turn, wherein fertility is increasingly treated as a future asset, thereby becoming financialized—valued in terms of investment, risk, and potential future returns. This shift links reproductive decisions to market logics, turning biological processes into speculative ventures shaped by corporate interests, insurance models, and venture capital (Van de Wiel, 2020). Finally, some scholars have also looked at failed institutionalization. Examining the aftermath of the Human Genome Project, Reardon (2019) showed how institutionalized genomics has often failed to deliver on promises of equity and medical breakthroughs, highlighting the ethical and social challenges of Big Data genomics.

The institutionalization of genomics has also occurred through technoscientific infrastructures, such as large-scale consortia and the establishment of biobanks, which order and institutionalize the value of biological samples and associated data (Mitchell & Waldby, 2010; Argudo-Portal & Domènech, 2020; Llorian et al., 2023; Helén & Lehtimäki, 2023). Following Callon's (1991) concept of innovation networks, Stemerding and Nahuis (2014) have attended to the institutionalization of medical genomics in the Netherlands, exploring valorization modes and the links between clinical genomics centers and patient care, or between universities and industry. Attending the institutionalization of genomics is crucial when delving into public-private partnerships or seeking a better understanding of the economic and knowledge flows between different poles. For instance, a formalization of the institutionalization of genomics can be found in national genomics strategies or the broader genomics agenda (Zimani et al., 2021; Onstwedder et al., 2022).

According to this vast literature, which we have only very briefly presented, the institutionalization of genomics has produced very relevant consequences, in terms of (a) the biologization of identity, risk, and responsibility; (b) the marginalization of social and structural determinants of health; (c) the creation of new inequalities in access to genetic knowledge and care; (d) the expansion of surveillance and governance over individuals based on genetic data; and (e) the commercialization of life itself, with genes treated as property or capital.

This volume, which addresses mundane aspects as a way in which genomics has become institutionalized, includes chapters that contribute to this literature and touch upon all these set consequences, unpacking how institutional genomics and its related modes of governance come to be activated in different locales—highlighting one more time that the mundane is indeed significant, analytically and politically.

The chapter by **Urquizas Campello** on forensic genetics and mass incarceration in Brazil, for instance, examines the introduction of genetic databases into the criminal system and the daily practices of sample collection in prisons, aiming to unpack the reactivation of historical racism and social selectivity through the quotidian sample-taking practices for genetic profiling. His work shows how Brazil's DNA database, driven by the institutional strategy to collect genetic material in prisons, is a clear example of the mundane as institutional. Mundane genomics, in this respect, addresses how genetic testing gets linked to a technical and institutional apparatus. More specifically, the chapter attends to the arrangements of judicial, police, and scientific institutions in Brazil, particularly looking at DNA databases for criminal purposes, which the author reads as part of the "contemporary universe of mundane genomics". Urquizas Campello pertinently notes that the socio-racial composition of DNA records can serve to feed a vicious circle of "anticipatory suspicion of criminal recidivism" against black and poor people. His chapter points out that the use of forensic genetics as a form of mundane genomics that tends to be overlooked in the prison routine can amplify long-standing violence and abuses.

In the case of **García Sancho**'s chapter, genomics is framed through a particular dimension of its mundanity: its entanglement with the infrastructures that enabled the systematic abduction of a specific kind of children in Francoist Spain. Neo-Lamarckian genetic discourse, used to justify the removal of children from lower-class or politically dissident families, became embedded in state infrastructures that persisted well beyond the dictatorship. Genomics was not modest in this context; rather, it was a central biopolitical tool for state-making and population control, that is, yet another example of mundane genomics as institutional. The tendency to overattribute power to genomics persists today, particularly in the expectation that DNA testing can uncover the truth and restore justice. A similar argument is made by García-Deister in her chapter when noting that the promises associated with genetic databases serve as necropolitical devices to administer hope in regard to human identification in Mexico—despite their modest role in human identification—with a strong presence in the institutional narrative through, for instance, DNA collection campaigns, while structural violence remains obscured.

Addressing yet another dimension of the mundane as defined in institutional terms, the work of **Tam and Raffaetá** unpacks the new sociopolitical normality enacted by the Hong Kong Government and the Chinese State by studying mundane aspects of microbiome research. Their work highlights how "the mundane is a set of knowledge practices that are both scientific and biopolitical", pointing out that the molecularization of biopolitical governance is crucial for the Chinese State following the 2019 Hong Kong protests and the COVID-19 pandemic. Drawing on insights from the Asian context, Tam and Raffaetá's work explores how postgenomics research becomes mundane in its intersections with the public domain and state-led objectives, highlighting the importance of attending to the institutional collaborations, incompatibilities, and hybrids that are configured. The mundane as institutional also recalls the normalization issue, as stated previously, but brings it to a totally different, and much bigger, scale. Tam and Raffaetá call our attention to how mundane, nation-based research can assemble biosocial groups with local and planetary connections. Or, to put it differently, the chapter shows that mundane (genomics) research can be tightly connected to national identity-building and geopolitical positioning processes. The distributed effects of mundane practices are then identified with this analytical reorientation.

Mundane as Modest and Situated

For a long time, feminist scholars have critically approached the noun modesty and the adjective modest as reinforcing oppressions against women, illustrated by the call for modest dress and behavior—an imposition for multiple oppressed collectives. Following its etymological meaning, modesty as a virtue, moderation, and measure has served as a mechanism to ensure women to remain in the background. In science studies, modesty as a virtue has been sharply addressed by Donna Haraway and Elisabeth Potter in an analysis revolving around the well-studied case of Boyle and the air pump; in such a context modesty appears as a masculine virtue which sustained modern science. In the modest witness of a scientific event, "this modesty was to be key to the gentleman-scientist's trustworthiness, he reported on the world, not on himself" (Haraway, 1997, p. 30). *Modestia* has also been part of clerical culture, transferred to the scientific realm, which Noble (1992) refers to as a "scientific asceticism" in which objectivity arises as "the culture of no culture" (Traweek in Haraway, 1997), as detachment. Despite the numerous critical readings of the modest witness in science, Haraway (1997) acknowledges that the modest witness had an encounter with the world and prompted an anti-heroic narrative that should not be dismissed; indeed, as noted at the end of her *Modest_Witness* book, what will count as modest is what is at stake.

For the study of mundane genomics, the mundane as modest recalls attention to humble practices overplayed in some hegemonic biomedical and scientific discourses, which are constantly fed by presumptuous promises. In this volume, we identified how pairing the mundane with modesty serves to relocate down-to-earthness, moderation, and attention to others as worth considering and productive for making care—and even genomics—doable. In reasoning on technoscience and the role of modest witness, Haraway (1997, p. 269) notes that "valid witness depends not only on modesty but also on nurturing and acknowledging alliances with a lively array of others". Thus, the mundane as modest can recall situatedness. In this volume, **Argudo-Portal** refers to *genomics modesty*, which is inspired by feminist scholars' work on knowledge politics; here, modesty cannot be disentangled from attention and alliances with others—in this case, decentering geneticists' and genomics' protagonism as detached from others' work (nurses, cardiologists, oncologists, lab technicians, bioinformaticians, etc.) and knowledge. In her work on labor division and genomics integration in the Spanish healthcare system, genomics modesty is posited as an ambivalent mode of care that requires attention to others while leaving room for frictions and dissent from different professional standpoints. Argudo-Portal's examination of the controversy around the specialty of clinical genetics in Spain recalls that mundane frictions and jurisdictional tensions also foster relationships and ways of caring for patients. As such, she considers cultivating *genomics modesty* vital to ensure the best possible care in a still uncertain field and fragile healthcare landscape. Mundane genomics, then, recalls modesty as a mode of care that does not necessarily reject tensions between professionals but, rather, considers them productive, particularly when approaching the integration of genomics into public healthcare settings.

García-Deister's chapter refers to the "modest place" of DNA analysis in the forensics of human identification in Mexico. This modesty requires including other tools, professionals, and objects in the picture. The work of García-Deister demonstrates that "DNA is not the central tool in human identification, but rather one of many important techniques". The role of DNA is, then, modest—understanding modesty as the acknowledgment of a broader network of actors and objects that requires resituating and de-centering genomics. As the genetics laboratory director in her chapter notes, genetic data is just another link in the quest for human identification, along with tattoos, clothing, and contextual information, among other practices. She argues that DNA analysis holds a modest place within forensic institutions; that is, the role of genetic databases in practice is not the definitive tool to address Mexico's forensic crisis. Thus, while in different contexts and with different topics, Argudo-Portal and García-Deister's chapters link the mundane and modest to recognize analytically the overlooked companions that make the work doable, resituating the significance of genomics in a broader network, be it in the healthcare system of Spain or the forensic realm in Mexico. Mundane as modest serves to restrain the de-contextualization of genomics and recall attention to those "others" that also do the work, whether other physicians, nurses, tattoos, clothes, or collective organization.

In this vein of decentering genomics protagonism, **García Sancho**'s chapter on the "stolen babies" in Spain serves to note that genomics alone cannot resolve the complex legacy of violence; it cannot trace coercion, expose manipulated consent, or fill archival voids left by forged documents and empty graves. In contrast, his chapter outlines a more modest role for genomics, one that supports, but does not determine, the process of restoring genealogical links and recovering historical memory. Genetic tools do not offer final truths; rather, they function as partial aids within broader efforts that include testimony, archives, legal processes, and activism. Thus, García Sancho enacts a form of mundane genomics: infrastructural, situated, and ethically modest. It rejects both the grand promises of past genomic imaginaries and the idea of genomics as a panacea. Instead, it offers a vision of genomics as a companion practice in the ongoing, collective effort toward justice.

Modesty in mundane genomics also highlights resistance or moderation (or the lack thereof). **Hilberg, Stermach, Kleinhout-Vliek, and Martin**'s work in this volume on media narrative structures associated with gene therapy shows how such narratives overlook persistent and complex challenges in favor of hope. However, such narrative structures, they argue, are increasingly becoming mundane, ordinary, and common in the portrayal of gene therapies, their mundanity making other topics invisible compared to alleviating suffering and offering hope. Cures are overpromised in contrast to calculations of affordability, which seem mundane, and overblown scientific hopes for those who suffer and their families. Their work discusses the shortcomings of heroic gene-therapy narratives that rely on "radical reversal of an otherwise fatal condition". In this case, the mundanity of such narratives makes a modest take improbable. Their chapter demonstrates how modesty, as a form of moderation, can be erased from narratives. Still, at the same time, they argue that heroic narratives of gene therapy can no longer neglect mundane considerations of economics, which would, in turn, require moderating media narratives and addressing questions that might seem mundane compared to the heroic promises. Such a genomics approach-linked modesty is not characteristic only of scholars in the social sciences; indeed, some geneticists advocate a call to "modesty, honesty, and realism" due to the learnings extracted from the Human Genome Project decades later (Emmer-Streib et al., 2017).

On the side of what could be considered modest technologies, **Giraud**'s chapter helps us see the mundane as a way to rethink technological progress in genomics and resistance to technological change. In the French context of preimplantation genetic testing (PGT), the persistence of older technologies, such as FISH (fluorescence in situ hybridization), is not due to inertia. Instead, it reflects moral and regulatory commitments to modesty and restraint. The mundane highlights the layered nature of genomic technologies, where innovation coexists with enduring tools shaped by ethical and institutional values. Rather than signaling backwardness, using so-called "outdated" methods reveals how transformative change in genomics can be slow, uneven, or deliberately constrained. Here, the mundane becomes a lens to de-spectacularize innovation and highlight how technological choices are socially and politically embedded. Mundane as modest calls attention to selective ignorance and the social life of genomics technologies.

Mundane as modest is also in dialogue with Gardner et al. (2015) sociology of low expectations, highlighting that modest future visions are crucial to recalibrating innovation in the biomedical realm. Indeed, they refer to *recalibration* as a form of clinical labor, showing in their case the limits of deep brain stimulation as a therapeutic technique for patients and families, in which such modesty serves explicitly to recalibrate expectations on a particular treatment or innovation. In this context, modest improvements also count as innovation and are worth pursuing. In this volume, mundane and modest refer to analytical attention to modesty as an ambivalent mode of care (**Argudo-Portal**, in this volume) that requires attending to others in multiple ways, as also illustrated by **Wahlberg et al.** in this volume, which repositions genomics within a broader field of care. Situatedness is, then, the result of analytically attending to the mundane as modest, paying attention to a broader social life of genomics made of hope but also multiple frictions, the relation of genomics' everyday relations to those of others, recognizing that genomics does not necessarily take the lead in everyday life—be it in forensics, hospital settings, or households—and where minor transformations are not seen as an innovation failure but constitutive of everyday technological shifts.

Mundane as Leaky and Unstable

One of the key features of mundane genomics is its semantic and ontological instability—its leakiness. Nowhere is this more apparent than in the gene itself. As science historians such as Evelyn Fox Keller (2000) and Rheinberger and Müller-Wille (2018) have shown, the gene's extraordinary cultural and scientific success has always depended on its malleability. From its early conceptualization as a unit of inheritance in Mendelian theory to its molecular reification as a sequence of nucleic acids, the gene has continuously shifted across epistemic frameworks, experimental practices, and metaphorical registers. This polysemantic character is not a flaw but precisely what has enabled the gene to circulate, perform, and endure.

What Keller dubbed *the century of the gene* was never defined by a single, stable referent. Instead, the gene has functioned as a placeholder: for risk, for identity, for disease, for destiny. Rheinberger and Müller-Wille emphasized its nature as a "transition object"—an entity whose meaning is never fixed but emerges through use. Rather than acting as a universal biological constant, the gene has proven to be a historically contingent and epistemically fluid construct. In the postgenomic age, this mutability has only intensified. Following the sequencing of the human genome, the central dogma of molecular biology—that DNA codes for RNA, which codes for protein—began to unravel. The realization that protein-coding genes constitute only a small fraction of the genome, and often fail to account for complex traits or diseases, ushered in a more relational and environmental view of genetic function. Postgenomics, as described by Richardson and Stevens (2015), marks a shift from deterministic to systemic understandings of biology. Epigenetics, microbiomics, and metagenomics now foreground the interdependence of genes, environments, and social worlds.

In this broader landscape, genomics is no longer primarily about discovering truths *in* DNA, but rather about working *with* DNA as a medium. These dynamics have already been observed in how users engage with direct-to-consumer (DTC) genetic tests. Rather than reinforcing genetic essentialism, such practices often enable new modes of self-exploration, reflection, and narrative construction (O'Riordan, 2010). Individuals use their genomic data not to anchor identity, but to experiment with it. Genomics, in this frame, becomes less a revelation than a resource—less about knowing who we are than about exploring who we might become (Harris et al., 2016).

Similarly, in popular culture, DNA no longer functions as a unified signifier. Instead, it leaks—across discourses, affects, and domains—stretching across identity claims, market logics, and kinship narratives. In popular music, for instance, DNA becomes a metaphor for authenticity, destiny, and rebellion. In songs like BTS's *DNA* or Kendrick Lamar's track of the same name, it serves as a symbol of both intimate personal bonds and collective historical memory. Sometimes it evokes love, other times trauma, resilience, injustice, or Afro-American cultural inheritance. The gene is, at once, sacred and parodic, essentialist and resistant to any biological reductionism.

This instability, as **Santoro and Merlo** argue, is not merely symbolic—it is central to how people engage with genomics in everyday life. In their qualitative study of DTC test users, they identify three dominant metaphorical clusters: *lottery* (genes as luck or chance), *network* (genes as connections between people), and *switch* (genes as activatable triggers). These metaphors often coexist—even within the same individual—highlighting how mundane genomics accommodates, and perhaps even thrives on, contradiction. Users oscillate between determinism and plasticity, between destiny and contingency.

This semantically fluid and emotionally charged terrain extends into clinical settings as well. **Brown**'s chapter demonstrates how PGT-A (preimplantation genetic testing for aneuploidies) is used in IVF clinics not only for identifying chromosomally "normal" embryos but also for broader reproductive purposes—managing twin pregnancies, reducing emotional distress, or optimizing cycle outcomes. The technology thus "leaks" beyond its genetic rationale and becomes part of a wider logic of care, control, and commercial reassurance.

A similar form of interpretive leakage is explored in **Martin**'s analysis of kinship among donor-conceived individuals. Here, DNA matches from DTC platforms rarely speak for themselves. They require interpretation, emotional labor, and social embedding. Kinship is not established through genomic data alone but through photographs, shared memories, perceived resemblance, and storytelling. In this context, genomics is not determinative but dialogic—its authority shaped by the practices and imaginaries that surround it.

What emerges across these cases is a genomics that resists closure. The gene persists—not as a fixed truth but as a traveling signifier. It becomes mundane not because it is stabilized, but because it is so widely available and loosely tethered. The gene has become "a semantic reservoir of the contemporary imagination": ubiquitous, intuitive, and open to shifting and even contradictory interpretations (Turrini, 2024). Thus, to speak of the mundane as leaky is to recognize how genomics embeds itself in everyday life not through conceptual coherence, but through semantic multiplicity. It circulates as metaphor, as affect, as infrastructure. It is evoked to express love, ancestry, illness, shame, pride, and potential. Its instability is not a weakness but what makes it usable—by clinics, consumers, activists, and artists alike. Rather than correcting misunderstanding or striving for coherence, a critical approach to mundane genomics must embrace this messiness. Leakiness is not an accident of genomics—it is its mode of life today.

* * *

It is in these variegated ways that we suggest genomics has become mundane over the past quarter of a century, in different domains, and in different parts of the world. In between the hype of the "DNA revolution" and the disappointment of the "DNA delusion", genomics has nonetheless made significant inroads within the fields of medicine, forensics, genealogy, and *-omics* research. It is with the normalization, routinization, and banalization of these inroads that chapters in this volume engage, showing how genomics has both generated novel ways of understanding, relating to, and acting on one's community, family, and self, all the while folding into existing socio-economic inequalities, forms of discrimination, and neglect.

References

Argudo-Portal, V., & Domènech, M. (2020). The reconfiguration of biobanks in Europe under the BBMRI-ERIC framework: Towards global sharing nodes? *Life Sciences, Society and Policy, 16*(1), 9.

Argudo-Portal, V., & Domènech, M. (2022). 'We came here to stay': Making biobanks worth maintaining in Spain. *Sociology of Health & Illness, 44*(2), 328–344.

Bloomberg Businessweek. (2024, May 15). *The DNA test delusion.* https://www.bloomberg.com/news/features/2024-05-15/23andme-s-dna-test-drama-shows-limits-of-what-genetics-can-do

Buch, E. D. (2015). Anthropology of aging and care. *Annual Review of Anthropology, 44*(1), 277–293.

Callon, M. (1991). Techno-economic networks and irreversibility. In J. Law (Ed.), *A sociology of monsters* (pp. 132–161). Routledge.

Cheung, K., Earp, B. D., Patch, K., & Yaden, D. B. (2025). Distinctive but not exceptional: The risks of psychedelic ethical exceptionalism. *The American Journal of Bioethics, 25*(1), 16–28.

Clarke, A. E., Shim, J. K., Mamo, L., Fosket, J. R., & Fishman, J. R. (Eds.). (2010). *Biomedicalization: Technoscience, health, and illness in the U.S.* Duke University Press.

Deejay, A., & Henne, K. (2024). Creating a new normal? Technosocial relations, mundane governance and pandemic-related disruption in everyday life. *Sociology, 58*(3), 699–716. https://doi.org/10.1177/00380385231205135

Denis, J., & Pontille, D. (2015). Material ordering and the care of things. *Science, Technology & Human Values, 40*(3), 338–367.

Domínguez Rubio, F., Denis, J., & Pontille, D. (2025). *Fragilities: Essays on the politics, ethics, and aesthetics of maintenance and repair.* The MIT Press.

Emmert-Streib, F., Dehmer, M., & Yli-Harja, O. (2017). Lessons from the human genome project: Modesty, honesty, and realism. *Frontiers in Genetics, 8*, 184. https://doi.org/10.3389/fgene.2017.00184

Enloe, C. (2011). The mundane matters. *International Political Sociology, 5*(4), 447–450.

Enloe, C. (2013). The mundane matters. Why feminists take daily life seriously. In *Seriously!: Investigating crashes and crises as if women mattered* (pp. 39–48). University of California Press.

Fortun, M. (2023). *Genomics with care: Minding the double binds of science.* Duke University Press.

Franklin, S. (2013). *Biological relatives: IVF, stem cells and the future of kinship*. Duke University Press.
Franklin, S., & Roberts, C. (2006). *Born and made: An ethnography of preimplantation genetic diagnosis*. Princeton University Press.
Froger-Lefebvre, J., Lade, Q., Vallier, E., & Bourgain, C. (2023). E-prescription and invisible work in genomics in France. *Frontiers in Sociology, 8*, 1152364. https://doi.org/10.3389/fsoc.2023.1152364
Fullwiley, D. (2011). *The enculturated gene: Sickle cell health politics and biological difference in West Africa*. Princeton University Press.
Gardner, J., Samuel, G., & Williams, C. (2015). Sociology of low expectations: Recalibration as innovation work in biomedicine. *Science, Technology & Human Values, 40*(6), 998–1021.
Geelen, E., Van Hoyweghen, I., & Horstman, K. (2015). Shaping the future and living in the present: Living a "good" life with a familial heart disease. *BioSocieties, 10*, 70–83.
Gibbon, S. (2007). *Breast cancer genes and the gendering of knowledge*. Palgrave Macmillan UK.
Gibbon, S., & Novas, C. (Eds.). (2008). *Biosocialities, genetics and the social sciences* (pp. 1–18). Routledge.
Haraway, D. (1997). *Modest_Witness@Second__Millennium.FemaleMan©_Meets_ OncoMouse™*. Routledge.
Harris, A., Kelly, S. E., & Wyatt, S. (2016). *Cybergenetics: Health genetics and new media*. Routledge.
Heinsen, L. L., Petersen, H. V., & Wahlberg, A. (2021). *Punctuated lives–staying healthy with Lynch syndrome*. https://somatosphere.com/2021/punctuated-lives.html/
Helén, I., & Lehtimäki, H. (2023). Valuation in emerging technoscience business: A case study of Finnish biobank research. *European Journal of Innovation Management, 26*(7), 611–634.
Hoeyer, K. (2010). After novelty: The mundane practices of maintaining a bone bank. *Science as Culture, 19*(3), 297–320. https://doi.org/10.1080/09505430903164937
Hollin, G. (2019). From the profound to the mundane: Questionnaires as emerging technologies in autism genetics. *Science as Culture, 28*(1), 93–116. https://doi.org/10.1177/0162243919841691
Hyde, S. T., & Denyer Willis, L. (2020). Balancing the quotidian: Precarity, care and pace in anthropology's storytelling. *Medical Anthropology, 39*(4), 297–304.

Isascon, W. (1999, January 11). *The biotech century*. TIME; nextgen. https://time.com/archive/6734363/the-biotech-century/

Jamali, L. (2025, May 19). *Struggling DNA testing firm 23andMe to be bought for $256m*. BBC. https://www.bbc.com/news/articles/c0ln0e5g6kgo

Keller, E. F. (2000). *The century of the gene*. Harvard University Press.

Kelly, S. E., Wyatt, S., & Harris, A. (2018). Mainstreaming genomics and personal genetic testing. In S. Gibbon, B. Prainsack, S. Hilgartner, & J. Lamoreaux (Eds.), *Routledge handbook of genomics, health and society* (pp. 32–38). Routledge.

Krupar, S., & Ehlers, N. (2020). Biocultures: A critical approach to mundane biomedical governance. *Culture, Theory and Critique, 61*(4), 440–456.

Latour, B. (1992). Where are the missing masses? The sociology of a few mundane artifacts. In W. E. Bijker & J. Law (Eds.), *Shaping technology/building society: Studies in sociotechnical change* (pp. 225–258). MIT Press.

Lemke, T. (2013). *Perspectives on genetic discrimination*. Routledge.

Levenson, M. (2025, May 29). 23andMe to be bought by biotech company for $256 million. *The New York Times*. https://www.nytimes.com/2025/05/19/business/regeneron-pharmaceuticals-23andme-data.html

Llorian, E. R., Kopac, N., Waliji, L. A., Borle, K., Dragojlovic, N., Elliott, A. M., & Lynd, L. D. (2023). A rapid review on the value of biobanks containing genetic information. *Value in Health, 26*(9), 1286–1295.

Lock, M. (2013). The epigenome and nature/nurture reunification: A challenge for anthropology. *Medical Anthropology, 32*(4), 291–308.

Löwy, I. (2017). *Imperfect pregnancies: A history of birth defects and prenatal diagnosis*. Johns Hopkins University Press.

Metzler, I. (2020). Imaginaries as infrastructures? The emergence of non-invasive prenatal testing in Austria. *BioSocieties, 15*(4), 601–626.

Michael, M. (2003). Between the mundane and the exotic: Time for a different sociotechnical stuff. *Time & Society, 12*(1), 127–143. https://doi.org/10.1177/0961463X03012001372

Michael, M. (2006). *Technoscience and everyday life: The complex simplicities of the mundane*. Open University Press.

Mitchell, R., & Waldby, C. (2010). National biobanks: Clinical labor, risk production, and the creation of biovalue. *Science, Technology & Human Values, 35*(3), 330–355.

Musée d'éthnographie & Mission Scientifique Dakar-Djibouti. (1931). *Instructions sommaires por les collecteurs d'objets ethnographiques*. Trocadéro Palace.

Nash, C. (2015). *Genetic geographies: The trouble with ancestry*. U. of Minnesota Press.
Nelkin, D., & Lindee, M. S. (1995). DNA mystique: The gene as a cultural icon. W.H. Freeman
Nelson, A. (2016). *The social life of DNA: Race, reparations, and reconciliation after the genome*. Beacon Press.
Noble, D. (1992). *A World Without Women: The Christian Clerical Culture of Western Science*. Oxford University Press.
O'Riordan, K. (2010). *The genome incorporated: Constructing biodigital identity*. Ashgate.
O'Riordan, K. (2011). Revisiting digital technologies: Envisioning biodigital bodies. *Communications, 36*(3), 291–312. https://doi.org/10.1515/comm.2011.015
Onstwedder, S. M., Jansen, M. E., Leonardo Alves, T., Cornel, M. C., & Rigter, T. (2022). Pursuing public health benefit within National Genomic Initiatives: Learning from different policies. *Frontiers in Genetics, 13*. https://doi.org/10.3389/fgene.2022.865799
Pickersgill, M., Cunningham-Burley, S., & Martin, P. (2011). Constituting neurologic subjects: Neuroscience, subjectivity and the mundane significance of the brain. *Subjectivity, 4*(3), 346–365. https://doi.org/10.1057/sub.2011.10
Pinel, C., Wyatt, D., & Faulkner-Gurstein, R. (2025). The hidden labor of translation: Introduction to special issue. *Science, Technology & Human Values, 50*(2), 242–250.
Pink, S., Sumartojo, S., Lupton, D., & Heyes La Bond, C. (2017). Mundane data: The routines, contingencies and accomplishments of digital living. *Big Data & Society, 4*(1), 1–12. https://doi.org/10.1177/2053951717700924
Pollock, A. (2012). *Medicating race: Heart disease and durable preoccupations with difference*. Duke University Press.
Pollock, A., M'charek, A., Ehlers, N., Creary, M., & García-Deister, V. (2021). Race and biomedicine beyond the lab: 21st century mobilisations of genetics—Introduction to the special issue. *BioSocieties, 16*(4), 433.
Puig de La Bellacasa, M. (2017). *Matters of care: Speculative ethics in more than human worlds*. University of Minnesota Press.
Rabinow, P. (1996). Artificiality and enlightenment: From sociobiology to biosociality. In *Essays on the anthropology of reason* (pp. 91–112). Princeton University Press.
Rabinow, P., & Rose, N. (2006). Biopower today. *BioSocieties, 1*(2), 195–217.

Rajan, K. S. (2006). *Biocapital: The constitution of postgenomic life*. Duke University Press.

Rapp, R. (1999). *Testing women, testing the fetus: The social impact of amniocentesis in America*. Routledge.

Reardon, J. (2019). *The postgenomic condition: Ethics, justice, and knowledge after the genome*. University of Chicago Press.

Rheinberger, H. J., & Müller-Wille, S. (2018). *The gene: From genetics to postgenomics*. University of Chicago Press.

Richardson, S. S., & Stevens, H. (Eds.). (2015). *Postgenomics: Perspectives on biology after the genome*. Duke University Press.

Rifkin, J. (1998). *The biotech century: Harnessing the gene and remaking the world*. Penguin Putnam.

Rose, N. (1999). *Powers of freedom: Reframing political thought*. Cambridge University Press.

Rose, N. (2007). *The politics of life itself: Biomedicine, power, and subjectivity in the twenty-first century*. Princeton University Press.

Rose, N., & Novas, C. (2005). Biological citizenship. In A. Ong & S.J. Collier (eds.), *Global assemblages: Technology, politics, and ethics as anthropological problems* (pp. 439–463). Blackwell.

Rothstein, M. A. (2007). Genetic exceptionalism and legislative pragmatism. *The Journal of Law, Medicine & Ethics, 35*(2_suppl), 59–65. https://doi.org/10.1111/j.1748-720X.2007.00154.x

Saldaña-Tejeda, A. (2018). Mitochondrial mothers of a fat nation: Race, gender and epigenetics in obesity research on Mexican mestizos. *BioSocieties, 13*(2), 434–452.

Skovgaard, A. L., Tjørnhøj-Thomsen, T., Johansson Jørgensen, M., & Høybye, M. T. (2024). Making it work: Everyday life and healthcare with multiple chronic illnesses in Denmark. *Medical Anthropology, 43*(5), 397–410.

Soulier, A., Leonard, S., & Cambon-Thomsen, A. (2016). From the arcane to the mundane: Engaging French publics in discussing clinical applications of genomic technology. *New Genetics and Society, 35*(1), 1–28.

Stemerding, D., & Nahuis, R. (2014). Implicit and explicit notions of valorization in genomics research. *New Genetics and Society, 33*(1), 79–95.

Taussig, K. S. (2009). *Ordinary genomes: Science, citizenship, and genetic identities*. Duke University Press.

Thomas, G. M., Rothman, B. K., Strange, H., & Latimer, J. E. (2019). Testing times: The social life of non-invasive prenatal testing (NIPT). *Science, Technology and Society, 26*(1), 81–97. https://doi.org/10.1177/0971721820960262

Turrini, M. (2018). Online genomes: Problematizing the disruptiveness of direct-to-consumer genetic tests. *Sociology Compass, 12*(11), e12633. https://doi.org/10.1111/soc4.12633

Turrini, M. (2024). Metáforas post-genómicas: el ADN en la música contemporánea. In M. F. Mancera Martínez & A. Márquez Carrascal (Eds.), *Arias Montano Humanismo en perspectiva Multidisciplinar para el Siglo XXI* (pp. 155–177). Centro Europeo del Humanismo Arias Montano, Fregenal de la Sierra.

Turrini, M., & Otín-Gavín, J. (2025). Let "Pregnant Women Choose the Destiny for Themselves and Their Child". How fertility clinic digital platforms frame preimplantation genetic testing (PGT) in Spain. *Sociology of Health & Illness, 47*(1), 1–12. https://doi.org/10.1111/1467-9566.13876

Van de Wiel, L. (2020). The speculative turn in IVF: Egg freezing and the financialization of fertility. *New Genetics and Society, 39*(3), 306–326.

Wahlberg, A. (2018). Exposed biologies and the banking of reproductive vitality in China. *Science, Technology and Society, 23*(2), 307–323.

Wahlberg, A., & Gammeltoft, T. M. (Eds.). (2018). *Selective reproduction in the 21st century*. Springer International Publishing.

Wailoo, K., Nelson, A., & Lee, C. (Eds.). (2012). *Genetics and the unsettled past: The collision of DNA, race, and history*. Rutgers University Press.

Weiner, K., Martin, P., Richards, M., & Tutton, R. (2017). Have we seen the geneticisation of society? Expectations and evidence. *Sociology of Health & Illness, 39*(7), 989–1004. https://doi.org/10.1111/1467-9566.12551

Zhu, J., & Dong, D. (2018). From quality control to informed choice: understanding "good births" and prenatal genetic testing in contemporary urban China. In *Routledge handbook of genomics, health and society* (pp. 47–54). Routledge.

Zimani, A. N., Peterlin, B., & Kovanda, A. (2021). Increasing genomic literacy through national genomic projects. *Frontiers in Genetics, 12*. https://doi.org/10.3389/fgene.2021.693253

Part I

Gametes and Embryos

2

My Genes, Your Genes: PGT-A in the Spanish Reproductive Bioeconomy

Vincenzo Pavone, Paula Bosch, and Mauro Turrini

Introduction

Over the past 15 years, reproductive bioeconomies, particularly those associated with assisted reproduction technologies (ARTs), have undergone significant change. This shift has been driven by expanding markets, rising parental age, and technological advances, especially in human genetics (Franasiak et al., 2022).

The increasing age of maternity and paternity has led to a substantial rise in the use of ARTs globally, particularly in Europe (Wyns et al., 2022). In developed countries, women, on average, have their first child at just over 30 years of age, with Spain reporting an average of 31.5 years (INE, 2023). Many women begin childbearing at 35 or older, and it is increasingly common to have children after 40 (Eijkemans et al., 2014). In Spain, men and women often have their first child later than they

V. Pavone (✉) • P. Bosch • M. Turrini
Instituto De Políticas y Bienes Públicos (IPP), Consejo Superior de Investigaciones Científicas (CSIC), Madrid, Spain
e-mail: vincenzo.pavone@csic.es; paula.bosch@csic.es; mauro.turrini@csic.es

V. Argudo-Portal et al. (eds.), *Mundane Genomics*,
https://doi.org/10.1007/978-981-95-4550-6_2

would prefer and have fewer children than desired (Merck Group, 2022), with 20% of women starting IVF treatment over the age of 40, a share that becomes 35% if we consider IVF cycles with egg donation (García, 2022).

Apart from the changing demographics, technological advances have expanded treatment options and the range of conditions that can now be addressed with ARTs (Fernandez et al., 2020). Additionally, increased affordability, due to improvements in private insurance and public health coverage, has made ARTs accessible to larger segments of the population (Chambers et al., 2014). According to the European Society of Human Reproduction and Embryology (ESHRE, 2024), over 1 million IVF cycles were performed in Europe in 2021, resulting in approximately 160,000 live births from ART procedures each year (Smeenk et al., 2023).

Another significant trend of reproductive economies is that ARTs are increasingly offered with "add-ons," additional techniques or treatments designed to enhance fertility cycle success rates or ensure the conception of healthy children (Glatthorn & Decherney, 2022). Common add-ons include Intracytoplasmic Sperm Injection (ICSI), used in cases of male factor infertility or previous IVF failure; assisted hatching, which facilitates embryo implantation; immune treatments, which imply the use of drugs or therapies to address immune-related issues that might interfere with implantation; and time-lapse imaging incubators for continuous embryo monitoring and assessment.

This chapter focuses on preimplantation genetic testing (PGT-A), a specific add-on technique that has been central to ongoing ethical, social, and economic debates for the past three decades. PGT-A involves screening embryos for chromosomal abnormalities before implantation during IVF cycles, and is only permitted in a few European countries, such as the UK, Spain, and Italy, but is routinely performed primarily in Spain (Smeenk et al., 2023). As of 2022, 20,502 genetic tests on embryos were performed in Spain, representing 12% of the total of IVF cycles, 22% of all IVF cycles susceptible to PGT, or 32% if we consider only IVF cycles with their own eggs. Of these 20,502 tests, 88.5% are PGT-A (SEF, 2022).

The growing numbers of PGT-A, however, need to be situated in the overall context of the Spanish reproductive bioeconomy. The Spanish reproductive market is among the largest globally, ranking first in Europe and third worldwide, behind Japan and the USA. In 2023, the aggregate turnover for the approximately 310 private centres specializing in these procedures was 630 million euros, with public hospital treatments adding an estimated value of 95 million euros (DBK Informa, 2024). In 2022, 198,830 ARTs cycles were performed nationwide, with 12% of all births in Spain resulting from such treatments, one of the highest percentages globally (SEF, 2024).

The considerable size of the Spanish reproductive market is due to several factors. The legal framework regulating ART is among the broadest and most permissive in the EU. Spain's 1988 Law on Assisted Reproductive Techniques, updated in 2006, was one of the first of its kind in Europe. The law does not specify a maximum age for accessing ARTs and, in addition to straight couples, permits single women and lesbian couples to use these techniques. It does not require an infertility diagnosis, and imposes no limit on the number of cycles a woman can undergo. Spain allows the use of all ART technologies, including PGT-A, excluding only surrogacy and non-medical sex selection.

Apart from a permissive legal environment, Spanish prices are internationally competitive, being considerably lower than in other economies like the United States. The high level of confidence among both national and international patients in the clinical practices and quality standards established by the Spanish legal and medical framework further supports the preference for the Spanish reproductive market (Vlasenko & Pavone, 2025).

While the permissive legal framework and the relatively affordable prices may explain why so many PGT cycles can be performed in Spain (Pavone & Arias, 2012), it does not explain why patients choose to have it performed and are willing to pay the extra costs it implies. Given that 60% of the PGT-A performed in the EU is actually conducted in Spain (Smeenk et al., 2023), Spain represents a formidable case study to address how the reprogenetics economy works and, more specifically, how the routinization of genetics add-ons like PGT is affecting reproductive care, practices, and relationships. Last but not least, it may cast light on how fertility patients make sense of this technique in the broader process of their reproductive journey.

Drawing on semi-structured interviews with ART patients in Spain, this chapter aims to deepen our understanding of how genomic techniques and knowledge have become part of everyday life. We use the term *mundane* to highlight the ambivalent process by which these technologies have been normalized and made personally meaningful to a growing number of people, both within and beyond medical settings. Our analysis focuses on the economic and medical ecosystems that have enabled the widespread adoption of genetic testing—particularly PGT-A—along with the roles played by key stakeholders and the social, cultural, and economic factors that motivate many patients to incorporate these tests into their reproductive journeys. Finally, we explore the broader implications of this routinization, with particular attention to its impact on patient experiences and reproductive healthcare policy.

The "Speculative Turn" of PGT-A: Commercialization and the Mobilization of Hope

Preimplantation genetic testing (PGT) is an advanced tool in reproductive health that may identify either hereditary genetic traits or chromosomal abnormalities in embryos fertilized *in vitro* between their third and seventh day. While there exist different types of PGT, here we focus on PGT-A, which specifically addresses aneuploidies: chromosomal abnormalities usually attributed to advanced maternal age, implantation failures, or recurrent miscarriages. For this reason, PGT-A is often offered as an *add-on* to in vitro fertilization (IVF) to couples or women with infertility in order to address these specific issues. However, it is not uncommon to see PGT-A routinely offered without a specific clinical indication, solely to improve the success rate of IVF. Offered simultaneously to enhance the chances of a successful pregnancy and avoid congenital diseases in future babies, PGT-A tends to blur the distinction between assisted reproduction and selective reproduction.

PGT-A, along with many other add-ons in assisted reproduction, raises significant concerns due to the lack of robust evidence supporting its efficacy and safety, as well as challenges related to informed consent and patient autonomy. Many add-ons lack sufficient clinical trial data to demonstrate their benefits (Braga et al., 2022; Glatthorn & Decherney, 2022, Harper et al., 2017), as the hyper-commercialization in this sector hinders the setup of randomized controlled trials and discourages patient participation (Perrotta & Geampana, 2020). In the case of PGT-A, several studies (including some randomized trials), have been conducted; however, this evidence has recently been classified as low quality (Cornelisse et al., 2020).

The cost of add-ons is, indeed, a major issue as it can significantly raise the expenses of assisted reproduction, burdening patients, and healthcare systems. Medical staff often feel pressured to offer all available treatments, even when evidence of their benefits is lacking (Perrotta & Geampana, 2021; Geampana & Perrotta, 2023). Public healthcare systems face challenges when subsidizing these costs (Perrotta & Hamper, 2021, 2023); thus, patients, often pay for add-ons that may not meaningfully improve outcomes, exacerbating the social inequalities associated with ARTs (Zhang et al., 2021).

The combination of inadequate evidence and rising costs is compounded by the false hope add-ons create. Their availability can lead patients to believe they guarantee success or significantly enhance conception chances (Bracewell-Milnes et al., 2021). IVF clinics often market these procedures to capitalize on patients' hopes, particularly among older couples seeking parenthood (Perrotta & Hamper, 2021). For example, in the UK, 80% of add-ons are offered routinely by clinics, while only 20% are specifically requested by patients (Cirkovic et al., 2023). This dynamic interaction between technological fixes and social pressure can lead to emotional distress and disappointment when outcomes fall short. Risks and side effects are frequently minimized, exposing patients to unnecessary complications (Armstrong et al., 2023). These concerns have sparked debates about the regulation and oversight of add-ons in ARTs. Current variability in regulatory frameworks across countries complicates efforts to ensure consistent quality and safety standards (ESHRE Add-ons working group et al., 2023). Informed consent and patient autonomy also remain challenging, as patients may not receive sufficient information about the evidence, risks, and alternatives, limiting their ability to make fully informed decisions in a situation where hope often outweighs considerations of safety and efficacy (Armstrong et al., 2023).

While all the add-ons are commercially available in Spain, the country stands out in the European context for its widespread implementation of PGT-A. This technique is considered by the HFEA to be one of the most controversial add-ons, especially in relation to its promises of delivering "healthy" babies, and its alleged capacity to increase the success rate.[1] The effectiveness of PGT-A remains fiercely debated, and the medical literature is divided on its impact on successful pregnancy outcomes (Rubio et al., 2013; Simopoulou et al., 2021). Studies evaluating the effectiveness of PGT-A have produced mixed results, with some suggesting it may be effective for specific populations, such as older women—generally over 38—with a good prognosis in terms of enough blastocysts to test, or those experiencing recurrent miscarriages (Murphy et al., 2019; Harris et al., 2025). While others question its overall value (Paulson, 2020), recent studies suggest not only that PGT-A could have a slightly lower success rate compared to general IVF (Roberts et al., 2022) but also that certain populations, such as those aged under 35, could be negatively affected (Harris et al., 2025). Despite these controversies, the Spanish reproductive industry has invested massively in this technique, without any sound evidence that PGT-A has improved the success rate (Hernandez-Nieto et al., 2018; Insogna et al., 2021).

Apart from its technical effectiveness, other concerns—ranging from equity and access to the psychological and social impact on patients' reproductive journey and medical care—suggest caution when it comes to its implementation. Access to PGT-A varies across Europe, influenced by healthcare systems, insurance coverage, and socio-economic status. As a result, scholars have voiced concerns about equity and disparities in access, particularly for disadvantaged populations who face barriers to accessing reproductive technologies. Some studies, for instance, suggest that PGT may exacerbate existing social inequalities, as it is expensive and may not be equally accessible to all socio-economic groups (Zhang et al., 2021). PGT-A potentially poses significant costs for individuals and healthcare systems, constituting a financial burden both for the direct costs of testing and related procedures and the indirect costs arising from subsequent decisions based on the results.[2] Overall, while PGT can provide critical insights into embryo viability and genetic health, the high financial cost requires families to weigh the economic burden against the potential benefits of avoiding genetic conditions or failed IVF cycles (Gualtieri et al., 2023).

Moreover, in many countries PGT-A is essentially an out-of-pocket expense, contributing to the general stratification process associated with add-ons. Policymakers across Europe have considered the economic implications of integrating PGT-A into clinical practice, balancing potential benefits with financial sustainability, but full implementation in public healthcare remains rare. Consequently, PGT-A is largely subject to market dynamics, technological advancements, competition among providers, and consumer demand. The ongoing debate also addresses the need for normative guidelines to manage access and implementation. The Preimplantation Genetic Diagnosis International Society regularly issues guidelines that elicit ambivalent reactions, ranging from trust to strong criticism (Gleicher et al., 2020). European researchers and professional organizations continue to strive for common guidelines and best practices for PGT-A in clinical settings.

Against this controversial background, a growing body of scholarly contributions has addressed how patients experience these techniques and how they impact on their reproductive journey. In 2017, Pavone and Lafuente showed that Spanish patients mainly considered PGT-A a tool to boost their chances to achieve a successful and healthy pregnancy. In the medicalization process involved, patients were willing to introduce PGT-A as a tool to target infertility while adopting a selection logic to discard "bad" embryos, understood as those which would not develop further, and select "healthy" embryos that are expected to be able to survive until birth. In this context, the importance of information was paramount and patients interviewed did not have enough information about the kind of selection to which their embryos were subjected, nor did they know precisely how these embryos were processed. PGT-A, in particular, extends the objective of ARTs from merely achieving pregnancy to ensuring the birth of a genetically healthy baby, thereby shifting the focus from fertility assistance to genetic selection and reproductive optimization (Pavone & Lafuente Funes, 2018).

In general, the information patients receive from the clinics is often too detailed and excessive and contains a large volume of medical terminology, which does not help them to make decisions (Gebhart et al., 2016). When specifically prompted and urged to elaborate on the implications and specificities of the testing, medical professionals do not refrain from providing more information, but many patients either do not have enough background information to be able to ask detailed and meaningful questions or shy away from doing so, due to their deep trust in the medical profession or to the actual lack of it (Quinn et al., 2018). When medical counselling is not available or fails to deliver, patients tend to turn to information extracted, unsupervised and unsupported, from social media, medical websites, or their inner circle of friends and acquaintances (Cheng et al., 2022). These findings seem to support the suggestion that patients' decisions to include add-ons in their treatment are allegedly due to a general lack of understanding of what counts as evidence or because they are not well-informed (Lensen et al., 2021). More recent work, however, has revealed a more positive attitude of patients towards PGT-A, with patients considering themselves sufficiently knowledgeable to make decisions, although some patients misinterpreted the actual benefits of the test (Bracewell-Milnes et al., 2021).

Information provided by private fertility centres on the Internet seeks to influence the perception of medical advancements by adopting a rhetoric that lies at the intersection of medicine and consumer culture (Swoboda, 2015). A recent study of the Spanish context found that information about PGT-A is severely biased in several respects: it presents PGT-A as excessively unproblematic and uncontroversial, and conflates preimplantation genetic diagnosis (PGD) or preimplantation genetic testing for monogenic disorders (PGT-M)—used primarily for families with rare genetic conditions—with PGT-A, and, in doing so, reinforces the notion of the "genetic responsibility" of mothers and couples for their prospective children. Finally, PGT-A is framed as a modulator of female fertility and a means to control maternal, age-related decline (Turrini & Otín-Gavín, 2025; Boe Hüttel et al., 2024).

Patients, however, are neither active and rational choosers nor defenceless victims open to manipulation by the health-care market. Rather, they are called to make complex treatment choices in highly uncertain conditions where biomedical knowledge is unable to provide explanations and there is a lack of robust evidence. Patients have different strategies to deal with uncertainty, medical evidence, and decision-making. Whilst some of them, indeed, decide to delegate the decision to the medical staff, others engage actively with the medical staff and the available evidence to make their decision and choose their own personal trajectory (Perrotta & Hamper, 2023).

Lamb et al. (2018) have conducted an in-depth study on the precise decision factors that lead IVF patients to choose or reject PGT-A. Whilst a relatively late age at their first pregnancy was a factor common to all patients interviewed, other important dimensions contributed to the decision-making process. For instance, some patients wanted to avoid therapeutic abortion during the pregnancy, or giving birth to a child with chromosomal abnormality. Other participants emphasized the importance of avoiding miscarriages and the emotional consequences of an unsuccessful embryo transfer. Interestingly, those who chose PGT-A had a more optimistic view of the ability of technological advancements in science to improve their chances of a successful pregnancy. Previous experiences also played a significant role and many of the individuals who opted for PGT-A stated that they had failed previous transfers and wanted more explanation of why they were miscarrying. Some patients, however, do express concerns that PGT-A could harm the embryos (Gebhart et al., 2016).

Last but not least, a number of scholars have emphasized the psychological consequences of undergoing PGT-A. Unexpected outcomes, uncertainty, unanticipated emotional consequences, doubts about embryo viability, and an over-emphasis on women's role and responsibility may give rise to a psychological burden that too often remains unattended. It is no surprise, therefore, that the emotional and psychological impacts of infertility during in vitro fertilization are often the primary concerns discussed (Rothwell et al., 2020). Resorting to PGT-A, however, may also absolve patients from the regret of not having tried everything, and help doctors feel reassured that they have done their best (Armstrong et al., 2023). Most importantly, the promotion of PGT-A in a highly commercialized sector as reproductive care reinforces and prolongs hope, fostering a narrative that keeps promising success at the next attempt and delegates treatment decisions to patients, increasing their psychological and financial burden (Perrotta & Hamper, 2021).

Data and Methods

This chapter draws from qualitative data gathered through fieldwork conducted in Spain in 2023, which involved semi-structured interviews with 20 female patients, mostly located in private clinics across five key regions. The objective of the fieldwork was to capture a wide range of perspectives on the different PGT-A practices adopted by clinics located in both touristy areas, such as the Mediterranean coast, and less touristy regions. The decision to interview only women was based on the well-documented notion in the literature that women bear the primary responsibility for reproductive decision-making, including the choice to undergo PGT-A and the potential consequences that follow. Research highlights that societal and medical discourses often place a disproportionate burden on women in matters of genetic testing, fertility, and reproductive choices, making their perspectives particularly relevant to this study.

The sample consisted of patients who were undergoing IVF with PGT-A, or had experienced it very recently. The fieldwork was conducted by three researchers—the principal investigator (PI), a senior researcher, and a research assistant—who carried out face-to-face, semi-structured interviews, each lasting between 45 and 90 minutes. Ethical approval for the project was granted in 2022 by the ethical committee of the research institution. After obtaining explicit informed consent from the interviewees, the semi-structured interviews, which included open-ended questions, were audio recorded when permitted by the participants and transcribed. The interviews focused on several key topics: (a) socio-economic background and motivations of patients; (b) choice of clinics and treatments; (c) information received and actively searched for; (d) medical protocols and informed consent on PGT-A; (e) medical and health risk procedures; (f) process of decision-making; (g) treatment implementation and outcomes; (h) regulation and the role of the state; and (i) suggestions and proposals to improve the actual reproductive healthcare system.

Interviews were conducted until saturation was reached. They were transcribed, coded, and analysed through thematic analysis (Marshall & Rossman, 2011) with the aim of systematically developing analytic categories and concepts (Silverman, 2015). Initially, the transcripts were analysed independently by the researchers involved. Subsequently, differences in the two coding systems were discussed, leading to the creation of a common set of codes for a final analysis of the interviews, until consensus was reached.

Information gathered from the interviews was cross-checked with secondary sources, such as clinic websites, as well as national and regional laws and reports. To protect the identity of the interviewees we just mention the coded province in which the interview took place, which includes both the main city and the villages nearby.

Results

a. *Bumpy Road Ahead: The Journey into PGT-A*

The patients interviewed show a variety of different routes into PGT-A, which, however, share some common characteristics. First, they come across PGT-A when they switch from the public healthcare system into the private one, after years of unsuccessful artificial insemination and failed IVF, often due to miscarriages. "It was like, if this doesn't work [standard IVF in the public clinic] we go [to a private clinic] and we decided to spend money [in contracting PGT] because it was like a plan. We have already discovered what happens [embryos with anomalies]" (ALICANTE 1).

When their journey in assisted reproduction begins, many of these patients come across the importance of genetics, which is highlighted through the information published online, through other patients' accounts, or through the physicians' counselling, but it is not until they cross the doorstep of a private clinic that they get to know of PGT-A in more detail. Sometimes they know of it in advance, at least in general terms, sometimes they are unaware of it, at other times, they actually move into private clinics precisely in order to get PGT-A. The connection between PGT-A and the risk of Down syndrome is one of the entry points of PGT-A when clinics perform their counselling. Fear of having a child with trisomy is a strong motivational factor:

> The clinics use the magic word, "Down Syndrome"; they always tell you, "Down Syndrome, Down Syndrome." They explain what it is: "No, this way we rule out chromosomal abnormalities that can cause miscarriages or genetic diseases such as Down's Syndrome," and they tell you like super … you know? […] "You don't want to have a child with Down's Syndrome, do you?" Well, of course, we are afraid and out of fear we do the PGD [PGT-A]. (VALENCIA 1)

The offer of PGT-A can place a huge burden of parental responsibility on patients experiencing fertility issues at the very moment of conception. This excerpt also shows a common tendency among such patients to confuse PGT-A with PGD (preimplantation genetic diagnosis—*diagnóstico genético preimplantacional* in Spanish), which is the former nomenclature to refer to what is today known as PGT-M. This misunderstanding reflects the way PGT-A is presented on private clinics' websites and social networks. Far from being a purely linguistic mistake, patients hardly differentiate between chromosomal and genetic conditions.

Patients are also very aware of the decrease in the quality of gametes as they age and its effect on the prevalence of miscarriages and chromosomal-related diseases. In this respect, the fear of having a child affected by chromosomal abnormalities is not the only motivational factor; the desire to avoid miscarriages is also quite important. At the same time, they consider PGT-A as a way to avoid abortion if the foetus turns up with chromosomal abnormalities at the triple test stage.

> So, you're going there at 42 or so, and you're not going to be able to do a PGT-A, so what are you, aborting every now and then or what? Sure. Or, well, the best thing that can happen to you is to have an abortion, because if you are going to have a baby that is not well, well, I don't know. (MADRID 1)
>
> If what had happened to us had not happened [miscarriages], maybe I would have taken the risk [of not doing PGT-A]. But of course, this in the

> end influences your decision; when it happens to you once, then it's like you are more afraid. (ALICANTE 1)

While fear of congenital anomalies, miscarriages, or voluntary abortion for medical reasons remains the main motivational factor, quite a few patients firmly believe that PGT-A will increase their chances of a successful pregnancy. Some patients, for instance, believed that PGT-A could compensate for the reduction of fertility due to age factors.

> We chose the clinic directly telling us that we needed a PGD, we were uncertain between two telling us exactly the same thing, which is that women are defined both by the quality of eggs and the number of eggs, and that of course after 35 years of age, due to the genetic damage we carry, and due to the number of eggs we produce, [this] is reduced. (MADRID 2)

An economic rationale is also an important factor to some patients, as paying for the test and discarding aneuploid embryos is seen as a saving strategy.

> Three thousand euros for the DGP [PGT-A]. But of course, he told us that each transfer is 1,600 euros, 1,600 times six, that's more than 3000 euros. So he told us, "It is only economically worthwhile for you to do DGP." [...] He said, "Once, you're going to do it anyway, but six times? Are you going to do all six transfers, considering that they might be wrong?" I was very convinced by his argument. (SEVILLA)

However, patients are acutely aware of the financial impact of PGT-A on their reproductive journey. How much are the patients willing to pay for the treatments? When and how do they decide to stop? PGT and other add-ons obviously make each step significantly more expensive.

> Indeed, very expensive. That is to say, it is absolutely disproportionate. Now, I do not know what real costs it covers, whether it is the machine that is expensive or what it is ... but, apart from that, I understand that it is also big business. I don't know the profit margins they have, but certainly for those who choose it, [PGT-A] is a lot of money. (MADRID 1)

The price of PGT-A is indeed high and represents around 40–60% of the costs of a cycle, thus considerably increasing the cost of a normal IVF cycle. There is a great range of prices, from 600 euro to 4,000 euro, depending on the clinic and the plan. Moreover, in some clinics there is a fixed price independent of the number of embryos created, but in others each embryo increases the price (making it more expensive for couples with a lot of embryos). If the couple contracts a plan with a "baby guarantee," where they pay a fixed price (around 20,000 euro) which includes all treatments, PGT-A is usually also included and the decision whether to implement it or not relies on the clinic.

> Each PGT-A is 500 euros. They charge you per single embryo analyzed and then there is a package starting with five or six embryos, I think. Since we had seven (embryos), we chose the package, which I think was 2,500 or 2,700 euros. (ALICANTE 2)
>
> In the first clinic, the cost to analyse the seven embryos was 2,500 euros. The genetic test is 50% of the total cost […] In the second clinic you could pay 2,500 if you had six, they would test up to six embryos. Now, if you had more than six embryos, each extra embryo testing cost 400 euros, which was a huge increase. But there was another option, which was 3,650 for unlimited embryo testing. Then it was up to you to make the calculations. (MADRID 1)

While some of the clinics offer the possibility to test only some of the embryos, making it more affordable, the majority insists on testing all the embryos: "That's exactly what we asked at that moment, if we could do DGP only to one or if it had to be to both embryos. But they told us that it was necessary to analyse all the embryos available in each cycle, that we could not choose one and not the other" (MADRID 3).

b. *Thirsty Even When It Rains: The Desperate Quest for Information*

The patients interviewed are very often not aware of the scientific controversy surrounding PGT-A, because physicians are not accustomed to disclosing it. In general terms, they find that the information provided by the clinics, when it comes to the informed consent, is too technical and difficult to understand:

> They can always get it right by saying, "I've already passed it on to you [the information] in written form." But it is not the same [as when they explain it to you in person]. Now I was asking myself, "If I, with a university education, and so on, could not really understand a lot of things, how is someone less educated going to understand it?" I mean, I don't know. (MADRID 1)
>
> It is the kind of information with [complex] arguments and with very technical terminology. That is, you read it once and you still don't get how they do it. (VALENCIA 1)

As a result, many patients end up searching for additional information on the web, starting from the clinic's webpage, and then moving on into fertility forums, and eventually into social media, such as Facebook, Instagram, or Twitter:

> The Internet and especially social media, where, in the end, you can find several embryologists or gynaecologists who do engage with users and explain everything quite well. Especially on Instagram. (MADRID 3).
>
> There is a very important community with the topic of fertility on Twitter, the hashtag infertilpandy, where women mention their processes, the tests that are available and those which are not, what is important and what is not. It is a supportive community, I actively follow it, I am part of it. I am also part of the National Infertility Network, where we talk about all these issues [...] sometimes, when talking about the subject of PGD, people say, "Well, it doesn't have enough consolidated scientific evidence to be used yet." Others say that it does, or, "That's just a way to extract more money." ... Yes, this debate is going on, isn't it? More on social media than elsewhere. (VITORIA)

However, the abundance of information available on the Internet is difficult to process and assess, and in the long run exposes patients to a burnout process:

Interviewer	What about social networks?
Patient	Yes, [I use them] quite a lot. I was getting a lot of information from many embryologists, endocrinologists, gynaecologists. But I got to the point of considering whether I should stop or

	not. I was offered psychological support by the clinic and they told me, "If you see that it does you more harm than good, at least you can try to limit your time on it." So I started to limit my time on social networks.
Interviewer	Did you participate with comments or forums?
Patient	Yes, I used to ask questions by direct message [Instagram], and read all the messages I received. So, I started my little master's degree in assisted reproduction, I guess like many others. (TERUEL)

In some exceptional cases, patients move further and delve directly into medical literature:

> Well, I try to look at reliable medical sites, right? Not on Facebook or similar things. I like to read a lot on the website called SciELO, Sci, where a lot of medical publications are available. When reading you get to know that there is a possibility of having problems with PGD, but in the end I think the cost-benefit is greater. (SEVILLA)

Patients do not usually come across the actual risks of the testing procedure—such as the risk of not finding any euploid embryos or the survival rate of embryos kept for six or even seven days in a petri dish—in the online forums and websites they use to consult or browse. These risks are, thus, often either downplayed or directly not communicated.

> And you know what happened? Well, in the end, none of them were genetically viable. And then that was a terrible blow. Why? First, because it is frustrating but, apart from that, nobody told me that there could be such a possibility. (MADRID 1)
>
> I really did not know that it could be dangerous for the embryo. (MADRID 4)

In fact, even the meaning of a "euploid embryo" is not always fully understood; some patients think that euploid equals healthy and *therefore more fit or more likely to implant*, ignoring that being euploid is not a predicting factor for future implantation and that euploid embryos may still have a genetic/polygenic alteration or disease.

> Well, when the embryos turned out to be healthy, I gradually transferred all of them but none of them implanted, so they told me, "Well, maybe they were not so healthy after all." And I replied, "Then, what have I done the PGD for? What was the purpose of it?" They answered, "Well, it [PGT-A] does not analyse everything." Well, why didn't they tell me this before? When I was told I had three healthy embryos, I assumed that it was nearly done. (ALICANTE 1)
>
> Well, that's it, a healthy baby at home. And I mean healthy in the sense that, once the PGD is done, it is assumed that the baby they put in you has been analysed and is fine. (ALICANTE 2)

c. *A "Compulsory Option": Ongoing Routinization of PGT-A*

Due to the promises of effectiveness and the new hopes fuelled by the implementation of add-ons, patients eventually give in to a process where they see PGT-A as something for which they are willing to cross their budget red line. "Given that we do not know how many [embryos] we are going to get, we are assailed by doubts: what if I only get one embryo and it is not healthy? In the end, you know, it is a lot of money, and you get the feeling that you are throwing money away, even if in reality you know you are not" (MADRID 4).

Mixing medical, economic, and psychological arguments, clinics, sometimes prompted by patients themselves, have made PGT-A something almost compulsory, especially for older patients with some history of miscarriages or chromosomal abnormalities resulting in previous abortions.

> Not only did they immediately recommend IVF, they also directly recommended preimplantation genetic diagnosis because of my age and in order to analyse the embryos chromosomally. (CASTELLÓ)
>
> In fact, I associate IVF with a mandatory PGT-A genetic analysis … I was never given any option about that. (MADRID 1)
>
> It was advised, well imposed really, because I was already 38. Since there is this drop in reproductive health, they say, "After 38 it is advisable because 50 or 60% …, "I don't know, they tell you this is the percentage of embryos that could be aneuploid, so it is advisable that you do it … I mean you really must do it. (VALENCIA 1)

In this process of medicalization and routinization, PGT-A has become an opportunity for clinics to increase the costs of IVF, and thus, their revenues, exacerbating the economic pressure and the inequality of access to ART. *Per se* this process would be less problematic if accompanied by an exhaustive provision of exact and reliable information to patients, but in Spain this does not often seem to be the case. In fact, some patients seem confused about what exactly is analysed with PGT-A, mixing genetic and chromosomal alterations.

> You need to sign the consent form and so on, and they explained what PGD is about. They said that they analysed the whole DNA sequence and so on. (ALICANTE 2)
>
> Interviewer: Did they tell you what type of PGD they used?
> Interviewee: No.
> Interviewer: Do you know that there are several types?
> Interviewee: No. I don't know that. (MADRID 5)

We also have anecdotal evidence that PGT-A has been implemented in cycles with donor's eggs even if the age of the woman is the main point the clinics make.

> But she [the doctor] told me, "Let's see, your uterus is not is not in a condition to be played with anymore." And she told me just like that, "I would also do PGD on the embryos from ovodonation so we can play it safe, because you can't endure so many abortions." (VALENCIA 2)
>
> Just in case the embryo comes wrong, because you never know these things, PGD is always done. […] Even when the donor is 18 years old, the child could still come with Down syndrome. (MADRID 6)

Confirming the main concerns and warnings of the existing debate on disability justice around genetic testing and screening (Powell et al., 2000), Spanish patients are routinely being denied the possibility to transfer aneuploid embryos, even when, after PGT-A, they wish to do so. While the vast majority of patients interviewed would not consent to such transfer, some were willing to go for it, but were prevented by the clinic:

> We ask, "For example, if it were an embryo with Down syndrome, would it be possible to transfer it?" They replied, "No, we don't transfer it. We discard it." [...] So, my husband and I talked about it, and we said, "Well, I wouldn't mind. You know, I would transfer it. I don't care." But, well, again they told us that it was their protocol and that they would not transfer it. (MADRID 3)

Another relevant outcome detected is that once PGT-A is introduced into the IVF treatment, there is no turning back, as new fears and self-limitations arise. As patients come to believe that only euploid embryos will develop into healthy babies, all the following cycles will include PGT-A. Some couples will even stop trying to conceive in the natural way because of the fear of having miscarriages or having an "unhealthy" baby.

> If I end up going through another in vitro fertilization treatment, I will choose to have another PGD, because it is true that I will be older and there are more possibilities of alterations. I do believe that PGD is necessary, at least, to be on the safe side. (MADRID 1)
>
> After all this [performing PGT-A], we became afraid to get pregnant naturally. We were a little bit scared, like "What if it happens?" We didn't want it to happen. Because sometimes we'd say, let's stop all this and try it [to conceive naturally]. I would talk to my husband and say, "Why don't we just keep trying?" And he would say, "No, because there's no control there, we don't know what might come out." (ALICANTE 1)

d. ***A Languishing Flame: Psychological Pressure, Disappointment, and Hopelessness.***

Once introduced into their journey, PGT-A is perceived by a lot of patients as the only option for keeping alive the languishing flame of hope—or, in other words, often after three or more years of failed attempts, as their last option to achieve the pregnancy for which they have been striving. PGT-A has often been presented to them as the "last resource" for the use and success of their own eggs, before either giving up on ARTs or moving into the egg donation programme. "I didn't want to have a treatment and then say, 'Let's see how it goes, let's see if it goes well and if not, we'll change this or we'll try another technique.' No. I wanted to go with all the options at once, with as much as I could, because I was already old, because I didn't have time" (CASTELLÓ).

They begin their PGT-A journey with enormous psychological pressure and high expectations, which are often misplaced. The outcomes, most of the time, lead to disappointment and hopelessness, especially when they see their last embryos failing to implant. If few or no euploid embryos are found or if the implantations are unsuccessful, the psychological ordeal the patients must go through (anxiety, depression …) is amplified by this false hope of success.

> The truth is that in the first one, perhaps I was a bit naive, I thought, "Well, we have obtained three embryos. Will all three be alright? We are not that old and we can't really have so many genetic alterations." When I realized that we only had one well one, I was very worried and very scared. I said, "There must be something else going on here, why didn't we get all of them fine?" So, in the following DGPs I was very worried. (TERUEL)
>
> Interviewer: If you could describe the feelings you have been going through?
>
> Interviewee: Anxiety, fear, frustration … (MADRID 5)
>
> When they told us that five embryos out of seven were wrong and one was a mosaic embryo, we were very disappointed […] and when we saw that we only had one viable embryo it was very hard. That's why, I imagine, waiting now for the outcomes of the transfer will be worse than for the previous two [cycles]. (ALICANTE 2)

Moreover, in the rare cases when a mosaic embryo is transferred, if it develops into a pregnancy the experience until birth is one filled with fear, anxiety, and feelings of uncertainty about the possible development of the embryo. In turn, this leads prospective parents to continue along the road of genetic testing through prenatal exams such as the triple test stage, amniocentesis, or morphological ultrasound.

> It was horrible. What made me very anxious was the great lack of knowledge about the topic. We were never really allowed to associate this alteration with any specific disease or syndrome. So, in the end, our greatest obsession was that the embryo would not prosper and that it would die

[...] We feared, "It is not going to implant," and, suddenly, it implanted. We thought, "We're not going to make it to week 12," and suddenly we got to week 12. Then we feared the worst: "The amniocentesis is going to go wrong," and eventually even the amniocentesis went right. We finally got past week 20 and we were worried about the morphological ultrasound, but we were told, "Chromosomally it's fine and it's developing." In the end, however, the fear didn't fade out until he was born. (MADRID 1)

Conclusion

Over the past 15 years, reproductive bioeconomies—especially those tied to ARTs—have evolved significantly due to rising parental age, socioeconomic shifts, and advances in genetics. With first-time motherhood increasingly occurring after age 30 in developed countries, the demand for ARTs has surged. This rise has also seen the expansion of genetic add-ons like PGT-A, marketed as improving treatment success. In Spain, where 20% of IVF patients are over 40 (35% with donors' eggs), ART usage is among the highest globally. A permissive legal framework, competitive pricing, and clinical quality have positioned Spain as a leader, performing over half of the EU's PGT and egg donation cycles by 2019. Its 630-million-euro reproductive market offers a key site to explore the impact on care of genetics-focused add-ons.

This chapter examines the growing routinization of these add-ons in Spain, focusing on PGT-A. Often promoted to improve pregnancy outcomes, it is commonly recommended for couples facing recurrent miscarriage or infertility; however, its clinical benefits remain contested, with limited and inconclusive evidence of its effectiveness. Despite this, PGT-A's use has risen, propelled by commercial incentives and marketing that cultivates hope and unrealistic expectations. Yet, the high cost of PGT-A deepens inequality in ART access, particularly in countries like Spain where it is not publicly funded. Patients frequently lack a full understanding of its value, making informed consent problematic. Many patients struggle with the complexity of medical information or receive insufficient explanations.

PGT-A also contributes to emotional distress—feelings of regret, uncertainty, and responsibility are common. Often seen as a last resort, it reinforces hope but also prolongs emotional and financial strain. These dynamics highlight the need for stronger regulation, better counselling, and clearer evidence to ensure its equitable and ethical use in reproductive care.

Drawing on interviews with ART patients, we have examined how Spanish medical, legal, and socio-economic systems have made it possible for PGT-A to be routinized, becoming a "mundane" technology: part of the everyday life of IVF patients and families in Spain. We have paid special attention not only to patient perspectives and experiences but also to the implications of PGT-A routinization for their reproductive journeys. Mundane as it is, PGT-A gives rise to some general trends which in many ways confirm the concerns debated in the literature. In Spain, many patients encounter PGT-A after transitioning from the public to the private healthcare system, typically following failed inseminations, IVF cycles, or miscarriages. Aware of age-related declines in fertility, they turn to PGT-A as a possible solution to improve success rates and avoid the emotional and physical toll of repeated failures. Although some are already familiar with the test through online sources, it is within private clinics that they fully engage with it, often drawn by concerns over chromosomal abnormalities like Down syndrome—concerns sometimes amplified by clinic messaging. For many, PGT-A becomes a last resort, perceived as increasing the genetic viability and implantation chances of embryos.

PGT-A is increasingly routine in private clinics, especially for older patients or those with miscarriage histories. It is sometimes offered even in donor egg cycles, where chromosomal risks are lower. Patients often feel it is a necessary step, with clinics presenting it as both a medical safeguard and a cost-saving investment. Yet, the financial burden is significant—ranging from 600 to 4,000 euros, with package deals reaching 20,000 euros—raising concerns about access and exacerbating inequalities in ART.

This routinization has led to new restrictions, such as clinics refusing to transfer embryos classified as aneuploid, materializing the worst concerns voiced by the literature on disability justice (Invalid, 2017; Kelly,

2009). Many patients, once introduced to PGT-A, feel compelled to continue its use in future cycles or avoid natural conception altogether. Although they seek clarity online, the information is often unreliable, adding to their stress. Some turn to psychological support as the emotional and financial toll mounts. Despite their efforts, most remain unaware of ongoing scientific debates—such as the limited predictive value of PGT-A or the potential of aneuploid embryos to develop into healthy blastocysts (Lee & Kiessling, 2017; Reshef et al., 2022). Many mistakenly believe that euploid embryos guarantee pregnancy, leading to confusion and disappointment when outcomes fall short.

Given these outcomes, a closer look at the social, regulatory, and economic ecosystem of PGT-A in Spain thus raises three main implications for policy-making. First of all, in a context where information is scattered, often unreliable, or overly technical, informed consent alone is not sufficient to regulate access to PGT-A. This growing reliance on PGT-A underscores the need for transparent, comprehensive information to help patients make informed decisions through a process that is not limited to reading documents and signing papers.

Second, the current divide between the public system, where PGT-A is almost never implemented, and the private system where it has become routine, is producing a very problematic situation. Patients who could benefit from this technique spend years on unsuccessful attempts in the public system before getting to access PGT-A in the private system, often too late and at a cost that they cannot afford, economically and psychologically. Conversely, in the private sector patients that could achieve a pregnancy without implementing genetic testing of embryos are routinely recommended PGT-A, leading to higher economic pressure and the risk of losing embryos.

As a result, and this is the third main issue, the National Commission for Assisted Human Reproduction (CNRHA) should make clear recommendations about the specific cases in which PGT-A is considered appropriate and, for those cases, such as patients of older age with a good prognosis or repeated miscarriages, its implementation in the public healthcare system should be taken into consideration instead of leaving it subjected completely to market dynamics. Nevertheless, randomized control trials should support this decision. The unfettered routinization of PGT-A in private healthcare is fuelling the social stratification dynamics already at work when it comes to assisted reproduction treatments. Most importantly, the marketing strategy used by the clinics presenting PGT-A as a last resource to overcome age-related limitations in reproductive processes is scientifically controversial and is capitalizing on the hope of thousands of patients every year, putting them under enormous economic and psychological pressure. This is, in turn, giving rise to widely experienced emotional and relational distress that only a few patients can handle through out-of-pocket psychological support.

Acknowledgments This work is the product of the IfGene research project, "Projecting the future of genetic prediction" (PID2020-115899GB-I00, PIs: Mauro Turrini and Rubén Blanco), funded by the Spanish Ministry of Science and Innovation (MICIU) through the Agencia Estatal de Investigación (AEI) (MICIU/AEI/10.13039/501100011033), with the support of the European Regional Development Fund (ERDF).

References

Armstrong, S. C., Vaughan, E., Lensen, S., Caughey, L., Farquhar, C. M., Pacey, A., Balen, A. H., Peate, M., & Wainwright, E. (2023). Patient and professional perspectives about using in vitro fertilisation add-ons in the UK and Australia: A qualitative study. *BMJ Open, 13*(7), e069146. https://doi.org/10.1136/bmjopen-2022-069146

Boe Hüttel, A., Haustedt Mossman, F., & Turrini, M. (2024). Drømmen om at blive mor. En antropologisk analyse af markedsføringen af genetisk testning af embryoner i Spanien. *Tidsskriftet Antropologi, 2024*, 90. https://tidsskrift.dk/tidsskriftetantropologi/article/view/153012

Bracewell-Milnes, T., Saso, S., Jones, B., Cato, S., Parikh, R., Thum, M.-Y., Johnson, M., Almeida, P., Norman-Taylor, J., & Nikolaou, D. (2021). A systematic review exploring the patient decision-making factors and attitudes towards pre-implantation genetic testing for aneuploidy and gender selection. *Acta Obstetricia et Gynecologica Scandinavica, 100*(1), 17–29. https://doi.org/10.1111/aogs.13973

Braga, D. P. d. A. F., Setti, A. S., & Borges, E., Jr. (2022). Ethics and IVF add-ons: We need to talk about it. *JBRA Assisted Reproduction, 26*(3), 371–373. https://doi.org/10.5935/1518-0557.20220030

Chambers, G. M., Hoang, V. P., Sullivan, E. A., Chapman, M. G., Ishihara, O., Zegers Hochschild, F., Nygren, K. G., & Adamson, G. D. (2014). The impact of consumer affordability on access to assisted reproductive technologies and embryo transfer practices: An international analysis. *Fertility and Sterility, 101*(*1*), 191–198.e4. https://doi.org/10.1016/j.fertnstert.2013.09.005

Cheng, L., Meiser, B., Kirk, E., Kennedy, D., Barlow-Stewart, K., & Kaur, R. (2022). Decisional needs of patients considering preimplantation genetic testing: A systematic review. *Reproductive Biomedicine Online, 44*(5), 839–852. https://doi.org/10.1016/j.rbmo.2021.12.011

Cirkovic, S., Wilkinson, J., Lensen, S., Jackson, E., Harper, J., Lindemann, K., & Costa-Font, J. (2023). Is the use of IVF add-on treatments driven by patients or clinics? Findings from a UK patient survey. *Human Fertility, 26*(2), 365–372. https://doi.org/10.1080/14647273.2023.2197628

Cornelisse, S., Zagers, M., Kostova, E., Fleischer, K., van Wely, M., & Mastenbroek, S. (2020). Preimplantation genetic testing for aneuploidies (abnormal number of chromosomes) in in vitro fertilisation. *Cochrane Database of Systematic Reviews*, (9). https://doi.org/10.1002/14651858.CD005291.pub3

DBK Informa. (2024). Informe Especial "Centros de Reproducción Asistida" 2023. *DBK Informa Observatorio sectorial.* https://www.dbk.es/es/detalle-nota/centros-reproduccion-asistida-2024

Eijkemans, M. J. C., van Poppel, F., Habbema, D. F., Smith, K. R., Leridon, H., & te Velde, E. R. (2014). Too old to have children? Lessons from natural fertility populations. *Human Reproduction, 29*(6), 1304–1312. https://doi.org/10.1093/humrep/deu056

ESHRE. (2024). *IVF and IUI treatment cycles increase across Europe, along with stable pregnancy rates.* European Society of Human Reproduction and Embryology. https://www.eshre.eu/ESHRE2024/Media/2024-Press-releases/EIM

ESHRE Add-ons working group, Lundin, K., Bentzen, J. G., Bozdag, G., Ebner, T., Harper, J., Le Clef, N., Moffett, A., Norcross, S., Polyzos, N. P., Rautakallio-Hokkanen, S., Sfontouris, I., Sermon, K., Vermeulen, N., & Pinborg, A. (2023). Good practice recommendations on add-ons in reproductive medicine†. *Human Reproduction, 38*(11), 2062–2104. https://doi.org/10.1093/humrep/dead184

Fernandez, E. I., Ferreira, A. S., Cecílio, M. H. M., et al. (2020). Artificial intelligence in the IVF laboratory: Overview through the application of different types of algorithms for the classification of reproductive data. *Journal of Assisted Reproduction and Genetics, 37*, 2359–2376. https://doi.org/10.1007/s10815-020-01881-9

Franasiak, J. M., Scott, K. L., & Scott, R. T. (2022). Preimplantation genetic testing. In T. Falcone & W. W. Hurd (Eds.), *Clinical reproductive medicine and surgery: A practical guide* (pp. 409–427). Springer International Publishing. https://doi.org/10.1007/978-3-030-99596-6_19

García, V. (2022). Informe estadístico de Técnicas de Reproducción Asistida 2022. *Sociedad Española de Fertilidad*. https://www.registrosef.com/public/docs/sef2022_IAFIV.pdf

Geampana, A., & Perrotta, M. (2023). Accounting for complexity in healthcare innovation debates: Professional views on the use of new IVF treatments. *Health, 27*(6), 907–923. https://doi.org/10.1177/13634593221074874

Gebhart, M. B., Hines, R. S., Penman, A., & Holland, A. C. (2016). How do patient perceived determinants influence the decision-making process to accept or decline preimplantation genetic screening? *Fertility and Sterility, 105*(1), 188–193. https://pubmed.ncbi.nlm.nih.gov/26474735/

Glatthorn, H. N., & Decherney, A. (2022). The efficacy of add-ons: Selected IVF "add-on" procedures and future directions. *Journal of Assisted Reproduction and Genetics, 39*, 581–589. https://doi.org/10.1007/s10815-022-02410-6

Gleicher, N., Albertini, D. F., Barad, D. H., Homer, H. D., Modi, M., Murtinger, D., Patrizio, P., Orvieto, R., Takahashi, S., Weghofer, A., Ziebe, S., & Noyes, N. (2020). The 2019 PGDIS position statement on transfer of mosaic embryos within a context of new information on PGT-A. *Reproductive Biology and Endocrinology, 18*(1), 1–8. https://doi.org/10.1186/s12958-020-00616-w

Gualtieri, M. R., Ory, S., Bazzi, M., Alburkat, Z., & Ali, B. (2023). The promise and challenges of preimplantation genetic testing for IVF. *Contemporary OB/GYN Journal, 68*(2) https://www.contemporaryobgyn.net/view/the-promise-and-challenges-of-preimplantation-genetic-testing-for-ivf

Harper, J., Jackson, E., Sermon, K., Aitken, R. J., Harbottle, S., Mocanu, E., Hardarson, T., Mathur, R., Viville, S., Vail, A., & Lundin, K. (2017). Adjuncts in the IVF laboratory: Where is the evidence for 'add-on' interventions? *Human Reproduction, 32*(3), 485–491. https://doi.org/10.1093/humrep/dex004

Harris, B. S., Acharya, K. S., Unnithan, S., Neal, S. A., Mebane, S., Truong, T., & Muasher, S. J. (2025). Success rates with preimplantation genetic testing for aneuploidy in good prognosis patients are dependent on age. *Fertility and Sterility, 123*(3), 428–438. https://doi.org/10.1016/j.fertnstert.2024.09.043

Hernandez-Nieto, C., Lee, J., Slifkin, R., Duke, M., Luna, M., Copperman, A. B., Sandler, B., & Flisser, E. (2018). Slow growing embryos biopsied on day 7: What are realistic expectations? *Fertility and Sterility, 110*(4), e72. https://doi.org/10.1016/j.fertnstert.2018.07.218

INE. (2023). *Indicadores demográficos básicos. Natalidad y fecundidad.* Instituto Nacional de Estadística. https://www.ine.es/jaxiT3/Datos.htm?t=1579

Insogna, I. G., Lanes, A., Ginsburg, E. S., & Racowsky, C. (2021). Quality of embryos on day 7 after medium refreshment on day 6: A prospective trial. *Human Reproduction, 36*(5), 1253–1259. https://doi.org/10.1093/humrep/deab038

Invalid, S. (2017). *Skin, tooth, and bone—The basis of movement is our people: A disability justice primer.* https://doi.org/10.1080/09688080.2017.1335999

Kelly, S. E. (2009). Choosing not to choose: Reproductive responses of parents of children with genetic conditions or impairments. *Sociology of Health & Illness, 31*(1), 81–97. https://doi.org/10.1111/j.1467-9566.2008.01110.x

Lamb, B., Johnson, E., Francis, L., Fagan, M., Riches, N., Canada, I., Wilson, A., Mathiesen, A., Sabatello, M., Gurtcheff, S., Johnstone, E., & Rothwell, E. (2018). Pre-implantation genetic testing: Decisional factors to accept or decline among in vitro fertilization patients. *Journal of Assisted Reproduction and Genetics, 35*(9), 1605–1612. https://doi.org/10.1007/s10815-018-1278-2

Lee, A., & Kiessling, A. A. (2017). Early human embryos are naturally aneuploid—Can that be corrected? *Journal of Assisted Reproduction and Genetics, 34*(1), 15–21. https://doi.org/10.1007/s10815-016-0845-7

Lensen, S., Hammarberg, K., Polyakov, A., Wilkinson, J., Whyte, S., Peate, M., & Hickey, M. (2021). How common is add-on use and how do patients decide whether to use them? A national survey of IVF patients. *Human Reproduction, 36*(7), 1854–1861. https://pubmed.ncbi.nlm.nih.gov/33942073/

Marshall, C., & Rossman, G. B. (2011). Managing, Analyzing, and interpreting Data. In Marshall, C., & Rossman, G. B., Eds., *Designing Qualitative Research* (pp. 205–227). Sage Publications.

Merck Group. (2022). *Las españolas tienen la mitad de los hijos que desean y son madres después de los 30 años, más tarde de lo que les gustaría.* https://www.merckgroup.com/es-es/pr/2022/Encuesta%20Fertifilidad%202022.pdf

Murphy, L. A., Seidler, E. A., Vaughan, D. A., Resetkova, N., Penzias, A. S., Toth, T. L., Thornton, K. L., & Sakkas, D. (2019). To test or not to test? A framework for counselling patients on preimplantation genetic testing for aneuploidy (PGT-A). *Human Reproduction, 34*(2), 268–275. https://doi.org/10.1093/humrep/dey346

Paulson, R. J. (2020). Hidden in plain sight: The overstated benefits and underestimated losses of potential implantations associated with advertised PGT-A success rates. *Human Reproduction, 35*(3), 490–493. https://doi.org/10.1093/humrep/dez280

Pavone, V., & Arias, F. (2012). Beyond the geneticization thesis: The political economy of PGD/PGS in Spain. *Science, Technology & Human Values, 37*(3), 235–261. https://doi.org/10.1177/0162243911411195

Pavone, V., & Lafuente Funes, S. (2018). Selecting what? Pre-implantation genetic diagnosis and screening trajectories in Spain. In A. Wahlberg & T. M. Gammeltoft (Eds.), *Selective reproduction in the 21st century* (pp. 123–148). Springer International Publishing. https://doi.org/10.1007/978-3-319-58220-7_6

Perrotta, M., & Hamper, J. (2021). The crafting of hope: Contextualising add-ons in the treatment trajectories of IVF patients. *Social Science & Medicine, 287*, 114317. https://doi.org/10.1016/j.socscimed.2021.114317

Perrotta, M., & Geampana, A. (2020). The trouble with IVF and randomised control trials: Professional legitimation narratives on time-lapse imaging and evidence-informed care. *Social Science & Medicine, 258*, 113115. https://doi.org/10.1016/j.socscimed.2020.113115

Perrotta, M., & Geampana, A. (2021). Enacting evidence-based medicine in fertility care: Tensions between commercialisation and knowledge standardisation. *Sociology of Health & Illness, 43*(9), 2015–2030. https://doi.org/10.1111/1467-9566.13381

Perrotta, M., & Hamper, J. (2023). Patient informed choice in the age of evidence-based medicine: IVF patients' approaches to biomedical evidence and fertility treatment add-ons. *Sociology of Health & Illness, 45*(2), 225–241. https://doi.org/10.1111/1467-9566.13581

Powell, C. M., Kent, D., Baily, M. A., Ferguson, P. M., Gartner, A., Lipsky, D. K., et al. (2000). *Prenatal testing and disability rights.* Georgetown University Press. https://books.google.it/books?id=RnDA0EKcMaEC&dq=Parens,+E.,+%26+Asch,+A.+(2000).+Prenatal+Testing+and+Disability+Rights.+Washington,+DC:+Georgetown+University+Press.&lr=&hl=es&source=gbs_navlinks_s

Quinn, M. M., Juarez-Hernandez, F., Dunn, M., Okamura, R. J., Cedars, M. I., & Rosen, M. P. (2018). Decision-making surrounding the use of preimplantation genetic testing for aneuploidy reveals misunderstanding regarding its benefit. *Journal of Assisted Reproduction and Genetics, 35*(12), 2155–2159. https://doi.org/10.1007/s10815-018-1337-8

Reshef, E. A., Robles, A., Hynes, J. S., Turocy, J. M., & Forman, E. J. (2022). A review of factors influencing the implantation of euploid blastocysts after in vitro fertilization. *Fertility & Sterility Reviews, 3*(2), 105–120. https://doi.org/10.1016/j.xfnr.2022.03.001

Roberts, S. A., Wilkinson, J., Vail, A., & Brison, D. R. (2022). Does PGT-A improve assisted reproduction treatment success rates: What can the UK Register data tell us? *Journal of Assisted Reproduction and Genetics, 39*(11), 2547–2554. https://doi.org/10.1007/s10815-022-02612-y

Rothwell, E., Lamb, B., Johnson, E., Gurtcheff, S., Riches, N., Fagan, M., Sabatello, M., & Johnstone, E. (2020). Patient perspectives and experiences with in vitro fertilization and genetic testing options. *Therapeutic Advances in Reproductive Health, 14*, 2633494119899942. https://doi.org/10.1177/2633494119899942

Rubio, C., Bellver, J., Rodrigo, L., Bosch, E., Mercader, A., Vidal, C., los Santos, M. J. D., Giles, J., Labarta, E., Domingo, J., Crespo, J., Remohí, J., Pellicer, A., & Simón, C. (2013). Preimplantation genetic screening using fluorescence in situ hybridization in patients with repetitive implantation failure and advanced maternal age: Two randomized trials. *Fertility and Sterility, 99*(5), 1400–1407. https://doi.org/10.1016/j.fertnstert.2012.11.041

Silverman, D. (2015). Interpreting Qualitative Data: A Guide to the Principles of Qualitative Research (5th ed.). Sage.

Simopoulou, M., Sfakianoudis, K., Maziotis, E., Tsioulou, P., Grigoriadis, S., Rapani, A., Giannelou, P., Asimakopoulou, M., Kokkali, G., Pantou, A., Nikolettos, K., Vlahos, N., & Pantos, K. (2021). PGT-A: Who and when? A systematic review and network meta-analysis of RCTs. *Journal of Assisted Reproduction and Genetics, 38*, 1939–1957. https://doi.org/10.1007/s10815-021-02227-9

Smeenk, J., Wyns, C., De Geyter, C., Kupka, M., Bergh, C., Cuevas Saiz, I., De Neubourg, D., Rezabek, K., Tandler-Schneider, A., Rugescu, I., & Goossens, V. (2023). ART in Europe, 2019: Results generated from European registries by ESHRE†. *Human Reproduction, 38*(12), 2321–2338. https://doi.org/10.1093/humrep/dead197

Sociedad Española de Fertilidad. (2022). Informe estadístico de Técnicas de Reproducción Asistida 2022. *Registro SEF.* https://www.registrosef.com/public/docs/sef2022_IAFIV.pdf

Sociedad Española de Fertilidad. (2024). 39.546 bebés nacieron tras las técnicas de reproducción llevadas a cabo en España en 2022. *Registro Nacional de actividad 2022 – Registro SEF.* https://www.sefertilidad.net/documentos/QrOPUTXSABq8_Y3Sf1X55jeBA9MZUk_KBKRIMfq5TH8.pdf

Swoboda, D. (2015). Frames of reference: Marketing the practice and ethics of PGD on fertility clinic websites. *Advances in Medical Sociology, 16*, 217–247. https://doi.org/10.1108/S1057-629020150000016008

Turrini, M., & Otín-Gavín, J. (2025). Let 'Pregnant Women Choose the Destiny for Themselves and Their Child'. How fertility clinic digital platforms frame preimplantation genetic testing (PGT) in Spain. *Sociology of Health & Illness, 47*(1), 1–12. https://doi.org/10.1111/1467-9566.13876

Vlasenko, P., & Pavone, V. (2025). Regulations and altruism as valuation mechanisms: A political economy of ova markets in Ukraine and Spain. *Science, Technology & Human Values*. forthcoming.

Wyns, C., De Geyter, C., Calhaz-Jorge, C., Kupka, M. S., Motrenko, T., Smeenk, J., Bergh, C., Tandler-Schneider, A., Rugescu, I. A., & Goossens, V. (2022). ART in Europe, 2018: Results generated from European registries by ESHRE†. *Human Reproduction Open, 2022*(3), hoac022. https://doi.org/10.1093/hropen/hoac022

Zhang, J., Rubin, L. R., Zierhut, H., & Pastore, L. M. (2021). Comparison of patients' ethical perspectives of preimplantation embryo genetic testing for aneuploidy (PGT-A) vs. monogenic disorders (PGT-M). *Reproductive Sciences, 28*(11), 3272–3281. https://doi.org/10.1007/s43032-021-00644-w

3

Driving a Ferrari on a Country Road, or When the Technique Becomes a Gatekeeper: The Case of Preimplantation Genetic Testing in France

Anne-Sophie Giraud

Since the sequencing of the entire human genome in 2003[1] and the development of increasingly efficient sequencers, many doctors and scientists have continually warned of the Pandora's box such analyses could open—or have already opened—whether these be medical or more "social" concerns: the inability to interpret genome variations or disease-predisposing mutations, or the revelation of infertility fraud (Saruwatari Zavala, 2023) and family secrets. Professionals often find themselves facing potentially sensitive information. Furthermore, technical advances—such as the advent of very high-throughput sequencing, called next-generation sequencing (NGS)—and the falling cost of genetic

[1] With the Human Genome Project, launched in 1988 and completed in 2003. The goal was to complete the full sequencing of human genome DNA.

A.-S. Giraud (✉)
LISST-Centre d'anthropologie sociale, CNRS, Toulouse, France
e-mail: anne-sophie.giraud@univ-tlse2.fr

V. Argudo-Portal et al. (eds.), *Mundane Genomics*,
https://doi.org/10.1007/978-981-95-4550-6_3

testing have made it more accessible, both to the general public with the development of so-called DTC (direct-to-consumer) tests, and in research and clinical practice.[2] As emphasised in the introduction to this book, genetic testing is becoming routinised in various medical and non-medical contexts, including the field of assisted reproductive technologies (ART) through the use of preimplantation genetic testing (PGT): a genetic diagnosis carried out on cells from embryos conceived through[3] *in vitro fertilisation* (IVF). It allows the birth of a child free of the targeted genetic disorders by avoiding the transfer of affected embryos into the uterus, and holds a unique place in the media and cultural landscape in France and in most European countries (Franklin & Roberts, 2006). PGT (-M and -SR[4]) was authorised in France from the very first bioethics laws governing all assisted productive techniques in 1994. However, it is only permitted on an "exceptional" basis, in the case of a specific severe disease (such as Huntington disease or cystic fibrosis) that has been previously identified in the applicants or their relatives. PGT-A (for aneuploidy), formerly known as preimplantation genetic screening (PGS), is strictly prohibited. It consists of the analysis of the entire karyotype of an embryo to detect possible chromosomal abnormalities, extra or missing chromosomes, or to identify structural chromosomal abnormalities caused by duplications or insertions of a chromosomal segment, without being previously known.

ART in general continue to constitute an "experiential system" (Knecht et al., 2012) in which what is right and wrong remains highly uncertain and contested. This is even more true for PGT as it is a selective reproductive technology (SRT) (Wahlberg & Gammeltoft, 2018), a technology "used to prevent or allow the birth of certain kinds of children" (p. 1) which can be viewed through the lens of the eugenic horrors of the Nazi past (Kerr, 2004). Many studies suggest that SRTs are merely a more "liberal", "gentle", or "voluntary" version of past eugenic policies (Taussig et al., 2005), and that they carry implicit value judgements about

[2] Source, website of the National Human Genome Research Institute: https://www.genome.gov/about-genomics/fact-sheets/Sequencing-Human-Genome-cost [link available the 12/04/2024].

[3] The procedure was former known as preimplantation genetic diagnosis (PGD).

[4] PGT-M for Monogenetic disorder and PGT-SR for structural rearrangement.

disabled people, particularly the notion that their lives are less worth living (Rapp, 1999; Shakespeare, 1998). Thus, PGT illustrates the ambivalence inherent in the so-called "mundanity" of genetics: even when implemented—and in some cases routinised—genetic testing is never used completely without resistance and fear.

PGT has developed in this context, first emerging as an experimental procedure in the 1990s in England using PCR (polymerase chain reaction) and Sanger sequencing (Handyside et al., 1990). Initially, it was used to detect X-linked and monogenic diseases. Only a few years later, the FISH (fluorescence in situ hybridisation) technique was introduced (Munné et al., 1995), and became the standard method for detecting chromosomal abnormalities (structural and numerical). In recent years, these methods have been largely replaced internationally by other sequencing techniques with the capacity to sequence whole genomes faster and at a lower cost: first CGH-Array,[5] then NGS.

This chapter originates in ethnographic observation conducted at a PGT centre in France. The medical field is generally seen as driven by a technological imperative to use the most complex (Spiegel, 1981; Timmermans, 2010), and above all, the most efficient technologies. Yet, in an era where fast whole genome sequencing is becoming dominant, during my research in 2020–2021, and still in 2025, French PGT genetics laboratories were using "older" technologies—FISH and Sanger sequencing—which professionals themselves labelled "archaic". As I demonstrate, this can be partly explained by the French legal framework, which acts as a "social lock" (Cresswell, 1996, p. 80), defined as ideological or political constraints hindering the evolution of a technical process that would otherwise develop solely according to technical principles. The techniques used allow professionals to work within the authorised framework of PGT in France; given their technical limitations, these technologies embody and support compliance with legal restrictions. They also help professionals to avoid morally uncomfortable situations and the burden of sensitive knowledge that may arise from the use of the latest sequencing technologies, thereby helping to prevent the "moral

[5] For Array comparative genomic hybridization. This is a chip-based cytogenetic technique that allows analysis of copy number variations in DNA.

tensions" (Heinsen, 2022) that can arise when using NGS. Indeed, whole-genome sequencing exposes professionals to burdensome knowledge, which they may not know how to handle and which could even lead them to break the law.

This chapter explores how French professionals navigate these ethical and legal frameworks, and how these frameworks shape the techniques used. It also tells a story of resistance and ambivalence towards the rapid development of sequencing technologies and the tendency to turn genetic testing into a routine procedure. It reflects the moral labour required of professionals as they adapt new technologies to existing legal frameworks. But moral concerns alone cannot explain this resistance, which is entangled with organisational and economic trade-offs. Many studies of ARTs and prenatal diagnostics have shown how professionals, within a specific legal context, navigate the legislation in their relationships with patients (Ehrich & Williams, 2010; Heinsen, 2023; Ivry & Teman, 2019; Rapp, 1999), especially how they frame information. In this work, I shift the focus from the professional-patient relationship to how professionals engage with the law through their technical practices, and how their technical choices help them maintain compliance.

Fieldwork and Methodology

This research was authorised by the INSERM[6] Ethics Evaluation Committee (CEEI/IRB) (IRB00003888, IORG0003254, FWA00005831) in February 2020 (opinion number 20-658, amended in 2022). It has been conducted in a French hospital.

The work draws from ethnographic observation and semi-structured interviews. Observation lasted for six months, from November 2020 to April 2021, in the three different departments forming the PGT centre: (1) the assisted reproductive medicine centre, including the reproductive biology laboratory; (2) the PGT genetics department; and (3) the clinical genetics department. In addition to observing numerous consultations with a range of professionals, I attended various meetings, including

[6] National Institute of Health and Medical Research.

those of the multidisciplinary PGT commission, the multidisciplinary prenatal diagnosis (PND) centre and commission, the clinical staff, and others. In total, 29 semi-structured interviews were conducted with professionals, including 12 in the assisted reproductive medicine team, 16 in genetics (PGT, PND, and clinical genetics), and one with a researcher working on the human embryo. I also conducted interviews with 13 couples, who were offered follow-up interviews. However, in this chapter, I primarily draw from the observation conducted in the PGT centre and the professionals' interviews.

I subsequently made several research visits to the same centre in 2024 and 2025 which gave me the opportunity to observe the evolution of practices. In 2024, I undertook further interviews (formal and informal) with the same PGT professionals. This new period of fieldwork allowed me to observe a profound transformation in clinical genetics diagnostic practices, particularly the increasing prescription of whole genome sequencing and the development of "genomic medicine". Several national plans aim to establish a strong national framework in the genetic field, improving patient health outcomes while positioning France competitively on the international stage. In addition, in 2025, I carried out an interview-based study for an ethics working group on PGT within the French Association of Human Genetics with other colleagues.[7] This ongoing research focuses on the introduction of next-generation sequencing (NGS) in PGT and PGT-A.

PGT: A Complex Organisation with Multiple Partners

The PGT Process

PGT combines IVF techniques with clinical genetics. After the co-fertilisation of oocyte and sperm, reproductive professionals (lab technicians, biologists and engineers, the biological team) biopsy the embryos. A few cells are removed under a microscope and sent to the PGT genetics

[7] These interviews were conducted mostly in pairs, with health professionals.

laboratory for analysis. The genetics team passes on the results of the genetic analysis to the reproductive biology laboratory, which is responsible for embryo selection and transfer. Only the "unaffected" embryos with good morphokinetic[8] characteristics (providing information about embryo quality, that is, their capacity to continue development) are selected for transfer into the uterus, thus enabling a pregnancy to begin with a foetus that is not affected by the targeted pathology.

The PGT process answers, then, to a "double imperative" (Ehrich & Williams, 2010), a two-stage embryo selection process that one biologist called a "double penalty". First, the embryos need to be evaluated by reproductive professionals as *viable*, sufficiently viable to be biopsied[9] and then to be transferred. This first selection is made using morphokinetic criteria, which provide an indication of an embryo's potential to develop, implant, and ultimately result in a live birth, even though the correlation between morphokinetics and viability is not guaranteed. This selection process is practiced daily in conventional IVF-ICSI procedures (Giraud, 2015; Helosvuori, 2019). But in PGT, embryos must also be diagnosed as unaffected or "healthy" by genetics professionals (Ehrich & Williams, 2010). Once the genetic results have been returned, the biologists, engineers, and technicians will discard all the affected embryos. Then they will make a new selection among the "healthy" embryos in order to transfer and freeze only the *viable* and *transferable* embryos, that is, those with morphokinetic characteristics deemed compatible with pregnancy.

Two Genetics Teams

PGT is, therefore, a complex procedure that requires the coordination and cooperation of three distinct teams: clinical, biological, and genetic (see Fig. 3.1). The clinical department deals with hormonal treatments, as well as the monitoring of women (gynaecology) and men (andrology).

[8] Morphokinetics combines the criteria of embryo appearance (morphology) with the knowledge of when and how the cellular processus which lead to that appearance occur (kinetic).

[9] Only embryos with good morphology are biopsied, not only because it is pointless to biopsy embryos that are not considered viable, but also because these embryos would not survive the biopsy.

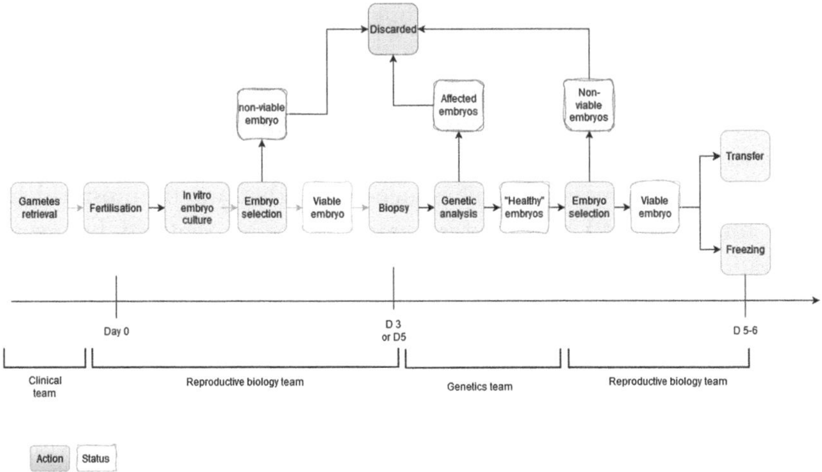

Fig. 3.1 PGT procedure

The reproductive biology department is responsible for all technical actions on gametes and embryos, from gamete processing to embryonic transfer and freezing. The genetics department carries out genetic analyses on embryonic cells. This department is itself composed of two teams, molecular biology and cytogenetics.

Molecular biology focuses on autosomal diseases, whether dominant or recessive. This process is called PGT-M, for monogenic hereditary disorder (in French, more commonly referred to as "DPI moléculaire"). Its scope is on a very micro level, analysing gene structures at the molecular level, specifically nucleic acids. For perspective, each chromosome contains several to thousands of genes. In the observed PGT genetics laboratory, the technique used is PCR, followed by the Sanger sequencing method. PCR, or polymerase chain reaction, aims to amplify a DNA sequence to obtain substantial amounts of DNA from a limited source—here, a single embryonic cell. This method appeared in the early 1990s, transforming molecular biology and becoming widely adopted. Some professionals, such as molecular technician David Dubois,[10] describe it as a "craft" technique: "These are old techniques that can seem a bit old

[10] All names have been anonymised.

school. PCR goes back 35–40 years, but it has evolved quite a bit". Before amplification, a reagent is added to the embryonic cell to specifically amplify the target gene(s) containing the mutation sought in PGT. Once amplified, DNA sequences are placed in a capillary sequencer, which sequences the DNA, allowing identification of chromosome copy numbers and specific mutations. This sequencing method was the most widely used for almost 40 years before gradually being replaced by next-generation sequencers.

Cytogenetics, on the other hand, works on a macro scale, focusing on chromosomes. This is referred to as PGT-SR, for structural chromosomal abnormalities (or "DPI-cytogénétique" in French). The technique used in the observed laboratory is FISH. The probes which are used are small DNA sequences with known locations on a chromosome, tagged with a fluorescent molecule detectable under a fluorescence microscope. These probes are applied to the embryonic cell, where they will hybridise, or bind, to complementary DNA in the cell. Once hybridised, the engineer or technician uses a fluorescence microscope to count the fluorescent dots within the embryonic cell, each indicating the location of marked chromosomes. This technology can detect numerical abnormalities (such as an extra or missing chromosome)—prohibited in France—or chromosomal rearrangements, also known as structural abnormalities. A decade ago, FISH was the gold standard for detecting chromosomal anomalies. But with the introduction of newer and more advanced techniques, it has gradually been replaced by genome-wide analysis methods (Committee et al., 2020); however, in France and in other countries, this method remains the primary choice in PGT genetics laboratories.

A Strict Legal Regulation

PGT is, then, a complex organisation, involving the coordination of medical and scientific knowledge, skills, techniques, and decisions, imbued with ethical, social, and cultural values and beliefs (Ehrich & Williams, 2010, p. 25): that is, *situated* (Haraway, 1988). In PGT centres, professionals have to deal simultaneously with political, social, technical, personal, and collective constraints. This study takes place in

France, where the state's responsibility in procreation is central, and where bioethics laws are underpinned by medical and traditional family norms.

Since PGT was authorised in France in 1994 with the first bioethics laws (although the practice only officially began in 1999 with the publication of implementing decrees), demand has grown steadily, from 742 requests in 2014 to 1016 in 2020. Since then, technical advances have made genetic diagnosis more accessible and efficient, broadening its possible applications: of the 366 genetic disorders covered by PGT in 2020, 19 were new.[11] However, the development of PGT in Europe has been much more gradual and limited (Franklin & Roberts, 2006, p. 95) than that of IVF, which spread rapidly from the late 1970s onwards because it was presented as a mere "aid" to nature (Franklin, 1997). Thus, 20,370 children were conceived using ARTs, but only 242 children were born after PGT in 2020 (including donor-assisted conception, ABM, 2020). Only five centres are authorised to perform PGT in the French metropolitan area (with no overseas centres). Significant also in terms of this desire to limit its use and unlike other European countries such as Spain or England, PGT in France is only performed in public centres. It is considered too sensitive an activity to be subject to financial logic (although the presence of pharmaceutical companies makes the situation more complex).[12]

This "slow progress" (Franklin & Roberts, 2006, p. 100) is partly due to the technical complexity of PGT, the potential for misinterpretation, even if now minimal (Harton et al., 2011, p. 6), low pregnancy rates (Cram & de Kreisler, 2002), and the significant cost and investment required, both in terms of highly qualified staff and specialised equipment housed in a single centre (Lavery et al., 1999). Additionally, PGT requires the complex coordination of three different teams—clinical, biological, and genetic—each with its own technical, organisational, and disciplinary identities, norms, and interaction regimes (Vertesi, 2012), which can make implementation challenging (Ehrich & Williams, 2010;

[11] Source: Medical and scientific report from the Agence de la Biomédecine (ABM), 2020.

[12] See, for instance, Linconstant (2020). However, financial considerations are, indeed, not absent from public centres, particularly when it comes to the purchase of machines, culture kits, freezing kits, or sequencing equipment.

Giraud, 2024). However, the limited uptake of PGT can also be explained by the reluctance expressed in the political debates and the media landscape since its development in the late 1990s (Franklin & Roberts, 2006), as well as the desire to restrict its use. Although PGT was quickly accepted for "compassionate" purposes, in order to alleviate human suffering (Comité Consultatif National D'Ethique, 2019, p. 8), critics argue that it leads to eugenics, aims to fulfil the fantasy of a "designer baby", and promotes genetic discrimination that could jeopardise the rights of people with disabilities (Kerr & Shakespeare, 2007).

In France, PGT is particularly controversial, and legislators fear its potential risks, meaning that strict safeguards are in place to regulate its use. As already noted, PGT (-M and -SR) is only authorised in "exceptional cases" (Art. L2131-4, Code of Public Health, CPH) and under very strict conditions: "The diagnosis may only be carried out when the anomaly or anomalies responsible for a serious, late-onset, life-threatening disease have been precisely identified in one of the parents or an immediate ancestor" (Art. L2131-4, CPH). There is no authorised list of diseases, with every situation being assessed on a case-by-case basis by a multidisciplinary panel. Severity is often evaluated based on several criteria: life expectancy, autonomy, quality of life, symptom variability, the availability of effective treatments, and the age of disease onset. Based on these criteria, some indications are almost systematically accepted (Huntington's disease or cystic fibrosis); others generate more debate. For instance, checking for the BRCA mutation, which increases the carrier's risk of developing hereditary breast-ovarian cancer syndrome, was prohibited for a long time because carrying this mutation only indicates a higher risk compared to the general population of developing the disease. Since then, however, some requests have been accepted, given the severity of the mutation's expression as well as the family history (Giraud, 2023).

By emphasising the "particular severity" of the indication, the law frames PGT within the broader principles that historically governed ARTs in France from 1994 to 2021, and still apply to prenatal diagnostic: ARTs should only address situations considered "pathological". Indications that do not meet strict "medical" criteria—such as sex selection for family balance[13]—are excluded. The professionals studied frequently use the argument that PGT is a "medical" practice aimed only to combat serious, debilitating diseases. They also emphasise the *proven* risk of transmitting a serious condition—excluding all diseases that could arise *de novo* during pregnancy—thus allowing couples to avoid transmitting to their child the disease from which they themselves suffer. This rhetoric is an integral part of the *moral labour* (Ivry & Teman, 2019) that professionals perform to make their practice acceptable (Giraud, 2024). The term "moral labour" refers to the various actions to justify, legitimise, and alleviate the moral and ethical tensions present in different contested social environments involving reproductive issues (Ehrich et al., 2007, 2008; Gammeltoft, 2014; Heinsen, 2022; Ivry & Teman, 2019; Rapp, 1999; Siegl, 2023; Svendsen & Koch, 2008): for example, as Sylvain Debré, a reproductive biologist, says, "I'm not convinced that it's eugenic to want to prevent someone from getting a serious, debilitating, disabling disease". It allows them to distance themselves from accusations of being part of a eugenic practice, to legitimise their work, and to create a practice and working environment that is not only morally acceptable for themselves but also for others. This conception of PGT as a medical procedure not "only" enabling the birth of a "healthy" child but also enabling individuals to avoid passing on the disease from which they themselves suffer would guarantee a non-eugenic practice.

The law also defines PGT as a "targeted" procedure that focuses on one or more previously identified pathologies. This excludes all techniques whose scope is too broad, which are prognostic rather than diagnosis, and are mainly aimed at improving fertility. This is the case for PGT-A. Although it is widely used in other European countries, such as Spain, its prohibition was reaffirmed in 2021 during the last revision of the bioethics laws. The critiques against PGT in general, and more specifically against PGT-A, are produced in particular by anti-choice lobbyists. Powerful in France, they play an important role in maintaining the ban on certain techniques, such as PGT-A. Their main argument is that this particular technique would contribute to the eradication of people with

[13] Sex-based selection is only authorised in the case of an X-linked disease (and even in this case, carrier female embryos are not automatically excluded).

disabilities, especially those with Down syndrome. All these elements create a "moral landscape" in which all the actors involved in PGT are enlisted, and which they continually update in their daily practice (Svendsen & Koch, 2008). It also affects the techniques used by the PGT professionals who are navigating a practice that they know to be highly sensitive in moral and ethical terms. Several situations—which I explore below—particularly concentrate moral dilemmas and tensions (Heinsen, 2022) because they are grey areas that are not covered by the law, putting professionals in difficult ethical and social situations.

Bulky Knowledges

In their practice, professionals encounter two particular situations that create moral tension, with one case placing them at the margins of French law.

The "healthy" Baby, or Something Fishy Is Going on[14]

The first situation relates to the ultimate goal of PGT: to produce a "healthy" child. Although this goal, as stated in the law, may seem simple at first glance, it turns out to be much more complex in daily practice (Ehrich et al., 2007; Ehrich & Williams, 2010). This complexity arises from the fact that the definition of a "healthy embryo" varies depending on the type of pathology, with each situation having distinct ethical and social implications.

In autosomal monogenic diseases,[15] there are both dominant and recessive forms. In dominant diseases, only one copy of a defective gene is needed for the person to be affected and to develop symptoms. For this reason, only homozygous embryos—those without the mutation—are transferred, as they are not affected. In the case of recessive diseases, two copies of the defective gene are required for the disease to manifest. An

[14] For a more detailed analysis of this question, see Giraud (2025).

[15] The term autosomal means that the transmission does not involve the sex chromosomes. X-linked may also be candidates for PGT. As they raise different ethical issues, I do not discuss them here.

individual with only one copy of the mutation is considered a "healthy carrier" or heterozygote. These individuals are unaffected, but when they are in a relationship with another healthy carrier as well,[16] they have a 50% chance of passing the mutation to their children, a 25% chance of not passing it on, and a 25% chance of having an affected child.

Structural chromosomal abnormalities occur when chromosomes break during the formation of gametes. If these abnormalities are "balanced" (all chromosomal material is present, but not in the correct location), they usually have no phenotypic effects on the carrier but can cause complications during procreation, such as infertility or miscarriage. Conversely, unbalanced chromosomes during embryo formation can result in foetal malformations and sick children, but most often, affected embryos do not develop, leading to miscarriages that occur at different stages. Only balanced embryos, both carriers and non-carriers,[17] are selected for transfer.

PGT's short-term goal of helping couples having a "healthy child" is achieved when non-carrier or healthy carrier embryos are transferred: the child-to-be will not develop any symptoms during their lifetime. However, the long-term consequences may differ for subsequent generations, depending on the different cases previously mentioned. In that of recessive diseases, even if the embryo is heterozygous, the child-to-be or their descendants will not necessarily face the same problems as the couple using PGT: it will depend on the partner and the rarity of the disease. "The rarer the disease, the lower the risk", a genetic counsellor explained to a couple consulting her to decide whether to proceed with genetic diagnosis. In the case of chromosomal abnormalities, by contrast, the transfer of balanced carrier embryos raises the problem of passing the same complications to the immediate next generation that the couple undergoing PGT encountered during procreation: "The children will have the same reproductive problems", says Rebecca Deligny, a cytogenetic engineer.

[16] Only couples in which both partners are healthy carriers are eligible for PGT.

[17] For simplicity, I use the term "carrier" to refer to both recessive diseases and chromosomal abnormalities.

The transfer of "healthy carrier" embryos, with all the consequences for future generations, raises for the professionals a number of questions and ethical dilemmas about the objectives of PGT. This situation creates a moral tension, as this is typically a situation "in which legally sanctioned decisions and actions [...] remain ethically unresolved for those involved" (Heinsen, 2022, p. 518). The law has opened up a social and ethical grey area (Ehrich et al., 2007): although it prohibits the destruction of "healthy carrier" embryos, nothing is said about the possibility of establishing a hierarchy wherein healthy homozygous embryos are transferred first, and only then healthy heterozygous embryos. But it is in these areas of interpretation that moral labour can take place, in the intersection between national guidelines and more individual procedures (here at the level of a centre), collective issues, but also more personal ones, depending on what the people at the heart of the practice believe to be "good", moral, and ethical. This is why some professionals wonder if the goal of PGT should not include, in addition to the health of the child itself, the health of their future children and future generations. However, for the majority of those interviewed, strict adherence to the legally defined purpose of PGT—"only" to facilitate the birth of "healthy" or "unaffected" children whether or not they are carriers—serves as a guarantee of maintaining morally acceptable practice: away from a drift towards eugenics, where any deviation from the norm, however small, would be eliminated for future generations as well as for the current one. The professionals I have met have, therefore, collectively decided not to establish a hierarchy among embryos once they have been diagnosed as "healthy", and patients are clearly informed about this policy. This decision appears to be shared by all the centres performing PGT in France.

As I demonstrate below, however, one technique allows some genetics professionals to avoid finding themselves in such situations.

A Targeted (un)Knowledge

The second situation that can create dilemmas for professionals concerns the scope of diagnosis in PGT, or in other words, what is permissible to know about the genetic status of the embryos being analysed.

In France, the law defines PGT as a "targeted" diagnostic: only mutations specifically identified in one of the patients or among their kin can be investigated.[18] The genetic analysis performed on embryos focuses exclusively on this specific mutation, and any other investigation is strictly prohibited. When everything goes well, "there are no problems of interpretation because we target the regions we want [...]. The law is designed to ensure that it is targeted", explains molecular geneticist Alexandre Wolf. Although accidental findings can occur during the diagnostic process, they usually pertain to the chromosomal regions being studied. These cases remain within the bounds of the law, and the protocol to follow in such instances is clearly established:

> Sometimes, chromosomal abnormalities are discovered accidentally. We target something but, for example, we find three alleles [...]. This could indicate a trisomy, which is an accidental, unexpected event, and we're not going to make [the embryo] transferable because the trisomy is not viable... and we've already established the desired embryo profile based on its alleles, and the fact of having three chromosomes is not transferable, so in that case we're going to make it non-transferable. (Alexandre Wolf, molecular biologist)

The management of diagnostic uncertainty and the challenges of interpreting results are integral to the work of clinical geneticists and biologists (Bourret & Rabeharisoa, 2008; Stivers & Timmermans, 2016; Timmermans et al., 2017). The same applies to the disclosure of incidental findings (Gaille & Horn, 2021). Accidental discoveries, known as "incidentalomas", are common in medicine. Initially, this term referred to a mass detected during an imaging test unrelated to the reason for the examination. Today, it includes any unexpected finding made during radiological, biological, or genetic tests. However, these discoveries are no longer entirely fortuitous but are secondary to the use of health technologies (Gesbert et al., 2020). Philippe Rousseau, an engineer specialising in

[18] In 2024, the laboratory discussed the possibility of performing "double indications", involving both molecular biology and cytogenetics: for example, when a chromosomal rearrangement is combined with a monogenic disease. However, mainly for technical reasons, they were not yet being implemented in this PGT centre.

prenatal diagnostics, highlights that this phenomenon is frequent in this field and becomes even more common with whole-genome sequencing.

At the time of my fieldwork in 2024, the main concern for clinical geneticists was how to inform patients about potential incidental findings, and whether they should be informed of all of them. This is why, during clinical genetics consultations and the consent process for analyses, patients are systematically asked whether they wish to be informed about mutations that might be discovered beyond the ones being sought. However, this is not the case with PGT. Due to the techniques used, their intrinsic limitations, and the legislative framework, such findings rarely occur. Outside the targeted regions, incidental discoveries are neither anticipated nor expected, whether during pre-PGT analyses of the couple and their relatives or during the analyses of the embryos. However, despite all these precautions, professionals may occasionally encounter unexpected information for which no protocol exists. It does happen, albeit rarely, that the probes used to identify a mutation reveal a different anomaly elsewhere than the targeted region. Sarah Meyer, a reproductive biology engineer, recalls such a case:

> It was annoying to think, "We're going to transfer something where we know …" Plus, I believe it was something that could have been pathological. This raises questions. I know there's something I shouldn't have seen, but I happened to see it … Should I pretend I didn't see it? But deep down, I know … That was what felt more disturbing.

In such situations, professionals face the question of whether to transfer an embryo known to carry an anomaly that is not the one originally sought. As I observed during my fieldwork in clinical genetics in 2024, professionals could see genetic variants other than those sought during the analysis of the results but do nothing about it because they are not part of the initial indication. However, most professionals involved in PGT in the centre studied agree that once an anomaly is known, it is impossible to ignore it. Transferring an embryo that is known to be affected does not align with the goal of PGT, which is to enable the birth of a "healthy" child. As Emmanuelle Delamarre, a molecular biologist, states, "When you know it's sick, you can't transfer it; that's not the 'deal'

". In such cases, the embryo is typically labelled "non-transferable" or "uninterpretable" even though this situation is a grey area in the eyes of the law. These discoveries place professionals in moral dilemmas, and require them to decide on their own how to proceed, on a case-by-case basis. While clinical geneticists are trained and used to dealing with such uncertainties as part of their daily work, this is not the case for those working in PGT; however, the machines in these laboratories do help to limit these situations.

Embody the Limitation

Cutting-Edge Technology at All Costs?

The technological imperative suggests that the medical field is driven to employ highly complex techniques and devices, particularly those that are the most effective. A strictly technical process governed by purely technical standards would necessarily evolve towards continuous improvement of the balance between effort and outcome. In this view, genetic analysis laboratories conducting PGT would be expected to adopt high-throughput sequencing technologies, as they reduce human resource demands and improve efficiency. This was a trend I observed in the clinical genetics field between 2020 and 2024, with a gradual shift from older, low-throughput sequencing machines to high-throughput sequencing equipment. The national "France Genomic Medicine 2025" (FMG2025) plan promoted this shift, fostering the development of "genomic medicine" and increasing the prescription of whole genome sequencing. In that scenario, analyses are no longer conducted on-site by the laboratories but are centralised on large sequencing platforms (Bourgain & Lade, 2022). This shift, as a geneticist shared with me in 2024, has led to a loss of technical skills (such as karyotype reading) and has also complicated the network of involved professionals (Bourgain, 2019).

However, the effectiveness or "cutting-edge" status of a technique does not guarantee its adoption (Cresswell, 1996; Lock, 2002; Timmermans, 2010). For instance, Lock has demonstrated that organ transplantation did not develop in Japan for cultural and structural reasons (Lock, 2002). This is something I observed in the genetic laboratory where I worked, and more broadly in all French PGT laboratories up to 2024.[19] The equipment used there is often described by professionals themselves as "archaic", "artisanal", or even "old school". A molecular technician explained that their sequencing method was "artisanal" not only due to its limited speed but especially because of its low data capacity compared to NGS: "It's still an artisanal technique compared to next-generation sequencers, which have much greater capacity for handling vast amounts of data". Professionals often described France as a global exception, one of the last countries still using these methods. As Rebecca Deligny, a cytogenetics engineer who worked for several years in a European country with much broader PGT indications and cutting-edge diagnostic technologies, told me:

> I feel like we're *has-beens* (both laugh). [...] In France, we have so many ethical laws that prevent us from doing a lot of things [...]. If you look at it purely from a technical point of view, it's extremely frustrating to be using methods that have been around for ... well, I'm really talking purely technically here—ethics aside ... to be doing things that have been done for 20 years without any improvement. It's all routine. It's not cutting-edge technology; it's a bit frustrating. [...] When you know there are techniques—I mean purely technical ones—that are much more advanced and state-of-the-art, and you can't access them because the laws hold you back, it's a bit ... well, you know. But then when you consider the eugenics angle and ethics, you think, well, in a way, it helps prevent any overreach. (Rebecca Deligny, cytogenetics engineer)

This archaic approach was highlighted again in January 2024 at the Paris Genetics Symposium, a conference gathering professionals in the field. During a presentation, one member of the French PGT Association (SFDPI) called for legislative change to authorise the use of broader and less "archaic" analysis techniques, in her words. The aim was to increase

[19] By 2025, only one centre in France had transitioned to a genome-wide technique—karyomapping—used in molecular biology for new indications. They still use the older technique for the most common diseases.

analysis capacities and reduce waiting times for PGT patients, which had reached an average of 18 months by 2024. However, an engineer I interviewed in 2021 downplayed the impact of using high-throughput sequencers in PGT. He argued that modernising genetic techniques would not necessarily shorten waiting lists. The delays were more due to bottlenecks in fertility clinics, as the teams had to handle not only PGT but also inseminations, IVF, ICSI, and third-party procreation.

Useful Technical Limitations

So, what is preventing these laboratories from using more advanced technologies? First, in a context where genome-wide technologies are developing more rapidly and at lower cost (Committee et al., 2020), the very decision to continue to use both the FISH technique in cytogenetics and the Sanger sequencer in molecular biology is partly an ethical choice and an element of moral labour. These techniques have limitations that allow the objectives of both French legislation and professionals to be met, that of transferring both healthy carrier and non-carrier embryos without discrimination and to make PGT a targeted technique.

The FISH technique does indeed have several limitations. It can only detect a limited number of chromosomes, mainly due to the limited number of available fluorochromes (the three most commonly used are green, red, and blue). In a single embryonic cell, the diagnosis usually involves just two pairs of chromosomes. The analysis relies on counting bright spots in the embryonic cell, corresponding to markers placed on the studied chromosomes. An even number of spots indicates that all studied chromosomes are present, while an odd number signals an imbalance in the studied region. However, the embryonic cells analysed are in the interphase stage, meaning the chromosomes are compacted and indistinguishable from each other. Not only is it impossible to detect other chromosomal abnormalities beyond those targeted by the fluorochromes, but it is also impossible to observe the exact chromosomal rearrangements—only whether the cell is balanced or unbalanced, that is, whether the entire chromosomal material in the studied regions is present or not. Only numerical abnormalities and imbalances can be identified:

three fluorochromes instead of four indicate an imbalance, or conversely, six fluorochromes will indicate a trisomy. The analysed embryos can therefore only be declared "balanced" or "unbalanced for the studied chromosomal regions" according to the categories used in cytogenetic reports for reproductive biology and patients:

> With our little probe, we looked at two chromosomes out of 23 and said, "Yeah, yeah, it's genetically good", but isn't the rest unbalanced? We don't know that […]. The only way to see a balanced carrier would be to do a karyotype. Except that, as you can see, when we fix the embryos [cells], it is at an interphase stage, so you don't see your chromosomes. So it's not that we don't want to, it's that we can't. (Rebecca Deligny, cytogenetic engineer)

The sequencing technique used in molecular biology, derived from Sanger sequencing, also has the advantage, in PGT, of being able to target only a limited number of specific genes. Similar to FISH, markers are used to identify the analysed regions. The studied laboratory only had four colours for these markers. The risk is that if the analysis area is expanded—thereby increasing the number of markers—the reliability of the results is not guaranteed. This is what Emmanuelle Delamarre (molecular biologist) explained to me when I asked her if it is possible to identify more than two diseases:

> The challenge is to have enough markers for the two diseases […], in order to reliably determine the embryo's status regarding the two diseases. […] Well, that increases the number of markers, and in a single tube, there can be reagent competition, which means that you can't multiply the number of markers. So you're somewhat limited in that sense. The technical challenge lies there: if we had to test for three conditions, we would end up losing informativeness and no longer have reliable results for one or more of the three diseases.

The professionals from the two teams (cytogenetics and molecular biology) do not face the same ethical dilemmas. Cytogeneticists are unable to differentiate between carrier and non-carrier embryos among the balanced ones. They do not face the ethical issues of a possible hierarchy among healthy embryos. Embryos are simply marked as "transferable genetic" in the analysis reports, without specifying carrier or non-carrier status. However, in molecular biology, the technique places professionals in a more complex situation, because they know the exact genetic status of the embryos (carrier or non-carrier). But they have made the decision to withhold such information to prevent themselves being tempted to prioritise non-carrier embryos. To do so, they actively produce un-knowledge by different means, through communication and organisation.[20]

Robust Techniques for Efficient Targeted Diagnosis

Techniques to Prevent Knowledge

Although their design was not originally intended to limit genetic reading, these machines still embody this intention through the preference for their use over more recent machines. Just as laboratory devices exemplify "reified theory", embodying the outcomes of scientific debates and controversies (Bachelard, 2020; Latour & Woolgar, 1979), they can also be seen to some extent as "reified legislation". Even more, these techniques are actors (Latour, 2007, p. 44) in the sense that they contribute to these limitations: they prevent the detection of mutations other than the targeted one, they prevent the distinction between non-carrier and carrier embryos among the healthy ones, and ultimately, they contribute to produce un-knowledge: un-knowledge about the "exact" genetic status of "healthy" diagnosed embryos and un-knowledge about the possible presence of other mutations than the targeted one. There is a moral delegation transferred to the technical objects (Latour, 2007, p. 44), but more, the techniques fully act as agents. They *contribute* to the moral labour carried out in PGT laboratories in order to reduce the moral

[20] This chapter focuses on the techniques used, I will not develop this part. But this is something that I have analysed in another paper (Giraud, 2025).

tensions professionals face in their practice. The techniques allow them to maintain and remain in ignorance or un-knowledge.

Maintaining "unknowing" (Geissler, 2013) may seem paradoxical in a context where the very purpose of the procedure is to make a diagnosis and determine the genetic status of the embryos analysed. More generally, science is often portrayed as generating, exchanging, and challenging knowledge (Latour & Woolgar, 1979), or as dominating other forms of knowledge with universal scientific understanding (Turnbull, 2000). This applies to medicine as well. However, in certain circumstances, ignorance or not-knowing can be a necessary condition for producing scientific knowledge: for instance, in the case of randomised control trials. Others studies have demonstrated that ignorance, or rather secrecy, helps maintain a particular social order, often sustaining inequalities (Taussig, 1999), or can be a source of power (McGoey, 2012). But in that case, the focus is not so much on preserving or affecting social relations, but rather on maintaining a moral order, staying in a situation in which people feel comfortable, and complying as far as possible with the law. Un-knowledge allows professionals to refrain from preferring healthy non-carriers to preserve an "ethical" practice and to adhere fully to the law, as well as to counteract the suspicion of eugenics that may be found in the debates. Un-knowledge allows them to avoid situations where knowing too much about the genetic status of embryo would place them in morally, socially, and ethically difficult situations, where they would have to make choices that would place them in borderline situations in relation to the law. A professional from another centre, who states that they would have no objection to transferring an embryo known to be affected by a disease other than the one for which it was tested, also raises another issue: sequencing the entire genome of an embryo, in itself, could be regarded as contrary to the law, even if professionals limit their analysis to the condition originally targeted.

This is, therefore, not a technical lock, that is, a limitation tied to technological constraints, an established technology, a patent, or a technical design that explains why PGT is not performed with NGS. NGS is used on a daily basis as part of genomic medicine, which the French Government has sought to promote through the France Genomic 2025 plan. However, the application of NGS to PGT faces resistance. The lock is of a social, ideological, and legislative nature, to use the terminology employed by Cresswell (1996, p. 83). This concept of socio-technical lock-in has also been studied in economics, for example, to show that suboptimal techniques in terms of efficiency are sometimes preferred because they stem from major historical institutional choices, making the transition to other options challenging and costly (Arthur, 1989).

An Effective but Inefficient Technique

Finally, the decision to continue using older sequencing techniques cannot be seen solely as a moral choice or one limited by legal constraints. This decision also addresses economic and organisational concerns. It involves a cost-benefit analysis, particularly in terms of the balance between investment and outcomes in a public centre that does not operate on a profit-driven model.

Using NGS for PGT does not appear to meet this requirement. "Why use such a machine if its full capacity won't be utilised anyway?" asks Alexandre Wolff, a molecular biologist:

Alexandre Wolf	It's a bit like using a bazooka to destroy something small. Do we need to use a super technology to target a single mutation when we could use a more basic technology that is much cheaper? I think that NGS could be very interesting, but only if we have the possibility of broadening our field of research. But I think it's simply because with our techniques, which are robust and simple, we can produce results that comply with French law, and if one day French law were to change, perhaps we could adapt to a more powerful technique that would give us access to this kind of information.
Anne-Sophie	OK. So, it doesn't make much sense then…
Alexandre	Yes, that's it. Yes, that's it. It's like if you want to drive a Ferrari on a shitty country road, you can't use it like you would on a racetrack! It's a bit like that (laughs)!

What matters here is the robustness of the technique used, along with the cost-benefit ratio. FISH and PCR are robust, proven methods that suffice to obtain a result where NGS would be overly powerful. It is the same principle as driving a Ferrari on a country road: it serves no purpose and the vehicle would not be used to its full potential. In this sense, the diagnostic techniques used in PGT are *effective* but not *efficient*. Effectiveness (*efficacité*) is defined as "the evaluation of the conformity of a result with a project" (Bensa & Cresswell, 1996, p. 122) or the ability to attribute an effect to actions. Efficiency, on the other hand, is the ability to deliver the best performance, the ratio between *input* and *output*.[21] However, this economic dimension was not the one most emphasised by the professionals interviewed. Rather, it was the moral dimension that stood out. I can only assume that, while organisational matters are not primarily their direct responsibility—but rather that of hospital management and administration—they are confronted by moral issues on a daily basis.

In addition to this aspect of NGS being inefficient, its use in the context of PGT cannot proceed without prior adjustments. This is due to the French legal framework, which makes PGT a targeted technique. It would require the application of restrictions or "filters" to mask the parts of the genome that are not supposed to be seen:

Anne-Sophie	Are there ways to adapt the technique?
Rebecca Deligny	Yes, there are. For example, with CGH Array, we could use custom-made software, what we call applying filters, to focus only on specific chromosomes. For instance, if there's a 3–4 translocation, we could just analyse chromosomes 3 and 4. That's feasible.

Or:

Alexandre Wolf	There are ways of not seeing. We only see what we're looking for. I'm not sure how this could be implemented in PGT. Apparently, in many European countries, it's already the case. But it would make our work much

[21] Presentation by Ludovic Coupaye (UCL) at the General Anthropology Seminar of the LISST-CAS laboratory.

> simpler, probably faster, and would require fewer people. We'd probably not need to go through all those probability steps.[22] We could search directly by launching the parents. I don't know how that would work, but it would certainly be a lot simpler.

These filters are already used in genome analyses when targeted analyses are requested. However, this solution has so far been deemed too costly, both technically and financially; it would require either purchasing kits from machine manufacturers to filter the information or hiring a bioinformatician to mask excluded zones and customise the filters for the couples and mutations handled at the centre. One biologist from another centre told me that they could not find a manufacturer willing to sell those kits because France is almost the only country requiring such filters, and it is not profitable. Concerning custom-filters, as the laboratory manager explained, they not only lack the resources for such recruitment, but it would also require a complete reorganisation of the laboratory, especially affecting the work of technicians responsible for technical procedures. Moreover, unlike basic sciences, acquiring a new instrument in the clinical and diagnostic field undergoes a long research and development process to ensure that the range of possibilities afforded by the technology complies with national legislation—particularly crucial in such a strictly regulated practice as PGT. Adherence to these rules and procedures allows professionals to maintain ethical practices in a highly monitored environment.

Conclusion

This chapter highlights the ambivalence surrounding the development of increasingly rapid sequencing technologies. Beyond fear or enthusiasm, professionals face the challenge of staying within the legal framework that governs their practice—a framework intertwined with economic and

[22] These refer to the steps prior to PGT itself to manufacture the probes adapted to the couple's mutation.

organisational constraints. An awareness of Europe's eugenic past also plays a central role in this process, reinforcing the desire to distinguish current practices from anything that might be suspected of being "eugenic". The professionals interviewed engage in moral labour aimed at remaining within the legal boundaries, which authorise PGT only for previously identified mutations. This moral work is delegated to and embedded in the technical procedures used to carry out genetic analyses.

However, in the context of "mundane" genomics—where genetic testing tends to become routine, particularly in ART—resisting the introduction of NGS appears increasingly difficult. The sense that this routinisation is inevitable was frequently expressed by professionals and reflects the broader shift already underway. In 2021, a molecular technician told me, speaking about "old-generation" sequencers: "In France, this is the only way to do PGT. But I think at some point, we'll get there—well, that's what I believe". Three years later, when I came back to the centre, the head of the PGT laboratory informed me that PGT genetic teams from various centres were discussing adapting NGS to PGT and its legislation. One centre had even announced that it had begun adjusting the setup. But the core dilemma for the professionals lies in the question of transitioning from current techniques to NGS, and how this shift might occur without exposing practitioners to unwanted knowledge. They were torn between the desire to adopt more up-to-date techniques and the concern to remain within an ethically "comfortable" practice.

It will be interesting to observe how professionals adapt this new technique to French regulations and economical constraints, and to see how they address the ethical and practical challenges that this change raises.

Acknowledgments This project was funded by the French National Research Agency (ANR) as part of the NORPRO project (Reproductive Norms) led by Anne-Sophie Giraud (ANR-22-CE41-0001).

References

Arthur, W. B. (1989). Competing technologies, increasing returns, and lock-in by historical events. *The Economic Journal, 99*(394), 116–131.

Bachelard, G. (2020). In V. Bontems (Ed.), *Le nouvel esprit scientifique*. PUF.

Bensa, A., & Cresswell, R. (1996). A propos de la technologie culturelle. Entretien avec Robert Cresswell. In: Genèses, 24. Trajectoires, sous la direction de Robert Salais. pp. 120–136.

Bourgain, C. (2019). De la génétique clinique à la médecine génomique: Enjeux d'une « démocratisation » de l'accès aux technologies génomiques en contexte de soin. *Cahiers Droit, Sciences & Technologies, 8*, 15–29. https://doi.org/10.4000/cdst.664

Bourgain, C., & Lade, Q. (2022). Jamais seuls face au génome! Produire un diagnostic génétique à l'ère du séquençage à très haut débit. *Médecine et philosophie, 7*, 25–31.

Bourret, P., & Rabeharisoa, V. (2008). Décision et jugement médicaux en situation de forte incertitude: L'exemple de deux pratiques cliniques à l'épreuve de la génétique. *Sciences sociales et santé, 26*(1), 33–66. https://doi.org/10.1684/sss.2008.0103

Comité Consultatif National D'Ethique. (2019). *Avis sur les problèmes éthiques liés aux diagnostics anténatals: Le diagnostic prénatal (DPN) et le diagnostic préimplantatoire (DPI)* (No. 107). Comité Consultatif National d'Ethique (CCNE).

Committee, E. P. C. S., Carvalho, F., Coonen, E., Goossens, V., Kokkali, G., Rubio, C., Meijer-Hoogeveen, M., Moutou, C., Vermeulen, N., & De Rycke, M. (2020). ESHRE PGT Consortium good practice recommendations for the organisation of PGT. *Human Reproduction Open, 2020*(3). https://doi.org/10.1093/hropen/hoaa021

Cram, D., & de Kreisler, D. (2002). Genetic diagnosis: The future. In I. C. J. Dejonge & C. L. R. Barratt (Eds.), *Assisted reproductive technology: Accomplishments and new horizons* (pp. 186–205). Cambridge University Press.

Cresswell, R. (1996). *PROMÉTHÉE OU PANDORE ?: Propos de technologie culturelle*. Editions Kimé. https://books.google.fr/books?id=RXc3DwAAQBAJ

Ehrich, K., & Williams, C. (2010). A 'healthy baby': The double imperative of preimplantation genetic diagnosis. *Health, 14*(1), 41–56. https://doi.org/10.1177/1363459309347477

Ehrich, K., Williams, C., & Farsides, B. (2008). The embryo as moral work object: PGD/IVF staff views and experiences. *Sociology of Health & Illness, 30*(5), 772–787. https://doi.org/10.1111/j.1467-9566.2008.01083.x

Ehrich, K., Williams, C., Farsides, B., Sandall, J., & Scott, R. (2007). Choosing embryos: Ethical complexity and relational autonomy in staff accounts of PGD: Choosing embryos: staff accounts of PGD. *Sociology of Health & Illness, 29*(7), 1091–1106. https://doi.org/10.1111/j.1467-9566.2007.01021.x

Franklin, S. (1997). *Embodied progress: A cultural account of assisted conception.* Routledge.

Franklin, S., & Roberts, C. (2006). *Born and made: An ethnography of preimplantation genetic diagnosis.* Princeton Univ. Press.

Gaille, M., & Horn, R. (2021). The ethics of genomic medicine: Redefining values and norms in the UK and France. *European Journal of Human Genetics, 29*(5), 780–788. https://doi.org/10.1038/s41431-020-00798-2

Gammeltoft, T. M. (2014). *Haunting images: A cultural account of selective reproduction in Vietnam* (1st ed.). University of California Press. https://www.jstor.org/stable/10.1525/j.ctt5vjzcm

Geissler, P. W. (2013). Public secrets in public health: Knowing not to know while making scientific knowledge. *American Ethnologist, 40*(1), 13–34. https://doi.org/10.1111/amet.12002

Gesbert, C., Torregrosa, C., & Mamzer, M.-F. (2020). De l'incidentalome à la découverte secondaire. *médecine/sciences, 36*(11), Article 11. https://doi.org/10.1051/medsci/2020188

Giraud, A.-S. (2015). *Les statuts de l'être anténatal: Un processus d'humanisation 'relationnel'. Assistance médicale à la procréation et mort périnatale.* Thèse de doctorat d'anthropologie sociale et ethnologie, EHESS.

Giraud, A.-S. (2023). Sous le régime de l'exception: Accéder à la parentalité grâce au diagnostic préimplantatoire en France. *Revue française des affaires sociales, 2*, 281–299. Cairn.info

Giraud, A.-S. (2024). *Silencing the embryos.* https://hal.science/hal-04687222

Giraud, A.-S. (2025). Un-knowing the embryo: The moral labor of doing preimplantation genetic testing in France. *Medical Anthropology, 44*(2), 185–201. https://doi.org/10.1080/01459740.2025.2479673

Handyside, A. H., Kontogianni, E. H., Hardy, K., & Winston, R. M. L. (1990). Pregnancies from biopsied human preimplantation embryos sexed by Y-specific DNA amplification. *Nature, 344*(6268), Article 6268. https://doi.org/10.1038/344768a0

Haraway, D. (1988). Situated knowledges: The science question in feminism and the privilege of partial perspective. *Feminist Studies, 14*(3), 575–599. https://doi.org/10.2307/3178066

Harton, G., Braude, P., Lashwood, A., Schmutzler, A., Traeger-Synodinos, J., Wilton, L., & Harper, J. C. (2011). ESHRE PGD consortium best practice guidelines for organization of a PGD centre for PGD/preimplantation genetic screening. *Human Reproduction, 26*(1), 14–24. https://doi.org/10.1093/humrep/deq229

Heinsen, L. L. (2022). Shouldering death: Moral tensions, ambiguity, and the unintended ramifications of state-sanctioned second-trimester selective abortion in Denmark. *Medical Anthropology Quarterly, 36*(4), 515–533. https://doi.org/10.1111/maq.12717

Heinsen, L. L. (2023). *The moral labor of making death: An ethnography of second-trimester selective abortion in welfare state Denmark*. Ph.D., Aalborg University. https://doi.org/10.54337/aau532692968

Helosvuori, E. (2019). Assembling viability: The art of mundane embryo selection in IVF. *BioSocieties, 14*(1), 1–22. https://doi.org/10.1057/s41292-018-0114-3

Ivry, T., & Teman, E. (2019). Shouldering moral responsibility: The division of moral labor among pregnant women, rabbis, and doctors. *American Anthropologist, 121*(4), 857–869. https://doi.org/10.1111/aman.13314

Kerr, A. (2004). *Genetics and society: A sociology of disease* (1st ed.). Routledge. https://doi.org/10.4324/9780203489512

Kerr, A., & Shakespeare, T. (2007). Genetic politics: From eugenics to genome. *Ethical Theory and Moral Practice, 10*(4), 409–418.

Knecht, M., Heibges, M., & Beck, S. (Eds.). (2012). *Reproductive technologies as global form: Ethnographies of knowledge, practices, and transnational encounters*. Campus-Verlag.

Latour, B. (2007). *Petites leçons de sociologie des sciences*. La Découverte; Cairn.info. https://www.cairn.info/petites-lecons-de-sociologie-des-sciences%2D%2D9782707150127.htm

Latour, B., & Woolgar, S. (1979). *Laboratory life: The construction of scientific facts*. Princeton University Press. https://books.google.co.uk/books?id=XTcjm0flPdYC

Lavery, S. A., Aurell, R., Turner, C., Taylor, D. M., & Winston, R. M. (1999). An analysis of the demand for and cost of preimplantation genetic diagnosis in the United Kingdom. *Prenatal Diagnosis, 19*(13), 1205–1208. https://doi.org/10.1002/(SICI)1097-0223(199912)19:13<1205::AID-PD727>3.0.CO;2-W

Linconstant, L. (2020). *L'assistance médicale à procréation comme technique relationnelle. Fabriquer des patients pour faire des parents (Lombardie, Italie).* Thèse de doctorat d'anthropologie sociale et ethnologie, Aix-Marseille Université.

Lock, M. M. (2002). *Twice dead: Organ transplants and the reinvention of death.* University of California Press.

McGoey, L. (2012). The logic of strategic ignorance. *The British Journal of Sociology, 63*(3), 533–576. https://doi.org/10.1111/j.1468-4446.2012.01424.x

Munné, S., Dailey, T., Sultan, K. M., Grifo, J., & Cohen, J. (1995). The use of first polar bodies for preimplantation diagnosis of aneuploidy. *Human Reproduction (Oxford, England), 10*(4), 1014–1020. https://doi.org/10.1093/oxfordjournals.humrep.a136027

Rapp, R. (1999). *Testing women, testing the fetus: The social impact of amniocentesis in America.* Routledge. http://books.google.fr/books?id=n_O0NdigZx4C

Saruwatari Zavala, G. (2023). The Pandora's box of transparency opened by direct-to-consumer genetic testing. *Frontiers in Genetics, 14*, 1175666. https://doi.org/10.3389/fgene.2023.1175666

Shakespeare, T. (1998). Choices and rights: Eugenics, genetics and disability equality. *Disability & Society, 13*(5), 665–681. https://doi.org/10.1080/09687599826452

Siegl, V. (2023). *Intimate strangers: Commercial surrogacy in Russia and Ukraine and the making of truth.* Cornell University Press.

Spiegel, A. D. (1981). In D. Rubin & S. B. Frost (Eds.), *Medical technology, health care, and the consumer.* Human Sciences Press.

Stivers, T., & Timmermans, S. (2016). Negotiating the diagnostic uncertainty of genomic test results. *Social Psychology Quarterly, 79*(3), 199–221. https://doi.org/10.1177/0190272516658770

Svendsen, M. N., & Koch, L. (2008). Unpacking the 'Spare Embryo': Facilitating stem cell research in a moral landscape. *Social Studies of Science, 38*(n°1), 93–110.

Taussig, M. T. (1999). *Defacement: Public secrecy and the labor of the negative (Original print).* Stanford University Press.

Taussig, K.-S., Rapp, R., & Heath, D. (2005). Flexible eugenics: Technologies of the self in the age of genetics. In J. X. Inda (Ed.), *Anthropologies of modernity* (pp. 194–212). Blackwell Publishing Ltd. https://doi.org/10.1002/9780470775875.ch8

Timmermans, S. (2010). Technology and medical practice. In I. C. E. Bird, P. Conrad, & A. M. Fremont (Eds.), *Handbook of medical sociology* (6th ed.). Vanderbilt University Press.

Timmermans, S., Tietbohl, C., & Skaperdas, E. (2017). Narrating uncertainty: Variants of uncertain significance (VUS) in clinical exome sequencing. *BioSocieties, 12*(3), 439–458. https://doi.org/10.1057/s41292-016-0020-5

Turnbull, D. (2000). Making malaria curable: Extending a knowledge space to create a vaccine. In *Masons, tricksters and cartographers*. Taylor & Francis.

Vertesi, J. (2012). Seeing like a Rover: Visualization, embodiment, and interaction on the Mars Exploration Rover Mission. *Social Studies of Science, 42*(3), 393–414.

Wahlberg, A., & Gammeltoft, T. M. (2018). *Selective reproduction in the 21st century*. http://search.ebscohost.com/login.aspx?direct=true&scope=site&db=nlebk&db=nlabk&AN=1575402

4

Promoting Genetic Testing to Prevent Non-Genetic Conditions: PGT-A and Solving the Problem of Twins in US Fertility Clinics

Eliza Brown

> Every single patient asks me, "Are you going to put two [embryos] in?" And I say, "Why?" And they say, "Well, we want to be one and done." That's the saying, right? And my response is, "Did you know that twins have this higher risk of cerebral palsy?" And by the time that conversation—and it only takes five minutes—is over, they never mention twins again.
>
> *United States fertility specialist on how to counsel patients about the risk of twins*

How do physicians promote genetic testing to patients? They might suggest genetic testing to diagnose disorders directly caused by genetic mutations, such as sickle cell anemia. Other times they recommend genetic testing to categorize patients at higher risk of diseases such as the genetic mutations that make breast cancer more likely. They might also

E. Brown (✉)
Department of Sociology, Berkeley University, Berkeley, CA, USA
e-mail: elizaclairebrown@berkeley.edu

V. Argudo-Portal et al. (eds.), *Mundane Genomics*,
https://doi.org/10.1007/978-981-95-4550-6_4

recommend genetic testing, especially to prospective parents, to ascertain their carrier status with regard to genetic mutations. Physicians might not always promote genetic testing for the management or prevention of genetic conditions, however. Instead, as this chapter illustrates, they might recommend genetic testing to patients as a means of indirectly preventing non-genetic conditions.

This is the case today in the United States in one of the central arenas for genetic testing: in vitro fertilization (IVF). IVF involves extracting eggs from ovaries and fertilizing them in a laboratory with sperm, some of which might then continue to develop into embryos. In many cases of IVF, some number of these embryos are transferred directly into the uterus; in other cases, they are subjected to preimplantation genetic testing (PGT). PGT involves removing cells from each embryo which are then shipped to an outside laboratory for analysis. Fertility specialists might recommend PGT to the small set of patients who are carriers of or affected by serious hereditable conditions, such as monogenic conditions (PGT-M) or structural rearrangements (PGT-SR). They might also offer a third kind of PGT—preimplantation genetic testing for aneuploidy, or PGT-A—to a much wider group of patients. In PGT-A, the laboratory technician looks for cells with 46 chromosomes, indicating a euploid or "normal" embryo, rather than those with some other number, which are called aneuploid or "abnormal." The patient would then be advised to only transfer euploid embryos to increase the chance of implantation in the uterus, avoid miscarriage, and to prevent conditions defined by chromosomal abnormalities, such as Trisomy 21, or Down syndrome. PGT-A is quickly becoming a feature of IVF cycles in the US (employed in nearly a third of cycles), particularly in clinics funded through private equity (Borsa & Bruch, 2022).

The routinization of PGT-A in the US fertility field means that genomic technologies and information have become normalized for an increasing number of people. And, in many ways, it seems that the growing popularity of PGT-A is just another sign of an increasingly geneticist medical logic and society (Lippman, 1991), particularly throughout reproduction. Pregnancy in the US has been marked by genetic testing for decades (Rapp, 2004; Rothman, 1986). Addressing this, anthropologists Tine Gammeltoft and Ayo Wahlberg (2014) coined the term *selective reproductive technologies* (SRT) to describe the fact that medically assisted reproduction is often not about creating any baby, but rather about making sure that only a certain kind of baby—in this case one that is genetically "normal"—is born. Still, I argue that even as PGT-M involves direct selection according to genetic criteria, the proliferation of PGT-A in the US actually has little to do with direct genetic selection.

In the US, as part of IVF, fertility doctors once transferred as many embryos as possible to increase the chances of pregnancy; sometimes these efforts led to pregnancies with twins and higher order multiples (triplets or more). Starting about 25 years ago, medical professional societies and public health agencies in the US called for decreasing the number of embryos transferred during IVF cycles to lower the incidence of multiples resulting from fertility treatment. Although nearly everyone agreed that higher order multiples were a problem, and practices were changed so that these outcomes became highly unlikely, the question of twins remained uncertain and ambiguous for much longer, in part because, even today, many patients express a preference for or an acceptance of a twin outcome (Barishansky et al., 2022). What is more, although twins constitute relatively higher health risks compared to singletons, most are ultimately healthy, and other risky pregnancies related to fertility treatment are uncontroversial in the US system.

Unlike many other countries, in the US there is not a central government authority dictating how physicians must treat fertility patients, including how many embryos to transfer in an IVF cycle. Besides not being bound by national legislation that dictates the course of care, fertility medicine in the US is also not supported by state funding; rather, it is a largely private, consumer-driven medical market (Conrad & Leiter, 2004). At most US clinics, a single cycle of IVF costs about $20,000 including medication. To date, 19 US states require private insurance companies to cover at least some fertility testing and treatment, but most fertility patients are still responsible for large out-of-pocket expenses, including that of PGT-A. Adding PGT-A to IVF cycles means adding about $5000 to already costly fertility treatment (van de Wiel, 2020).

Consequently, and as part of a US consumer medicine model, fertility providers often find themselves in situations in which they need to dissuade fertility patients who are also paying customers from transferring more than one embryo, even if doing so might decrease their chance of getting pregnant. One of the ways in which they do so is by emphasizing the health problems associated with twin pregnancy and birth, in particular its propensity for producing children with lifelong impairments and disabilities.

Extant reproductive scholarship emphasizes that even though it presents moral dilemmas related to sex selection, the sanctity of embryos, and questions about life itself (Ehrich et al., 2007; Franklin & Roberts, 2006; Raspberry, 2009), clinicians and patients are attracted to PGT because of its promise to improve prospective children's health before conception through the selection against "unhealthy" (or at least affected) embryos (Franklin & Roberts, 2006; Turrini & Otín-Gavín, 2025). PGT-A, I argue, is not so much promoted by US fertility specialists as a way of directly selecting against genetic abnormalities or diseases but, rather, is presented as a means of choosing single embryo transfer and thereby reducing health risks associated with twins. In this way, PGT-A continues to operate as an SRT that prevents births to children with impairments and disabilities, albeit one that selects against impairments that are not genetically transmitted but instead are often the product of difficult pregnancies and births. Thus, PGT-A is routinized as a tool to improve the functionality of IVF.

This routinization of PGT-A in US fertility clinics marks it as part of a shift to *mundane genomics*, which underlines the normalization, multiplicity, and commercialization of genetic data in contemporary society. The meaning of PGT-A in fertility clinics is flexible and layered; it can at one moment be framed as a way of preventing genetically transmitted conditions, at another be discussed as a means of averting miscarriages, and throughout be framed as an odds-upping approach that obviates the need for multiple embryo transfer. This flexible framing is related to its normalization; fertility providers can recommend PGT-A to nearly any patient as part of improving the chance of success per transfer. Still, the continued framing of PGT-A as a tool for preventing congenital anomalies underscores how deeply entrenched ableist and eugenic assumptions remain in medical practice for both provider and patient-consumers.

This chapter is primarily based on my observation in 2019 of an academic fertility practice ("University Clinic") in a large city in the US that promoted PGT-A to most patients, where I observed 67 patient-provider consultations. University Clinic had two sites in large indistinct urban medical buildings, and was co-operated by the university, the physicians at the practice, and a fertility company with several centers in its network. While I was there, there were seven physicians who conducted consults; it was standard-setter in the field whose doctors regularly published research and spoke at conferences. There were some patients who chose to go to University Clinic because of its expertise in PGT-M.

University Clinic's twin birth rate, at about 5% for women under the age of 35, was well below the national average for fertility clinics. When I asked fertility specialists at University Clinic about the clinic's view of twins, they responded that it was a problem that had already been solved through medical advances such as PGT-A. This allowed patients to raise their chances of getting pregnant without increasing their chance of multiples—something that was a part of the history of the fertility field, but that was not currently an issue. Nevertheless, the question of multiple embryo transfer and twins regularly came up in the consults I observed.

In addition to my observation of University Clinic, I observed two other fertility clinics in the same state, and attended conferences and government advocacy sessions, which serve as background evidence for this chapter. Moreover, I conducted in-depth interviews with 27 fertility doctors and 78 fertility patients living in the US, data from which serve as supplementary evidence. Two of the 27 doctors worked at University Clinic while the other 25 worked at other clinics in the US. Of the 79 fertility patients interviewed, 11 happened to have received care at University Clinic.

A Brief History of Managing Multiples in the United States

From the beginning of fertility medicine, increasing the chances of pregnancy also meant increasing the chance of multiples. In the 1960s, oral and injected medications were introduced to treat fertility by inducing ovulation. But they did not just induce the release of one egg, as would typically happen with unmedicated ovulation, but often several. For the first time there were widely available medications that increased the chance of pregnancies and births to multiples. Fertility doctors needed to strike a careful balance between prescribing enough medication so that at least one egg follicle developed, but not so much that too many follicles developed, creating the conditions for many eggs to be fertilized. In this way, fertility medications were promising but offered limited control.

IVF offered the promise of that control, first resulting in a live birth in 1978 to Louise Brown in England (Kolata, 1978). While that birth resulted from an unmedicated cycle (Edwards & Steptoe, 1980), US fertility specialists were soon using hormones to stimulate the production of multiple eggs with IVF to increase the chance of success (Macklon et al., 2006). Unlike combining injected medications with insemination or intercourse, with IVF physicians could select the number of embryos to transfer and therefore manage the multiples resulting from treatment.

Still, for all its promise, success rates from IVF were extremely low in the 1980s. Consequently, most fertility specialists chose to transfer several embryos per cycle. At first, most fertility specialists were not particularly concerned about pregnancies and births to multiples from these practices because the success rates were so low that the chance of multiples was also highly unlikely. What is more, in cases in which multiple pregnancies resulted, patients might opt for multifetal reduction, a type of abortion in which only one fetus in a multiple pregnancy is terminated, allowing for the rest of the pregnancy to continue with the remaining fetus(es).

Then, in 1992, then Representative (now Senator) Ron Wyden (D-Oregon) wrote a law called the Fertility Clinic Success Rate and Certification Act (FCSRCA). This law was motivated by concerns that fertility clinics were exploiting infertile couples by not being transparent about their low success rates. The act requires the Center for Disease Control and Prevention (CDC) to collect data from clinics and submit an annual report to Congress on Assisted Reproductive Technology (ART) success rates. The data collected include patients' diagnosis, type of ART, clinical information pertaining to the ART procedure, and information on pregnancy outcomes.

The law as originally implemented, however, contained within it an unintended accelerant to the crisis of multiple births. The initial reports did not include categories differentiating the births of singletons and multiples: any pregnancy, including one to higher order multiples, counted as a successful outcome. One reproductive endocrinologist described the problem, from his perspective, with the FCSRCA, in an interview:

> So essentially what he [Wyden] wrote was a law that I call the High Order Multiple Pregnancy Act of the United States because we have the highest multiple pregnancy rate in the United States because the United States government was publishing a registry that graded clinics on their success rate and didn't account for twins and triplets. They just looked at baby rate, pregnancy rate. And you want to get a high pregnancy rate, throw a bunch of embryos in. You'll get a high pregnancy rate, but you'll also make a lot of twins and triplets.

A law that had the intention of protecting patients from nefarious marketing had, to use Robert Merton's (1936) concept, "the perverse result" of encouraging aggressive treatment approaches more likely to result in multiples. Indeed, by 1998, three or more embryos were transferred in 80% of IVF cycles, with four or more embryos in 47% of cycles (Katz et al., 2002). Incredibly, the incidence of births to higher order multiples increased by a factor of 6.7 from 1971 to 1998 (Kulkarni et al. 2013). This was concerning because multiple pregnancies present higher relative risks compared to singleton pregnancies for both the pregnant person (e.g., gestational diabetes, C-section) and resulting children (e.g., preterm birth, low birthweight).

In the late 1990s, professional fertility organizations began to implement changes. In 1998, the first practice guidelines were published in the US recommending maximum numbers of embryos per transfer (Stillman et al., 2013). While there were still incentives to produce a high rate of pregnancies regardless of the birth outcome because of the cycle reporting system, now clinics could face professional censure for maximalist treatment approaches. Rates of higher order multiples began to decline, but there was no agreement about whether twins were also a problem. Indeed, in these years many doctors categorized singletons and twins together, framing both as a positive outcome in comparison to higher order multiples (e.g., Goldfarb et al., 1996). At this time, transferring a single embryo was seen as reducing pregnancy rates too much to be the best practice (Elsner et al., 1997).

In the 2000s, technological improvements, such as maturing embryos longer and better embryo freezing techniques, allowed fertility specialists to transfer fewer embryos without lowering success rates (Kuleshova et al., 1999; Yoon et al., 2000), but it was still common to transfer more than one embryo, especially for poor prognosis patients. The rates of higher order multiple births continued to decline, and success rates continued to improve.

It is important to note that fertility specialists, although trained as OBGYNs, stop treating patients in the first trimester. Consequently, they are not tasked with managing multiple pregnancies and births directly. What is more, although twins do present relatively higher risks compared to singleton pregnancies, there was not the same degree of consternation in the US fertility field about IVF and ICSI themselves increasing the chance of complicated pregnancies and births. Nevertheless, twins became a site of moral complexity for fertility specialists (Ehrich et al., 2007; Heinsen, 2018): although not prohibited, twins made them feel as though they were creating "pathological" pregnancies, and potentially, the birth of children with disabilities.

Still, in the first half of the 2010s, the twin rate stayed flat in the US. Then, in the second half of the decade, for the first time since the advent of fertility treatments, twinning began to decrease. Another major medical development facilitated the success of the transition to universal single embryo transfer in the US in the mid-2010s: PGT-A. Although first developed in the early 1990s, it was not until some key tinkering with the technique that made it more effective in the early 2010s that PGT-A started to proliferate rapidly (Parikh et al., 2018). PGT-A allows for embryonic attrition to occur before the embryos are transferred to the uterus. Accordingly, it does not increase the rate of pregnancy from each egg retrieval but does for each embryo transfer. Performing PGT-A is the only way single embryo transfer can be as successful as double embryo transfer (Scott Jr et al., 2013). As one physician explained in an interview,

"So you want to incorporate really a single embryo transfer cycle into your practice. The only way that you can do that and not have a 25 percent chance of pregnancy per treatment is to really use PGT-A as a tool." Single embryo transfer combined with PGT-A became the standard of care at many fertility centers, and twin rates began to decrease.

For all its promises, PGT-A presents many problems. Among poor prognosis patients there might not be enough embryos to mature to Day 5, a requirement to perform PGT-A today at most US clinics, without losing all of them to attrition in the laboratory and consequently having a failed IVF cycle without an embryo transfer. Although it is possible that embryos that fail to mature in the lab would also not survive in the uterine environment, many fertility scientists theorize the opposite. Additionally, some research suggests that for younger women PGT-A offers no additional benefits because of the lower rate of aneuploid embryos for that group. Some specialists take issue with the accuracy of PGT-A, arguing that it does not adequately account for mosaicism and cellular change. What is more, as egg freezing soars in popularity, it is of note that even those physicians who normally recommend PGT-A might hesitate to recommend it for patients who freeze their eggs and return to fertilize and implant them because PGT-A requires an additional freeze and defrosting of the embryos, which might lead to even more attrition for egg-freezing patients. In short, the single embryo transfer with PGT-A IVF model works best for patients who can produce enough embryos to tolerate attrition and with money to spend on additional treatments.

What kind of "tool," then, is PGT-A? PGT-A became the "right tool" for the job (Casper & Clarke, 1998) not because if its infallibility, but because it allowed fertility doctors to bypass disagreements with patients about how many embryos to transfer, as least with those who could afford it. As one fertility doctor put it in an interview:

> I really believe that that's the biggest benefit of PGS [preimplantation genetic screening, now called PGT-A] and even though it's not perfect and there could be misdiagnosis and false positives and the biopsy doesn't help them, you know, I think the ability to like get [patient] acceptance to do single embryo transfer has been a huge shift.

PGT-A allows patients to understand single embryo transfer not as losing out on a higher chance of success but as part of a high-success strategy, and subsequently eases potential conflict between physicians and patients. Still, selling high success rates is not necessarily enough to convince patients to pursue single embryo transfer, particularly for those patients who see twins as a benefit rather than a tolerable risk. In these cases, fertility specialists must convince patients that twins themselves are an undesirable outcome that can be solved through PGT-A. And as we will see, this involves connecting twins with life-long impairments and disabilities.

How Twins Became Connected to Prospective Children's Disabilities

In explaining the health problems more likely to occur with a twin pregnancy and birth to patients, fertility specialists can emphasize any number of conditions from increased back pain to a high chance of a C-section delivery. However, even if pregnancy and delivery risks are the most likely to occur, they might not be the most salient for patients. As one fertility specialist said in an interview, "What patients don't care about is the increased maternal risks of pregnancy-induced hypertension and increased risk of gestational diabetes. Those are maternal risks and those were very quickly discounted by patients. So what really needs to be emphasized in my experiences is the risks to offspring." Indeed, some patients' reactions to hearing the most likely risks related to twin pregnancies and births supported this physician's view. As one man who decided with his wife to transfer two embryos during IVF explained to me in an interview about his experience in a consult when his fertility specialist listed some of the health risks associated with twins:

> I remember [the fertility specialist] talking about elevated risk of pre-eclampsia, gestational diabetes … I remember going through the full litany of things that could go wrong. But then we looked at him and said, "We hear about all our friends with single babies going through gestational dia-

betes, pre-eclampsia, weight issues, and everything else." So this kind of sounds the same.

Many pregnancy risks seem so normal as to not be worth avoiding, particularly in the US context where interventions like C-sections are the norm (Morris, 2013), and where diagnoses of conditions like gestational diabetes are rising (CDC, 2023). Fertility specialists who are highly motivated to convince patients to pursue treatment approaches less likely to result in twin pregnancies, therefore, might turn to a risk that patients are much more likely to find untenable: children's disabilities.

Disability scholars have written critically about prenatal and antenatal genetic testing for the past several decades (Asch, 1989, 2002; Ouellette, 2015; Saxton, 1984; Shakespeare, 2013). Some of the central concerns include resurgent eugenics, how screening expresses negative messages about disability, the fact that tests are not always accurate, and the lack of accounts about disability that are not medical or pathological for prospective parents, among others.

One key question among disability scholars like Tom Shakespeare around PGT and other prenatal testing includes that of "predictable disablement": "Many congenital conditions have moved from being unfortunate examples of bad luck to blameworthy failures of surveillance and control" (Shakespeare, 2006, p. 130). That is, having a disabled child is transformed from something that one could not have known in advance to something that could have been known and altered, and consequently there is a difference in public expectations for prospective parents. This sense of responsibility is situated in unique contexts: In China and Taiwan, prospective parents are coerced to engage in screening to maintain population "quality" and to avoid burdening society (Zhu, 2013; Shih, 2018); in Vietnam, prospective parents are concerned that having a disabled child would "undo" their family's position in local communities (Gammeltoft, 2014); and, in Denmark, pregnant women delegate moral responsibility for selecting against disability to the state (Heinsen, 2018). Still, no matter the context, with prenatal testing, children's disability moves from a question of chance to one of selectivity.

Genes have been central to questions about "predictable disablement" but, especially with medically assisted reproduction, there are other ways of thinking about it. Indeed, the higher rate of disabilities among twins has next to nothing to do with genes, given that the primary causal pathway is pregnancy and birth complications, but still might represent a sort of "predictable disablement." Prospective parents who do not take steps to avoid twin births, then, might be considered blameworthy if their children experience any disability like blindness or cerebral palsy. The following sections illustrate how fertility specialists and patients discuss reducing the risk of prospective children's impairment and disability through reducing the chance of twins by employing PGT-A as part of the IVF process.

Avoiding Disability by Avoiding Twins

PGT-A "solves a lot of the problems that we have with fertility," as I observed Dr. Ricci, one of the fertility specialists at University Clinic, put it to a patient. As stated by other physicians, PGT-A not only solved the problem of needing to transfer multiple embryos to achieve a high success rate, but also the issue of needing to convince patients to choose single embryo transfer and avoid twins.

When I interviewed Dr. Doyle, he explained that "people like statistics but they also like stories" and that is why he told them about patients who ended up having twins with cerebral palsy and other disabilities. While I did not observe him doing so, he also told me that in consults, he would "do an imitation of cerebral palsy," holding his hand awkwardly to imitate someone with a physical impairment as he sat across his desk from me, and then patients would respond, "Oh, I've seen someone with that."

I did sit with Dr. Doyle several times when he purposefully scared fertility patients with frightening statistics and vivid horror stories about twin birth outcomes. For example, I sat in on a consult with Dr. Doyle and a patient in which he explained to her that he would not transfer two embryos in her IVF cycle, and not just because her fibroids would make a twin pregnancy difficult. The following is an excerpt of field notes taken at the time:

> Twins are bad, he says seriously, if God gives you twins you don't give them back, but they come early, their lungs aren't strong, their eyes can be weak, and they have a 20-fold increase risk of "being slow," of having movement problems. He says that he would recommend the genetic testing to save her doing more cycles, but they almost eliminate twins by doing the genetic testing and putting one embryo in, and very rarely they split.

In Dr. Doyle's telling, the main health problem associated with twins is their chance of lifelong disability ("being slow"). As Dr. Doyle made the problem of twins clear to the patient, he also immediately provided the solution in "genetic testing," namely PGT-A. For Dr. Doyle and the rest of the University Clinic physicians, PGT-A sufficiently removes the uncertainty about attrition after embryo transfer ("save her doing more cycles") to avoid risking the impairments more likely to occur with twins.

Another time that I observed Dr. Doyle with patients he moved away from statistics and more toward tragic narrative to underline his point about the problem with twins. He told the story of a patient who did not listen to him about the dangers of twins and transferred to another academic fertility clinic that allowed her to transfer two embryos. Then, 24 weeks pregnant with twins, she called him from the emergency room with one ruptured membrane, he said. He recounted that she ended up delivering at 26 weeks to twins who are both blind, one with permanent pulmonary problems. In his telling, this former patient told him that she loves her babies, but her life has never been the same. In this story, long-anticipated births are "spoiled" (see Rothman, 1986) and lives ruined by having blind children. This story and others like it work as moral fables, warning patients that if they pursue multiple embryo transfer then they will be responsible for the "predictable disablement" (Shakespeare, 2006) of their children.

To be clear, because University Clinic fertility specialists did not usually maintain relationships with patients beyond their early weeks of pregnancy, their main points of encounter with people pregnant with twins or with newborn twins was in their obstetrics residencies before their fellowships focusing on fertility medicine. Consequently, most horror stories that they had based on personal experience were often decades old and not reflective of state-of-the-art neonatal intensive care units. On cerebral palsy, which encompasses many different movement disorders and associated conditions with a wide range in severity, fertility specialists were not providing an informative perspective but rather a persuasive and stigmatizing one. Just as Rapp demonstrated how "miscommunication as well as communication, silence as well as conversation" (2004, p. 63) might guide genetic counseling sessions, so too can these elisions guide conversations about the heightened chance of disability associated with twins in fertility clinics. Educating patients about cerebral palsy had nothing to do with the reality of people living with cerebral palsy, and everything to do with convincing patients to transfer one embryo while maintaining success rates with PGT-A.

Medical sociologists and disability scholars have found that ableism is endemic to medical education, clinical interactions, and medical decision-making (Mauldin & Brown, 2021). I found this to be the case at University Clinic: there were no physicians who talked about the risk of disability in anything but disparaging and sometimes outright offensive terms. In University Clinic consults, physicians made patients anticipate the possibility of their pregnancy being "spoiled" by disability (Rothman, 1986). As with the doctor at the beginning of the chapter, the threat of disability functioned as a trump card; whatever the patient might say about wanting twins, doctors could steer them away by scaring them about twins' higher chance of being disabled.

Connecting with Patients' Aversion to Disabilities

These scare tactics worked because nearly all patients were scared about having a child with any impairments. This was especially the case for University Clinic patients, most of whom were wealthy and could afford to purchase PGT-A, which cost several thousand dollars and was often not covered by even high-quality insurance. Barbara Katz Rothman and Rayna Rapp's influential studies of amniocentesis underline how social class mediates prospective parents' relationship to disability in the US. As Rothman writes, "the rich are able to buy their way out of disability"

(1986, p. 8). Wealthy people avoid some of the extreme stresses of caring for children with disabilities through their access to high-quality healthcare and education (Bixby, 2023). Yet, upper middle-class and wealthy individuals might be more inclined to make, in their own view, the "selfish" decision to avoid raising a disabled child through testing and abortion to prioritize other aspects of their lives besides caregiving (Rapp, 2004). Thus, while in countries like Denmark or China public systems and government authority lead to a universal response to avoiding children's impairments (Heinsen, 2018; Zhu, 2013), in the US, those with the resources to pay for medical care and education for children with disabilities are precisely those who are the most likely to avoid having these children through the use of reproductive technologies.

While most patients did not enter University Clinic knowledgeable about the risks of pregnancies and births to multiples, once patients were engaged on the topic they sometimes remembered cases of twin disability or death. One day, I observed a follow-up consult with Dr. Ciccone and a couple ready to pursue IVF after receiving their test results. The patient, Iva, was 38 years old and had never been pregnant. Her husband, Eric, was in his late forties and had two children from a previous marriage, but for unknown reasons his sperm count had decreased substantially since conceiving them, making it necessary to pursue IVF to have more children. Iva told Dr. Ciccone that she wanted to transfer two embryos. He told her that she should only transfer one. He explained that nearly everyone who is pregnant with twins has a C-section, but more than that, twins are much more likely to be born prematurely and have "abnormalities." He told her that she could have one child and then another.

After noting that they can talk more about how many embryos to transfer later, Dr. Ciccone asked the couple if they might be interested in genetic testing. Eric replied that they were, and Dr. Ciccone explained that most University Clinic patients do it. When explaining how PGT-A works, Dr. Ciccone emphasized that the success rate for a "normal" embryo making a baby is "relatively high." They also discussed the strategy of trying to get as many "normal" embryos "in the freezer" as possible to increase the chance that they might be able to implant one now and another in a few years. Note that University Clinic providers sometimes mentioned to patients that their embryos might not survive the additional attrition of PGT-A, but they emphasized the higher chance of success per transfer. From the get-go, Dr. Ciccone used PGT-A to convince Iva and Eric that it is preferable to transfer one embryo.

Later in the consult, Dr. Ciccone asked the couple if they had any questions. Iva returned to the question of multiple embryo transfer, asking if putting in two embryos is dangerous. Dr. Ciccone responded that it is, primarily because of preterm birth in twins. Eric nodded along and told his wife that they end up in the ICU [intensive care unit] a lot of times. Dr. Ciccone added that twins might survive the NICU [neonatal intensive care unit] but then they have more learning disabilities. While earlier Dr. Ciccone might have promoted PGT-A as a way of not needing to transfer multiple embryos while keeping success rates high, when the conversation returns, he used the threat of disability to underline his recommendation for single embryo transfer.

Upon hearing Dr. Ciccone's remarks about the potential for disability, Eric became his ally, working with him to illustrate to his wife all the potential problems that come with pregnancy and birth to multiples. Even though Eric did not seem to enter the room knowing about the health problems that are more likely to transpire with twins, he quickly remembered and connected different cases of problematic multiples. He mentioned that one of his daughter's friends was a triplet and she had a lot of trouble seeing, and all three children had "some kind of problem."

Eric did not just mention recent cases of multiples but extended further into his family history. Eric said that his father was a twin in the 1930s but it was natural—his grandmother produced two eggs. Dr. Ciccone added that "it sounds horrible but bad ones died." Eric agreed, remembering that his ex-mother-in-law had a twin, and the twin died at birth. Although none of these examples were of twins born in the late 2010s with the top medical care of University Hospital, and indeed in his father's case at least there was seemingly no impairment, they were nevertheless brought to bear on this case as proof that twins must be avoided as a means of avoiding disability.

In the end, Eric became the one to convince his wife Iva that they would be better off spacing the children by a few years through a combination of PGT-A, single embryo transfer, and embryo banking than trying to have twins. Iva did not seem to harbor the same memories or concerns as her husband but was convinced by his argument to attempt to space out their children. From this case we can see how University Clinic physicians' argument about avoiding twins to avoid disability was often compatible with their patients' perception and fear of disability.

Not "Worth" the Risk of Disabilities

Not all patients, however, quickly acquiesced to University Clinic physicians' perspective on the health risks of twins, namely the potential for lifelong disabilities. On rare occasions, patients held firm in their preference for transferring multiple embryos and a twin birth. In these cases, physicians escalated their depiction of the health risks of twins. For example, I observed a consult with Dr. Ciccone and a couple starting fertility treatment at the University Clinic. Avani, 41-years-old, said to Dr. Ciccone that she did not have a problem with twins. Laughing lightly, she explained that she was getting old and maybe could not become pregnant later. She thought that she might have only one shot, and that way she could have two children. Rohan, her partner, agreed, telling Dr. Ciccone that he wanted to move fast because time was running out.

Rohan asked Dr. Ciccone if he has a philosophy about how many embryos to put back in. Dr. Ciccone said that he has a "very biased philosophy." He explained to Avani and Rohan that they test the embryos and then freeze them individually, and then they can transfer them one at a time. He told them that Avani might get pregnant now and then they could transfer another embryo in a year. He said, "Why are they resistant to putting in two? Twins can actually be very dangerous." He went on, "You just don't have access, you're misinformed, it's not your fault, you just don't have access to the information of really what happens to many patients who have twins." After going into risks related to pregnancy, Dr. Ciccone went into detail about the problems with preterm birth, as captured in my field notes:

> It's a huge problem if you deliver at 26 weeks, because then you could have children that are mentally and physically disabled and not only that, but you have two of them … the rate of cerebral palsy is six or seven times the

> rate with twins … some of that is not related to prematurity, the babies get stuck, even if you do a C-section the babies get stuck, and it takes a long time to get it out, and it's not always but if it happens to you it's not so good.

Nonetheless, after going through all these risks, Dr. Ciccone told Rohan and Avani that if they "really demand" a double embryo transfer that could be rearranged. In response, Rohan told Dr. Ciccone that they could always do a fetal reduction if both embryos progressed, and he was more worried about not even one embryo sticking. Dr. Ciccone said that it was not worth it and that the chance of each embryo sticking after it has undergone preimplantation genetic testing is very high.

In this case, simply describing the chance of causing disability through twins was not enough to dissuade patients from their preference for transferring more than one embryo. Dr. Ciccone also needed to be clear with Avani and Rohan that multiple embryo transfer was "not worth it" because PGT-A would still allow them to keep a high success rate—and because the potential for disabilities was so serious. Note too that PGT-A was also framed as solving the problem of multifetal reduction, even in a case in which the patients did not seem perturbed by its possibility. In its higher success rates at transfer, and avoidance of having children with disabilities, genetic testing was critical in convincing patients to avoid multiple embryo transfer.

One of the potential problems with bringing up serious impairments related to twins is that most twins, related to IVF or otherwise, are ultimately healthy. Some physicians at University Clinic were upfront about the relative and absolute risk of conditions like cerebral palsy with patients but framed the relative risk as more important because of defining the absolute risk as wholly avoidable. For example, I observed a consult with Dr. Ricci and patients in which he explained the IVF process to them, focusing on the problems with fresh embryo transfers and his work on PGT. As he explained, "There is seven-fold increased risk for cerebral palsy. It's still a tiny risk at 14 per thousand, but it's significant. The good news is all of the numbers are low, but seven times a low number—why take the risk, you have other embryos in the fridge; you can put them back another time." Dr. Ricci used similar language in many of his consults, describing twins as seven times more dangerous than singletons based on the relative risk of cerebral palsy, but also underlining that it was still a small risk overall. What was key to this framing was that transfer of frozen genetically tested embryos was provided as an easy solution. In this way, the health risks associated with twins—articulated as the main ones related to fertility treatment—are not just permanent: they are wholly avoidable.

Framing the lifelong disabilities related to twins as avoidable is key. The condition that PGT-A is most known for preventing—Down syndrome—is very different from the one that most fertility specialists described to patients as being prevented by PGT-A coupled with single embryo transfer—cerebral palsy. Down syndrome is a naturally occurring chromosomal abnormality "solved" by medical intervention, whereas cerebral palsy brought on by ART-related multiples, in this framing, is iatrogenic, a disease caused by medical intervention. Due to genetic testing during pregnancy, Down syndrome has become a nearly wholly avoidable condition: the majority of pregnant people in the US test for and elect to abort Down syndrome fetuses (Natoli et al., 2012). Consequently, for many fertility patients, it is not so much a question of *whether* to engage with this type of genetic selection, but rather, *when* to do so. Thus, it is more powerful for fertility specialists to bring up cerebral palsy as a possibility if patients do not heed their recommendations to transfer a single embryo, as a condition that might not be easily avoided at other stages of the pregnancy because of the lack of conversation and acceptance of fetal reductions from twins to singletons. With prospective parents socially obligated to avoid "predictable disablement" (Shakespeare, 2006), the specific threat of cerebral palsy worked as a powerful tool to promote single embryo transfers and avoid twins.

All Unhealthy Twins Means All Healthy Singletons

In the absence of national legislation directing care, and the need to appeal to patient-consumers in the US fertility field, University Clinic physicians promoted single embryo transfer by underlining lifelong and irreversible disabilities associated with twins. As much as we have entered the geneticist era of reproductive medicine (Raz, 2009), at University Clinic, physicians connected disability as much to twins as to genetics. PGT was framed as a way of avoiding twins, more so than directly preventing genetic disease, save for cases in which patients were known carriers for a condition. In this way, while testing for a known genetic mutation through PGT-M or PGT-SR directly prevented some genetic conditions for a small group of people, by preventing twins, the much more common PGT-A was a way of preventing disability for everyone (see Turrini & Otín-Gavín, 2025 on the blending of PGT-M, PGT-SR, and PGT-A).

More generally, shuffling health risks onto multiples then allows singletons from IVF to be "healthy" in comparison. Fertility specialists strategically foreground some chances of health problems related to a twin pregnancy and birth, but they also elide others, including the chance of a complicated singleton IVF birth and the chance of a worthwhile life with cerebral palsy. For example, on the patient education fact sheet produced by the American Society for Reproductive Medicine (ASRM 2023), the question is posed: "If I conceive with IVF, will my pregnancy be more complicated (than if I conceived on my own)?" The printed response is as follows:

- Having a multiple pregnancy (pregnancy with more than one baby) is more likely with IVF, particularly when more than one embryo is transferred. These pregnancies carry significant risks, including:
- Preterm labor and/or delivery: premature babies (regardless of whether or not they were conceived naturally or with IVF) are at higher risk for health complications such as lung development problems, intestinal infections, cerebral palsy, learning disabilities, language delay, and behavior problems
- Maternal hemorrhage
- Delivery by cesarean section (C-section)
- Pregnancy-related high blood pressure
- Gestational diabetes

At no point does this factsheet answer the question about increased pregnancy risks related to IVF without referencing multiples, even though there is a body of scientific research to suggest that these are in fact more complicated pregnancies, even when they are with singletons (Pandey et al., 2012). Both institutionally and interactionally, punting health risks to a type of fertility treatment saves fertility treatment as a whole from being implicated as a health risk to avoid.

This chapter highlights the normalization, multiplicity, and commercialization that define mundane genomics. Genetic testing is a tool (Casper & Clarke, 1998) that can be repurposed toward different ends depending on the needs of practitioners in particular clinical settings (Mol et al., 2010). In the context of US fertility medicine as a consumer-driven private medical market that is not governed by a central authority, PGT-A works as a tool to help fertility specialists avoid getting mired in arguments with patients who would prefer to transfer multiple embryos to improve their chances of conception. Put differently, PGT-A eases the mundane task of patient counseling.

Moreover, we can see how the selective aspect of selective reproductive technologies (SRTs) (Wahlberg & Gammeltoft, 2014) can sometimes be obscured or indirect. Fertility patients and physicians might aim to produce some children and not others not by directly rejecting certain embryos (though of course that is a part of PGT) but really by valorizing singleton birth and maligning twins. Prospective parents, then, are to some extent responsibilized to avoid twins and made blameworthy for any impairments or disabilities that they might experience (Shakespeare 2006).

Finally, this analysis highlights how in a biocapitalist society like the US, fertility clinics harness the ablism of wealthy patient-consumers to sell an expensive and highly profitable product like PGT-A (Borsa & Bruch, 2022; van de Wiel, 2020). Although fertility specialists might have sold PGT-A to patients partially based on its ability to avoid compromising success rates when transferring fewer embryos, they also did so by expounding on how PGT-A paired with single embryo transfer would prevent disabilities by preventing twins. This medical communication can only prove convincing to patients who agree that it is worth spending thousands of dollars to avoid even slightly increasing the chance of disabilities like cerebral palsy.

References

American Society for Reproductive Medicine. (2023). *Reproductive facts: Patient fact sheet developed by the American Society for Reproductive Medicine.*

Asch, A. (1989). Reproductive technology and disability. In S. Cohen & N. Taub (Eds.), *Reproductive laws for the 1990s* (pp. 69–124). Humana Press. https://doi.org/10.1007/978-1-4612-3710-5_4

Asch, A. (2002). Disability equality and prenatal testing: Contradictory or compatible. *Florida State University Law Review, 30*(2), 315.

Barishansky, S. J., Hutchinson, A. P., Lawson, A. K., & Pavone, M. E. (2022). The parent trap: Desire for multifetal gestation among patients treated for infertility. *Journal of Assisted Reproduction and Genetics, 39*(6), 1399–1407. https://doi.org/10.1007/s10815-022-02534-5

Bixby, L. E. (2023). Disability is not a burden: The relationship between early childhood disability and maternal health depends on family socioeconomic status. *Journal of Health and Social Behavior, 64*(3), 354–369.

Borsa, A., & Bruch, J. D. (2022). Prevalence and performance of private equity–affiliated fertility practices in the United States. *Fertility and Sterility, 117*(1), 124–130.

Casper, M. J., & Clarke, A. E. (1998). Making the pap smear into the right tool for the job: Cervical cancer screening in the USA, circa 1940–95. *Social Studies of Science, 28*(2), 255–290.

Centers for Disease Control and Prevention. (2023). QuickStats: Percentage of mothers with gestational diabetes, by maternal age—National Vital Statistics System, United States, 2016 and 2021. *MMWR. Morbidity and Mortality Weekly Report, 72*, 16. https://doi.org/10.15585/mmwr.mm7201a4

Conrad, P., & Leiter, V. (2004). Medicalization, markets and consumers. *Journal of Health and Social Behavior, 45*, 158–176.

Edwards, R., & Steptoe, P. (1980). *A matter of life: The story of a medical breakthrough.* Morrow.

Ehrich, K., Williams, C., Farsades, B., Sandall, J., & Scott, R. (2007). Choosing embryos: Ethical complexity and relational autonomy in staff accounts of PGD. *Sociology of Health & Illness, 29*(7), 1091–1106. https://doi.org/10.1111/j.1467-9566.2007.01025.x

Elsner, C. W., Tucker, M. J., Sweitzer, C. L., Brockman, W. D. W., Morton, P. C., Wright, G., & Toledo, A. A. (1997). Multiple pregnancy rate and embryo number transferred during in vitro fertilization. *American Journal of*

Obstetrics and Gynecology, 177(2), 350–357. https://doi.org/10.1016/S0002-9378(97)70194-9

Franklin, S., & Roberts, C. (2006). *Born and made: An ethnography of preimplantation genetic diagnosis*. Princeton University Press.

Gammeltoft, T. (2014). *Haunting images: A cultural account of selective reproduction in Vietnam*. University of California Press.

Gammeltoft, T. M., & Wahlberg, A. (2014). Selective reproductive technologies. *Annual Review of Anthropology, 43*(1), 201–216. https://doi.org/10.1146/annurev-anthro-102313-030351

Goldfarb, J. M., Austin, C., Lisbona, H., Peskin, B., & Clapp, M. (1996). Cost-effectiveness of in vitro fertilization. *Obstetrics & Gynecology, 87*(1), 18–21. https://doi.org/10.1016/0029-7844(95)00385-9

Heinsen, L. L. (2018). Moral adherers: Pregnant women undergoing routine prenatal screening in Denmark. In A. Wahlberg & T. Gammeltoft (Eds.), *Selective reproduction in the 21st century* (pp. 69–95). Palgrave Macmillan.

Katz, P., Nachtigall, R., & Showstack, J. (2002). The economic impact of the assisted reproductive technologies. *Nature Medicine, 8*(10), S29–S32. https://doi.org/10.1038/nm1002-29

Kolata, G. B. (1978). In vitro fertilization: Is it safe and repeatable? *Science, 201*(4357), 698–699. https://doi.org/10.1126/science.684419

Kuleshova, L., Gianaroli, L., Magli, C., Ferraretti, A., & Trounson, A. (1999). Birth following vitrification of a small number of human oocytes: Case report. *Human Reproduction, 14*(12), 3077–3079. https://doi.org/10.1093/humrep/14.12.3077

Kulkarni, A. D., Jamieson, D. J., Jones, H. W., Kissin, D. M., Gallo, M. F., Macaluso, M., & Adashi, E. Y. (2013). Fertility treatments and multiple births in the United States. *New England Journal of Medicine, 369*, 2218–2225. https://doi.org/10.1056/NEJMoa1301467

Lippman, A. (1991). Prenatal genetic testing and screening: Constructing needs and reinforcing inequities. *American Journal of Law & Medicine, 17*(1–2), 15–50. https://doi.org/10.1017/S0098858800010489

Macklon, N. S., Stouffer, R. L., Giudice, L. C., & Fauser, B. C. J. M. (2006). The science behind 25 years of ovarian stimulation for in vitro fertilization. *Endocrine Reviews, 27*(2), 170–207. https://doi.org/10.1210/er.2005-0015

Mauldin, L., & Brown, R. L. (2021). Missing pieces: Engaging sociology of disability in medical sociology. *Journal of Health and Social Behavior, 62*(4), 477–492. https://doi.org/10.1177/00221465211038458

Merton, R. K. (1936). The unanticipated consequences of purposive social action. *American Sociological Review, 1*(6), 894–904. https://doi.org/10.2307/2084615

Mol, A., Moser, I., & Pols, J. (Eds.). (2010). *Care in practice: On tinkering in clinics, homes and farms*. Transcript Verlag.

Morris, T. (2013). *Cut it out: The C-section epidemic in America*. New York University Press.

Natoli, J. L., Ackerman, D. L., McDermott, S., & Edwards, J. G. (2012). Prenatal diagnosis of down syndrome: A systematic review of termination rates (1995–2011). *Prenatal Diagnosis, 32*(2), 142–153. https://doi.org/10.1002/pd.2910

Ouellette, A. (2015). Selection against disability: Abortion, ART, and access. *Journal of Law, Medicine & Ethics, 43*(2), 211–223. https://doi.org/10.1111/jlme.12236

Pandey, S., Shetty, A., Hamilton, M., Bhattacharya, S., & Maheshwari, A. (2012). Obstetric and perinatal outcomes in singleton pregnancies resulting from IVF/ICSI: A systematic review and meta-analysis. *Human Reproduction Update, 18*(5), 485–503. https://doi.org/10.1093/humupd/dms018

Parikh, F. R., Athalye, A. S., Naik, N. J., Naik, D. J., Sanap, R. R., & Madon, P. F. (2018). Preimplantation genetic testing: Its evolution, where are we today? *Journal of Human Reproductive Sciences, 11*(4), 306–314. https://doi.org/10.4103/jhrs.JHRS_122_18

Rapp, R. (2004). *Testing women, testing the fetus: The social impact of amniocentesis in America*. Routledge.

Raspberry, K. (2009). The genesis of embryos and ethics in vitro: Practicing preimplantation genetic diagnosis in Argentina. In D. Birenbaum-Carmeli & M. C. Inhorn (Eds.), *Assisted reproduction, testing genes: Global encounters with the new biotechnologies* (pp. 213–238). Berghahn Books.

Raz, A. E. (2009). Eugenic utopias/dystopias, reprogenetics, and community genetics. *Sociology of Health & Illness, 31*(4), 602–616. https://doi.org/10.1111/j.1467-9566.2009.01165.x

Rothman, B. K. (1986). *The tentative pregnancy: How amniocentesis changes the experience of motherhood*. Viking.

Saxton, M. (1984). Born and unborn: The implications of reproductive technologies for people with disabilities. In R. Arditti, R. D. Klein, & S. Minden (Eds.), *Test-tube women: What future for motherhood?* (pp. 298–312). Routledge.

Scott, R. T., Upham, K. M., Forman, E. J., Hong, K. H., Scott, K. L., Taylor, D., Tao, X., & Treff, N. R. (2013). Blastocyst biopsy with comprehensive chromosome screening and fresh embryo transfer significantly increases in vitro fertilization implantation and delivery rates: A randomized controlled trial. *Fertility and Sterility, 100*(3), 697–703. https://doi.org/10.1016/j.fertnstert.2013.04.035

Shakespeare, T. (2006). The dilemma of predictable disablement: A challenge for families and society. In E. Parsons (Ed.), *The nature of difference: Science, society, and human biology* (pp. 213–228). Routledge.

Shakespeare, T. (2013). *Disability rights and wrongs revisited* (2nd ed.). Routledge.

Shih, L.-W. (2018). Moral bearing: The paradox of choice, anxiety and responsibility in Taiwan. In A. Wahlberg & T. Gammeltoft (Eds.), *Selective reproduction in the 21st century* (pp. 97–122). Palgrave Macmillan.

Stillman, R. J., Richter, K. S., & Jones, H. W., Jr. (2013). Refuting a misguided campaign against the goal of single-embryo transfer and singleton birth in assisted reproduction. *Human Reproduction, 28*(10), 2599–2607. https://doi.org/10.1093/humrep/det291

Turrini, M., & Otín-Gavín, J. (2025). Let 'pregnant women choose the destiny for themselves and their child'. How fertility clinic digital platforms frame preimplantation genetic testing (PGT) in Spain. *Sociology of Health & Illness, 47*(1), e13876. https://doi.org/10.1111/1467-9566.13876

van de Wiel, L. (2020). The speculative turn in IVF: Egg freezing and the financialization of fertility. *New Genetics and Society, 39*(3), 306–326. https://doi.org/10.1080/14636778.2020.1783699

Wu, C.-L. (2023). *Making multiple babies: Anticipatory regimes of assisted reproduction* (Vol. 52). Berghahn Books..

Yoon, T. K., Chung, H. M., Lim, J. M., Han, S. Y., Ko, J. J., & Cha, K. Y. (2000). Pregnancy and delivery of healthy infants developed from vitrified oocytes in a stimulated in vitro fertilization–embryo transfer program. *Fertility and Sterility, 74*(1), 180–181. https://doi.org/10.1016/S0015-0282(00)00557-1

Zhu, J. (2013). Projecting potentiality: Understanding maternal serum screening in contemporary China. *Current Anthropology, 54*(S7), S36–S44. https://doi.org/10.1086/673386

5

Genetic Testing in Assisted Reproduction Treatments with Donated Gametes at the Spanish Legislation

Flor Arias Aparicio

Introduction

Genetic medicine raises expectations with regard to the prevention of many diseases, particularly in the field of reproductive medicine, where genetic studies make it possible to obtain a pre-clinical diagnosis of patients that allows them to discover or rule out certain hereditary diseases. Reproduction in itself, whether through the use of one's own or donated gametes, involves a risk, hence the important role that genetic screening tests play in the prevention of recessive hereditary diseases in offspring. The treatments carried out in reproductive medicine are preceded by genetic studies of patients and gamete donors; however, even after genetic testing, disease-free offspring cannot be guaranteed. The clinical results of carrier testing (preconception genetic testing that tests for monogenic diseases that follow a recessive or X-linked pattern of inheritance) and genetic matching (study of the genetic data of donor

F. A. Aparicio (✉)
Public Law Department, Universidad de Extremadura, Cáceres, Spain
e-mail: florarias@unex.es

V. Argudo-Portal et al. (eds.), *Mundane Genomics*,
https://doi.org/10.1007/978-981-95-4550-6_5

and recipient, or the couple, to ensure genetic compatibility between the two to avoid the transmission of a recessive disease) have demonstrated their efficacy in minimising the risks of the diseases studied in these tests (SEF, 2019). Nonetheless, there is no such thing as zero risk: that is, it is not possible to guarantee offspring free of all congenital alterations, diseases, or defects. All human beings carry pathogenic variants in one or more genes associated with recessive or X-linked diseases and, when they reproduce, there is a 2–5% risk of having offspring with severe hereditary diseases.

The use of donated gametes is currently one of the most common resources for assisted human reproduction technique treatments to achieve the desired pregnancy. The contribution of donors, which is necessary in many assisted reproduction treatments, requires the application of protocols that guarantee compatibility and safety in the selection and application of these donations. Technological advances make it possible to detect mutations for recessive diseases, opening up the possibility of genetic matching between donors and recipients, thereby reducing the risk of diseases that can be transmitted to offspring. In this context, providing adequate information to patients about the genetic testing and its risks is essential given the likelihood that patients will not understand the complex information obtained from these tests; thus, medical teams play an important role in counselling and decision-making.

Spanish legislation contains provisions establishing the requirements to be met by the institutions and persons who carry out genetic analysis, as well as the purpose or type of genetic tests that can be performed. The proliferation of genetic testing in the field of assisted human reproduction has highlighted the inadequacy of the current regulation in this area in relation to issues such as the tests to be carried out on gamete donors, as the regulations do not establish which specific genetic tests should be included in a genetic screening programme. Moreover, the information to be provided to patients and donors regarding the genetic studies carried out on them is also not sufficiently addressed by the law. These problems have been pointed out by the doctrine that highlights the deficiencies and shortcomings of the current legislation in relation to innovative technology (Abellán-García, 2019; Martínez Otero, 2017; Reguera Cabe, 2021).

In this context, the reproductive decisions of genetic testing users are based on the counselling provided by the fertility clinics that disseminate the proposed use of genetic tests as a means of responding to reproductive problems—as well as on a positive perception of these tests. Meanwhile, the lack of specific Spanish legislation regarding the performance of certain genetic tests facilitates the normalisation and routinisation of genomic technologies and information in IVF cycles, both among users—to the extent that they trust them and consider them increasingly relevant—and by fertility clinics themselves, which recommend their use. This lack of attention on the part of the legislator, in turn, has favoured the spread of this new technology across the private sector, which compensates, in practice, for the legislative neglect by setting the rules on how and when to use the tests.

Specifically, this chapter analyses the legislation in force that regulates genetic testing in reproductive matters, mainly Law 14/2007, of 3 July, on Biomedical Research,[1] and Law 14/2006, of 26 May, on Assisted Human Reproduction Techniques.[2] As can be seen, these regulations are almost two decades old and do not take into account new scientific advances in the field, nor the problems (ethical and social) that they can generate. Other regulations that complement and detail the requirements and conditions are also analysed in relation to specific aspects foreseen in both laws. These rules must be read and interpreted within the framework of the international agreements signed by Spain, insofar as they form part of its legal system, as well as in accordance with the court rulings that have been handed down and in which the application of the laws under study is specified. Over the last ten years, a wealth of case law has been generated on the subject. The decided cases highlight the importance of adequately informing patients about the need for genetic testing and its risks, and also warn of the inability of users of this technology to understand the complex information obtained from these tests, and stress the decisive role—the responsibility—of medical teams and reproduction clinics in decision-making.

---

[1] «BOE» núm. 159, de 4 de julio de 200. https://www.boe.es/eli/es/l/2007/07/03/14.

[2] «BOE» núm. 126, de 27/05/2006. https://www.boe.es/eli/es/l/2006/05/26/14.

The following analysis of the existing legislation highlights its limitations, as it is slow to adapt to scientific developments and has shortcomings regarding the use of genetic testing (its scope and limits; the need for its application; its selective or assisted nature, etc.). In this respect, it is a commonplace that legal regulation rarely keeps pace with scientific progress; however, to consider that the regulatory response *always* lags behind scientific progress is to assume a linear and deterministic view of innovation (Parthasarathy, 2025). This time lag between law and science can be overcome through contemporary collaboration between technological advances and regulation, in a mutually supportive way, and by both spheres integrating socio-political dynamics through co-production.

The Regulatory Framework for Genetic Studies in Reproductive Medicine

The 1997 Convention for the Protection of Human Rights and Dignity of the Human Being with regard to the Application of Biology and Medicine (known as the Convention on Human Rights and Biomedicine or the Oviedo Convention) establishes provisions in relation to the application of biomedical research to human beings (Sánchez Patrón, 2024). This international agreement allows, specifically, the carrying out of analyses both for predictive purposes (the aim of which is to determine the predisposition or detection of a genetic disease of the individual), and for diagnostic purposes (the aim of which is to determine the genetic characteristics of a person). In this sense, Article 12 of the Oviedo Convention states:

> Tests that are predictive of genetic diseases or that make it possible to identify the subject as a carrier of a gene responsible for a disease, or to detect a genetic predisposition or susceptibility to a disease, may only be carried out for medical purposes or medical research and with appropriate genetic counselling.

Intervention involving a variation in the genome has to be justified for "preventive, diagnostic or therapeutic reasons" (Article 13). This condition constitutes a restriction on the scope of intervention in the genome, meaning that any modification of the genome has to be related to a previous disease or condition.[3] The alteration of the genome, therefore, constitutes the barrier beyond which it must be justified.

In this context, the Spanish law that regulates research in the field of biomedicine, both basic and clinical, is Law 14/2007, of 3 July, on Biomedical Research (hereinafter LBR). Specifically, this law establishes the conditions and guarantees for carrying out genetic analyses for research or diagnostic purposes (Abellán-García, 2009). Its Article 9 limits the performance of tests on present or foreseeable genetic diseases with a research or therapeutic objective in line with the Oviedo Convention. Genetic analyses of pre-embryos (embryos formed *in vitro* up to 14 days of their development), embryos, and foetuses in the mother's womb require the written consent of the pregnant woman (Article 48.4 LBR). Interventions carried out on embryos or foetuses in the mother's womb may only have a diagnostic or therapeutic purpose in the embryo's own interest (Article 30 LBR). In the case of pre-embryos not yet transferred to the future pregnant woman, the performance of these analyses, for diagnostic or therapeutic purposes, is regulated by the provisions of Law 14/2006, of 26 May, on Assisted Human Reproduction Techniques (hereinafter, LAHRT).

These two laws, therefore, constitute the two fundamental rules governing genetic testing in Spain. Other regulations, however, complement and detail the requirements and conditions foreseen in both laws in relation to aspects such as patient autonomy and their rights and obligations in terms of information and clinical documentation (Law 41/2002, of 14 November); the compulsory protocols for the study of donors and users related to assisted human reproduction techniques (Royal Decree 412/1996, of 1 March); the quality and safety standards for donation (Royal Decree-Law 9/2014, of 4 July); or the regulation of an advisory and consultative body, the National Commission on Assisted Human

[3] Rapport explicatif de la Convention pour la protection des Droits de l'Homme et de la dignité de l'être humain à l'égard des applications de la biologie et de la médecine: Convention sur les Droits de l'Homme et la biomédecine, Oviedo, 4 April 1997, p. 15, par. 90. https://rm.coe.int/16800cce7e.

Reproduction, in the use of assisted human reproduction techniques (Royal Decree 42/2010, of 15 January).

Genetic Testing in the Assisted Reproduction Treatment of Patients and Gamete Donors

In the field of reproductive medicine, genetic testing can take place at different times or stages: (i) genetic testing of the couple or woman who wish to have healthy children; (ii) genetic testing of gamete donors; (iii) testing of *in vitro* embryos, prior to their transfer to the mother's uterus, to detect and rule out a disease; (iv) genetic testing of born children. When carrying out one or other of these tests, it must be taken into account that the relevant legal regulations are different: while the genetic analysis of parents and donors is regulated by the Law on Biomedical Research (LBR), the analysis of *in vitro* embryos is governed by the regulations on assisted reproduction, more specifically, by the precepts that regulate preimplantation genetic diagnosis contained in the Law on Assisted Human Reproduction Techniques (LAHRT).

When those who wish to have offspring are aware of the existence in their medical history, or in their family, of a disease that they can transmit, they can carry out genetic studies to confirm or rule out this risk. Indeed, it is becoming increasingly common for a couple with a family history who wish to have a healthy child to undergo genetic testing to confirm or rule out their status as carriers of a particular inherited genetic mutation. The Biomedical Research Act establishes the types, requirements, and conditions of the genetic tests that can be carried out: (1) predictive tests for genetic diseases; (2) tests to identify the subject as a carrier of a gene responsible for a disease; and (3) tests to detect a predisposition to develop a disease or the condition of carrier of a genetic variant that may condition the response to a specific treatment.

The results of these tests will help potential users make the decision about whether to have children by natural methods (if the risk of transmission is ruled out) or to resort to assisted reproduction techniques to select pre-embryos that are not affected or carriers of the disease in question (preimplantation genetic diagnosis). Indeed, the finding of a hereditary disease can be a determining factor in deciding on embryonic genetic diagnosis in order to detect and rule out the disease in offspring (Pavone & Arias, 2012).

The tests that focus on embryos (known as preimplantation genetic testing) are aimed at selecting those that are free of a genetic or chromosomal alteration, and their regulation is found in the Law on Assisted Human Reproduction Techniques. Specifically, this law regulates the different variants of these tests according to the objective pursued: PGT-A (preimplantation genetic test for aneuploidy), PGT-SR (preimplantation genetic test for structural anomalies), and PGT-M (preimplantation genetic test for monogenic diseases). In addition, the LAHRT allows preimplantation genetic testing in combination with histocompatibility antigen determination (PGT + HLA) for therapeutic purposes for third parties ("saviour babies").

The LAHRT establishes the cases that can be carried out directly by the clinics (detection of serious, early onset, hereditary disease without postnatal curative treatment at the present time,[4] and detection of other alterations that may compromise the viability of the embryo, Article 12 LAHRT),[5] as well as others that require administrative authorisation, and a prior favourable report from a commission of experts (the National Commission for Assisted Human Reproduction [CNRHA]).[6]

The use of donated gametes is one of the most widely used assisted reproduction techniques today. The contribution of donors in the

[4] Vid, Comisión Nacional de Reproducción Humana Asistida (2025a). *Recomendaciones para la realización de técnicas de PGT en casos de Enfermedad de Huntington.* https://cnrha.sanidad.gob.es/documentacion/home.htm.

[5] Vid. Comisión Nacional de Reproducción Humana Asistida (2025b). *Recomendaciones en relación con Enfermedades susceptibles de ser consideradas entre los supuestos del artículo 12.1 de la Ley 14/2006, de 26 de mayo, sobre técnicas de reproducción humana asistida.* https://cnrha.sanidad.gob.es/documentacion/home.htm.

[6] Article 20 of the LAHRT establishes the National Commission on Assisted Human Reproduction as a "collegiate body, of a permanent and consultative nature, aimed at advising and guiding on the use of assisted human reproduction techniques, contributing to the updating and dissemination of scientific and technical knowledge in this field, as well as to the development of functional and structural criteria of the centres and services where these techniques are carried out". The purpose, composition, and functions of this Commission are developed in Royal Decree 42/2010, of 15 January, which regulates the National Commission on Assisted Human Reproduction (2025c). The activity reports are available at: https://cnrha.sanidad.gob.es/documentacion/home.htm.

reproductive process requires ensuring the selection and application of donations in terms of compatibility and safety; it is, therefore, essential to be able to identify whether gamete donors are carriers of recessive diseases. The current legislation on gamete donation does not define the specific genetic tests to be performed, merely establishing the criteria for the selection and clinical evaluation of gamete donors. In relation to genetic studies, it is indicated that genetic screening for recessive genes prevalent in the donor's ethnicity should be carried out on the basis of international scientific data, as well as an assessment of the risk of transmission of known hereditary diseases present in the family.

Advances in scientific knowledge of the human genome and improvements in genomic techniques (in particular mass sequencing) have led to the emergence and implementation of numerous genetic tests in the clinical practice of assisted human reproduction centres and in gamete donation programmes, without there being any legal provisions in this respect. This is the case with carrier screening tests, which detect thousands of mutations related to hundreds of serious genetic diseases that respond to an autosomal recessive or X-linked inheritance. Carrier screening tests for recessive mutations in patients and donors in the field of assisted reproduction techniques are aimed at checking the presence of genetic mutations for severe recessive diseases in gametes in order to establish the existing reproductive risk and assess the possible options that would reduce the risk of having a child affected by these diseases (Reguera, 2021). This practice can be accompanied by the use of genetic matching techniques. So-called genetic matching consists of matching genetic studies—the genetic tests of the donor and the patient who provides his or her own gametes—in order to determine the genetic risk in the offspring.

The large number of known genetic conditions means that there is a wide variety of test types that differ in the range of diseases they study. A distinction is made between basic tests and extended tests, which differ in the number of diseases included; however, there is no normative provision determining the prevalence cut-off point (Reguera, 2021). The practical consequence is that it is up to the reproduction centres to decide which genetic tests will be performed, and it is up to these centres to provide adequate information on both the possible advantages and limitations of the use of these tests (Abellán-García, 2019). This lack of regulation has led to the development of documents and guidelines by national and international scientific associations and societies containing recommendations and criteria to be followed for the selection of diseases. This was undertaken at the time by the Spanish Fertility Society (SEF) and the Association for Reproductive Biology Studies (ASEBIR), which in 2012 published a joint document of recommendations for the application of the relevant regulations.[7] In this regard, the disorders recommended for inclusion in the expanded carrier screening tests must be severe, early onset, with associated pathogenic variants with high penetrance and with a clear genotype-phenotype relationship (ACOG, 2017; Urbano et al., 2024).

Gamete Donation in Assisted Human Reproduction: Donor Selection and Genetic Screening

As noted, gamete donation is one of the most widely used assisted reproduction techniques today. Spanish legislation establishes the criteria to be followed by the centres to select gamete donors suitable for the donation process (Arias et al., 2018). The Law on Assisted Reproduction Techniques establishes that donors must be of legal age and in good physical and mental health (Article 5.6 LAHRT). In order to determine the donor's health condition, Spanish legislation provides for a series of medical tests to be carried out, as set out in Royal Decree 412/1996, of 1 March, which establishes the compulsory protocols for the study of donors and users related to assisted human reproduction techniques. When selecting donors, and with the aim of avoiding the appearance of chromosomal malformations, a legal upper age limit of 35 is set for the donation of female gametes, while there is no age limit for the donation of semen (SEF, 2007).

[7] More recently, the position of these entities is contained in ASEBIR's position on genetic screening studies for recessive diseases in donation programmes, 2016; SEF, Genetic screening of gamete donation, 2019.

With regard to the health of donors, Article 3 of Royal Decree 412/1996 establishes that "donors will undergo a medical examination, which will be reflected in a medical history, including personal and family history as well as a physical examination, which must contain at least the data listed in the annex to this Royal Decree". In addition to the "basic protocol for the study of donors", Royal Decree 412/1996 establishes the studies that, in all cases and as a minimum, must be carried out on donors:[8] blood group, Rh factor, VDRL or a similar test to detect syphilis, hepatitis screening, HIV marker detection test, a clinical study for the detection of clinical stages of toxoplasmosis, rubella, herpes virus, and cytomegalovirus, and clinical study for the detection of neisseria gonorrhoeae and chlamydia trachomatis.[9] Likewise, it is established that persons with a family history of malformations linked to chromosomopathies, genopathies, or metabolopathies cannot be admitted as gamete donors, and likewise those with genetic, hereditary, or congenital transmissible diseases. Royal Decree-Law 9/2014 also establishes the quality and safety standards for donation[10] in relation to the use of reproductive cells from donors other than the usual partner, and determines the criteria that must be met:[11] (a) Donors will be selected on the basis of their clinical history, which must be taken by the responsible physician. This assessment will include any factors that may be relevant in the identification and selection of those whose donation may pose a health risk to others, such as the possibility of transmitting a disease, or to themselves.

[8] Vid. Article 4 and Annex of Royal Decree 412/1996 establishing the compulsory protocols for the study of donors and users related to assisted human reproduction techniques and regulating the creation and organisation of the National Registry of Gamete and Pre-embryo Donors for human reproduction purposes.

[9] Vid. Article 6 of Royal Decree 412/1996 establishing the compulsory protocols for the study of donors and users related to assisted human reproduction techniques and regulating the creation and organisation of the National Registry of Gamete and Pre-embryo Donors for human reproduction purposes.

[10] Royal Decree-Law 9/2014 of 4 July 2014, establishing quality and safety standards for the donation, procurement, testing, processing, preservation, storage, and distribution of human tissues and cells, and approving the coordination and operational rules for their use in humans, transposes Directive 2004/23/EC of the European Parliament and of the Council of 31 March 2004 on setting standards of quality and safety for the donation, procurement, testing, processing, preservation, storage, and distribution of human tissues and cells.

[11] Vid. Point 3 of Annex IV: "Selection and evaluation of the donor of reproductive cells" of Royal Decree-Law 9/2014.

(b) Donors must have negative serological markers for HIV 1 and 2, HVC and HVB and syphilis. (c) Donors who live or come from areas with a high incidence of disease or whose sexual partners or parents live in or come from areas with a high incidence of the disease will be tested for anti-HTLV I and II antibodies. (d) In some circumstances additional tests may be required depending on the donor's medical history or cell/tissue characteristics (i.e. malaria—CMV—Tripanosoma cruzi, RhD). f) A genetic load assessment will be carried out in relation to the existence of autosomal recessive genes according to scientific knowledge and known prevalence in the donor's ethnicity. (g) An assessment of the risk of transmission of known inherited diseases present in the family will be carried out.

The lack of legal provision regarding the specific genetic tests that should be included in a genetic screening programme for gamete donors, which is left to the discretion of the centres, has been compensated for by the recommendations and guidelines provided by professional associations that are in favour of including them in the clinical practice of assisted reproduction techniques (ASEBIR, 2016). Consequently, most scientific societies have published their positions on which diseases should be included in genetic tests for donor selection, and carrier testing has been incorporated into gamete donation programmes in order to determine the carrier status of donors (SEF, 2019). The performance of these tests opens up the possibility of genetic matching of donors and recipients to reduce the risk of transmission of the genetic diseases studied in them. Although legislation does not require this test to be performed, fertility centres endorse genetic matching as a tool to avoid the coincidence of two mutations of the same gene and thus reduce the risk of transmission.

The vast majority of assisted human reproduction centres or units incorporate screening for recessive diseases into their gamete donation programmes (Capalbo et al., 2024), which involves decision-making by the biomedical teams responsible for the procedure in relation to issues such as the use of donors and their assignment to recipient couples, the weighing of the risks and benefits involved in genetic testing, the feasibility of the programme, and the responsibility for the treatment of the information thus obtained (SEF, 2019).

Finally, Spanish legislation provides that donors must be informed of the results obtained in the tests carried out on them.[12] This information must be as complete as possible in relation to the associated risks and the measures adopted or that may be adopted, and must be clearly transmitted and explained to the recipient. The following section is devoted to the correct transmission of information to all parties involved in gamete donation.

The Scope of Information Related to Genetic Testing

Tests that are predictive of genetic diseases, or that make it possible to identify the subject as a carrier of a gene responsible for a disease or to detect a genetic predisposition or susceptibility to a disease, require appropriate genetic counselling (Article 12 of Oviedo Convention). In the field of assisted reproduction treatments, all genetic tests carried out on both patients and donors must be accompanied by express consent for their performance, genetic counselling, and the possibility of knowing the results of the tests.

The generic obligation to inform patients and donors about treatments is included in the LAHRT (Article 3, sections 3 and 4), while the information and consent relating to specific tests to ascertain the condition of being a sufferer or carrier of diseases are regulated in the LBR (Articles 44–57). Thus, the LBR establishes that when a genetic analysis is carried out for health purposes, the interested party must be guaranteed appropriate genetic counselling, in a manner to be determined by regulation, respecting in all cases the criteria of the person concerned (Article 55 LBR). The LBR defines genetic counselling as a procedure aimed at, firstly, informing a person about the possible consequences for him/her or his/her offspring of the results of a genetic analysis or screening and its advantages and risks, and, secondly, where appropriate, advising him/her on the possible alternatives resulting from the analysis (Article 3 e] LBR). This transmission of relevant information takes place both before and

[12] Vid. Article 45 LBR.

after a genetic test or screening and even in their absence. The professional carrying out or coordinating the genetic counselling must provide adequate information and advice, both on the significance of the resulting genetic diagnosis and on the possible alternatives that the subject may choose in view of it (Article 56 LBR).

In the field of assisted reproduction, information and counselling on the possible techniques to be used must be provided both to those who wish to use them and to those who are going to be donors (Article 3 LAHRT). This information, which must be provided by the heads of the medical teams carrying them out, includes the biological, legal, and ethical aspects of the practice of the tests, as well as the economic conditions of the treatment. At this point, it is important to draw attention to the work that the digital platforms of fertility clinics play in relation to the information they provide on the different techniques, insofar as in many cases it is the main source of a patient's knowledge and the basis for their making a decision (Turrini & Otin-Gavin, 2025).

The content of the information to be transmitted to the patient includes both the explanation of the consequences of the resulting genetic diagnosis for the affected person and their offspring, the possible alternative tests, and a warning about the possibility of unexpected discoveries and their repercussions for the subject (Articles 3 e] and 55. 2 LBR). Genetic counselling is, therefore, a process in which the patient is informed and advised so that he/she can make a free and informed decision, aware of its implications, and even has the right to decide not to receive the data (Article 4.5 LBR). The exponential increase in the amount of information obtained from genetic testing, mainly from the increased use of next generation sequencing (NGS) tests, has dramatically increased the assessment of the implications and transcendence of these tests (Abellán-García, 2019). In the context of assisted reproduction, the complexity of the information obtained from the application of carrier testing in combination with the practice of genetic matching makes it difficult for professionals to transmit this information to patients.

In order for the couple and donors to be able to make an informed decision, genetic counselling about the tests should take place before and after the study, so that they are aware of all the implications of the results for themselves and their relatives. This communication should include information on the technical limitations of the tests that do not allow the identification of some types of mutations—that is, that are not 100% effective.

In the case of donor involvement, the genetic information that may be of interest relates to the donor as the biological parent; for example, it should be borne in mind that in gamete donation the rule of donor anonymity applies. The Law on Assisted Human Reproduction Techniques recognises the right of the offspring and the recipient to obtain general information on the donors that does not include their identity; this information includes those data with clinical significance requested by a recipient, where appropriate, provided that it is justified and does not compromise the identity of the donor, especially in cases in which a child is born with a serious health problem (Article 5.5 LAHRD).[13] The donor's medical records, without revealing his or her identity, may also be requested by the offspring once they have reached the age of majority (Article 18.3 LAHRH), and scientific societies have expressed their support for providing information of clinical significance to recipients and their offspring if they request it. In general, however, they do not recommend informing recipients of the genetic profile of donors with regard to the presence of pathogenic variants not involved in the matching, as these are data that belong to the donor and could jeopardise their anonymity (SEF, 2019).

On the Right Versus the Obligation to Report on Genetic Studies of Donors

Genetic testing of gamete donors raises the dilemma of the relevance and appropriateness of informing them about the results of their tests when unexpected findings appear, insofar as this information may be useful for their own reproductive project. The LBR recognises the right of the

[13] Article 5.5 LAHRT states: "Only exceptionally, in extraordinary circumstances involving a certain danger to the life or health of the child or when appropriate under criminal procedural law, may the identity of the donors be disclosed, provided that such disclosure is indispensable to avoid the danger or to achieve the proposed legal purpose. Such disclosure shall be restricted in nature and shall in no case entail publicity of the identity of the donors".

participants both to know and not to receive information on the results of the genetic tests carried out, a situation which is expressly reflected in the formula of informed consent (Article 4 LBR). However, the legislator leaves this question open by leaving it up to the doctor and his or her discretion to make a decision in this regard. In this sense, the right of the person to decide not to be informed of genetic data, including any unexpected discoveries that may occur, is recognised. However, when this information, according to the criteria of the doctor responsible, is necessary to avoid serious harm to the donor's health or that of his or her biological relatives, a close relative or a representative shall be informed, after consulting the care committee, if there is one. In all cases, the communication must be limited to the data necessary for these purposes only (Article 4.5 LBR).

This is, in any case, a controversial issue on which there is no consensus at the biomedical and scientific level on the appropriateness of reporting in the event of the appearance of such unexpected findings. Ultimately it is a decision that depends to a large extent on the importance of such information in terms of clinical repercussions, pathogenicity, and severity, and, in short, on the overall usefulness of its being known for the establishment of the relevant preventive or therapeutic actions. Again, some scientific societies have published their positions on this matter, such as the American College of Medical Genetics and Genomics (ACMG) (ESHRE, 2022).

On the Legal Consequences of Not Informing Donors

The Law on Assisted Human Reproduction Techniques lists a series of actions—those that violate the obligations related to the appropriate communication with and informing of donors/recipients of the results of genetic tests—regarded as serious infringements that carry the corresponding sanction. These include the following:[14] (i) the violation by the work teams of their legal obligations in the treatment of the users of these techniques; (ii) the omission of the necessary information or prior studies

[14] Vid. Article 26 LAHRT.

that might avoid harm to the interests of donors or users, or transmission of congenital or hereditary diseases; (iii) the omission of data, consents, and references required by the LAHRT, as well as the failure to take the clinical history in each case. Similarly, the LBR considers the omission of data, consents, and references the law requires, or the breach of the conditions of confidentiality of donor data it establishes, to be serious offences.[15] The absence or defective practice of these obligations may, in turn, give rise to the corresponding liability for compensation insofar as the medical action has caused harm in violation of the general duty not to harm another (Aviles & Ramón Fernández, 2019; Díaz Martínez, 2014).

Legal Liability Arising from Failure to Comply with the Information Obligation

The number of disputes involving claims for compensation for damages suffered in assisted reproduction treatments has increased in the last decade—both in number and in the grounds for the dispute—as a result of the broadening use of these methods of procreation. In this sense, the doctrine maintains that, despite the multiple scenarios and the impossibility of imagining all potential compensable damages, there are solid jurisprudential criteria, as well as a specific norm (LAHRT) that contains clear rules to be applied in the solution of conflicts with cause in the relationships that it regulates (Cervilla Garzón, 2021). Liability for damages in assisted reproduction treatments falls within the scope of what is known as "voluntary" medicine. In contrast to curative, therapeutic, or assistance medicine—which obliges the doctor or those responsible for health care "not to cure the patient, but to provide them with the care they require according to the current state of medical science"—in voluntary medicine (satisfactory or satisfaction medicine) the patient undergoes an intervention not because there is a pathology but to improve a physical or aesthetic aspect or to transform a biological activity (Elizari Urtasun, 2021). This latter type of medicine has the peculiarity of both a greater obligation on the part of the practitioner to provide information,

[15] Vid. Article 74 LBR.

to the level of atypical risks, and a higher degree of diligence, bringing the obligation assumed closer to a result. The physician must inform patients both of the possible risks of the intervention and of the possibility that the desired objective will not be achieved. The information to be provided to the patient must be greater, since the patient has a greater margin of freedom to choose to accept or reject the intervention when assessing the possible risks. Indeed, case law has focused on the information on risks, stating that in voluntary medicine it must be exhaustive and comprehensive, including atypical or infrequent risks and excluding only those unknown to medical science (Elizari Urtasun, 2021).

The right of users of assisted reproduction techniques to be informed extends throughout the entire reproductive process. The Patient Autonomy Act stipulates that the information must extend to "the purpose and nature of each intervention, its risks and consequences".[16] For its part, the LAHRT specifies and adds some aspects to this general right to information by indicating the obligatory content of the information due, which includes biological, legal, and ethical aspects, as well as the economic conditions of the treatment (Article 3.3 LAHRT).

The information due is necessary for the issuing of informed consent, a requirement prior to the commencement of the medical act which this matter presents. The minimum content of this information includes, according to the Patient Autonomy Act, the following risks: (a) the consequences relevant that the intervention safely originates; (b) the risks related to the patient's personal or professional circumstances; (c) the probable risks under normal conditions, in accordance with experience and the state of science or directly related to the type of intervention; (d) the contraindications.[17] To this information must be added the clarifications of the LAHRT, which establishes that the information must include "the possible risks, for the woman herself during treatment and pregnancy and for the offspring, that may arise from childbearing at a clinically inappropriate age" (Article 6.2). Moreover, insofar as assisted reproduction techniques constitute a non-curative medical act, the scope

[16] Vid. Article 4.1 of Law 41/2002, of 14 November, regulating patient autonomy, and rights and obligations regarding clinical documentary information.

[17] Vid. Article 10.1 of Law 41/2002, of 14 November, regulating patient autonomy and rights and obligations regarding clinical documentary information.

of the information must be wider, including both typical risks (those that are likely to occur as a consequence of the medical act) and atypical risks (understood as those that occur very occasionally).

Relevant to these conditions, the Judgment of the Provincial Court of Valencia of 30 March 2016,[18] in relation to a child born with haemophilia in a heterologous insemination with genetic material from a donor, condemns the medical centres for breach of the obligation to provide information. This was despite the fact that the form signed by the users had been drawn up in accordance with the models published by the Spanish Fertility Society, a circumstance that does not provide exemption from the obligation to pay compensation, as this fact does not guarantee compliance. The Court stated:

> From the text transcribed as informed consent, we conclude that it is not communicated to the recipients, that it is not transmitted to the people who are going to undergo this reproduction technique—who are not experts in medicine or in its terminology—the reality of the tests and analyses to which the oocytes have been subjected. In our opinion, the person who reads this informed consent assumes that all the necessary tests have been carried out to discover and exclude genetic, hereditary, or infectious diseases transmissible to the offspring that can be detected according to the state of science. Furthermore, in the documentation given to the recipient parents and in the informed consent, the future parents are not warned of the existence of many detectable diseases for which technical tests are not carried out and of the existence of many others that cannot be detected, a matter of special importance in cases such as the present one.
>
> In view of the above, we must conclude that the risk of transmission of diseases must be clearly and precisely stated in the informed consent signed by the parents, from the very first moment, and, given its transcendence, such information cannot be considered to be fulfilled by mere statements relating to [the claim] that they were informed verbally, and by not doing so the defendant has violated the recipients' right to information, incurring the corresponding civil liability.

[18] SAP Valencia, 30 March 2016, ECLI: ES:APV:2016:2303.

On the other hand, the Murcia Provincial Court Judgement of 13 March 2018[19] understands that there has not been a breach of the obligation to provide information by the fertility clinic that advertises its services with a guaranteed 100% success rate for its procedures (with a refund of the amount paid if not successful), as the informed consent signed was not expressed in the same way.

In the case of preimplantation genetic diagnosis, which is regulated by Article 12 of the LAHRT, if the conditions required by the regulation are met, and after the diagnosis of the disease that the pre-embryo presents, treatment will be carried out if possible or, alternatively, the parents may decide to reduce the embryo or destroy the defective pre-embryo. It is also obligatory to provide the necessary information in this scenario, and failure to comply with the required content, detail, and accessibility will generate the relevant liability of the medical centre and the intervening doctors (Mediavilla, 2024).

Some Final Considerations

Scientific projects such as the Human Genome Project, aimed at decoding the human genome, and its technological applications (such as genetic sequencing), have made it possible to prevent certain genetic diseases by modifying the responsible gene. Advances in genetic medicine continue, and at the same time, the scope of new technologies is expanding. However, these discoveries are not always accompanied by specific regulations or by updates to current legislation that governs new advances and protects individual rights. Indeed, both international and national legislation establishes principles and standards that allow us to address the new challenges related to genetic testing by setting limits on biomedical research. However, current regulations are not without limitations, as their adaptation to scientific advances is often either slow or narrowly focused compared to the dynamic pace of scientific research in fields such as genomics. As a result, grey or unregulated areas emerge, with significant shortcomings regarding the use of genetic testing: its scope and

[19] SAP Valencia, 30 March 2016, ECLI: ES:APV:2016:2303.

limits, the rationale for its implementation, its selective or assisted nature, and so on.

It is, thus, essential to move beyond the notion that legal regulation inevitably lags behind scientific progress, as this reflects a linear and deterministic view of innovation. Lawmakers must certainly act proactively to avoid a regulatory vacuum in which emerging technologies become normalised and routinised without critical reflection on their implications and boundaries. However, the relationship between law and innovation should be grounded in a collaborative, contemporaneous dialogue, one wherein regulation and technological development inform and reinforce one another. This integration always responds to socio-political dynamics through a process of co-production of norms and technologies, which include ethical, social, and cultural dimensions.

The widespread use of genetic testing in assisted reproduction treatments is driven by several closely related circumstances: both technological advances in the tests, and the information and dissemination provided by the centres that apply them, have generated confidence among the patients who seek them. While it is true that the available genetic tests do not prevent diseases, and only include a limited number of genes and diseases, their use can increase the likelihood of having a healthy child; however, a risk still exists and the result may be unsatisfactory. The complex information obtained from these tests overburdens patients with multiple scenarios and places them in a vulnerable situation, as they are sometimes unable to, or have serious difficulty in understanding it. In this highly charged context, medical teams play a key role in the decision-making process.

Genetic counselling, in particular, is a key element before, during, and after testing. It is doctors who present the woman/partner with the different available options, possible outcomes related to specific variants, risks, and limitations, and the meanings of each result. This minimal information is intended to ensure that the patient understands the clinical, technical, and ethical implications of the procedure to be followed. However, for a test such as the one considered here—expanded carrier screening in the context of gamete donation—there is no clear consensus on which variants should be investigated or disclosed to recipients or donors based on their clinical utility. Moreover, the inclusion of extremely rare variants often seems to be in response to commercial strategies more than to a clearly medical logic.

Fertility centres also tend to present both carrier screening and preimplantation genetic testing—the two main genomic testing modalities in assisted reproduction—as a way to ensure the avoidance of hereditary or spontaneous congenital conditions. From a legal point of view—in accordance with legislation and jurisprudence—the focus has so far been merely on informed consent. Those executing these procedures must clearly and intelligibly inform patients that genetic testing does not guarantee that their offspring will be free of diseases, malformations, or chromosomal abnormalities. While fertility clinics have the legal obligation to incorporate information on available genetic guarantees into the provision of their treatments, their specific way of complying with this obligation has allowed them to broaden their scope of action to the extent of offering all available tests, even when there is no consensus on their limits and clinical usefulness. In this way, technologies that have been developed and adopted by the private sector lie outside the scope of current legislation, and are thus normalised. This is how they have become mundane.

In this context, once a technology has become normalised, the legislator's scope for regulating it is reduced, and future legislation tends to accept and endorse already existing protocols. Effectively, this is how the creation of standards, of laws and regulations, becomes one of the ways, perhaps the most institutional way, of making a new technology mundane, with all the consequences that this entails. These consequences, outlined and discussed in detail elsewhere in this volume, call for much more precise and articulated regulation, especially in relation to the specific cases and target groups for which carrier screening, particularly in the form of genetic matching, may be offered.

Such regulation must guarantee the rights of those who use these techniques, mainly the right to get comprehensible information allowing for an autonomous, unconditioned decision. This includes not only recipients, but also donors, whose right to access understandable results often appears less recognised than that of recipients to be informed. Moreover, individuals born through ART should also be protected by law. These improvements in the regulatory framework must be preceded and accompanied by a necessary ethical and political debate in which all variables (social, cultural, economic, etc.) are assessed, and in which the contours of decisions on the use of genetic tests for reproductive purposes, as well as on the handling of the genetic information resulting from them, are clearly delimited.

Acknowledgements This work is the product of the IfGene research project, "Projecting the future of genetic prediction" (PID2020-115899GB-I00, PI: Mauro Turrini and Ruben Blanco), funded by the Spanish Ministry of Science and Innovation (MICIU) through the Agencia Estatal de Investigación (AEI) (MICIU/AEI/10.13039/501100011033) with the support of the European Regional Development Fund (ERDF).

References

Abellán-García Sánchez, F. (2009). Los análisis genéticos dentro de la ley de investigación biomédica. *Revista de la Escuela de Medicina Legal, 11*, 22–42.

Abellán-García Sánchez, F. (2019). Pruebas genéticas en los tratamientos de reproducción asistida, con especial referencia a los test de portadores y matching genético. Controversias en materia de información. *Revista de Derecho y Genoma Humano, n. extraordinario*, Nº Extra 1, 325–348.

ACOG. (2017). Carrier screening in the age genomic medicine. *Committee Opinion, Number 690*. https://www.acog.org/clinical/clinical-guidance/committee-%20opinion/articles/2017/03/carrier-screening-in-the-age-of-genomic-medicine.

Arias Aparicio, F., Nicolás Jiménez, P., & Romeo Malanda, S. (2018). Aspectos jurídicos de la obtención, utilización y circulación de los gametos humanos. *InDret, Revista para el Análisis del Derecho*, 3.

ASEBIR. (2016). Position on genetic screening studies for recessive diseases in donation programmes. https://asebir.com/documentos-asebir/

Aviles Alepuz, B., & Ramón Fernández, F. (2019). La donación de óvulos en la reproducción asistida: riesgos y responsabilidad. *Ius et Scientia, 5*(2), 11–54.

Capalbo, A., Pla, J., Janssens, S., Accoe, D., Pennings, G., & Mertes, H. (2024). Should we use expanded carrier screening in gamete donation? *Fertility and Sterility, 122*(2), 220–227. https://doi.org/10.1016/j.fertnstert.2024.05.157

Cervilla Garzón, M. D. (2021). La responsabilidad civil derivada de los daños ocasionados en el uso de las técnicas de reproducción asistida. In *Cuestiones clásicas y actuales del Derecho de daños: Estudios en Homenaje al Profesor Dr. Roca Guillamón* (pp. 1481–1512). Aranzadi.

Comisión Nacional de Reproducción Humana Asistida. (2025a). *Recomendaciones para la realización de técnicas de PGT en casos de Enfermedad de Huntington.* https://cnrha.sanidad.gob.es/documentacion/home.htm.

Comisión Nacional de Reproducción Humana Asistida. (2025b). *Recomendaciones en relación con Enfermedades susceptibles de ser consideradas entre los supuestos del artículo 12.1 de la Ley 14/2006, de 26 de mayo, sobre técnicas de reproducción humana asistida.* https://cnrha.sanidad.gob.es/documentacion/home.htm.

Comisión Nacional de Reproducción Humana Asistida. (2025c). *Memoria de actividades (2024–2025).* https://cnrha.sanidad.gob.es/documentacion/comision/home.htm

Díaz Martínez, A. (2014). Pruebas y consejo genético ante nuestros tribunales de justicia. (Más allá del diagnóstico prenatal erróneo). *Revista Doctrinal Aranzadi Civil-Mercantil,* n. 8.

Elizari Urtasun, L. (2021). Reflexiones en torno a la distinción entre medicina satisfactiva y medicina curativa y sus consecuencias. In *Cuestiones clásicas y actuales del Derecho de daños: Estudios en Homenaje al Profesor Dr. Roca Guillamón*, pp. 283–326. Aranzadi.

ESHRE. European Society of Human Reproduction Embryology. (2022). *Good practice recommendations for information provision for those involved in reproductive donation, Human Reproduction Open.* https://doi.org/10.1093/hropen/hoac001

ESHRE. Add-ons working group, Lundin, K., Bentzen, J. G., Bozdag, G., Ebner, T., Harper, J., Le Clef, N., Moffett, A., Norcross, S., Polyzos, N. P., Rautakallio-Hokkanen, S., Sfontouris, I., Sermon, K., Vermeulen, N., & Pinborg, A. (2023). Good practice recommendations on add-ons in reproductive medicine. *Human Reproduction*, *38*(11), 2062–2104. https://doi.org/10.1093/humrep/dead184

ESHRE. European Society of Human Reproduction Embryology. (2024). IVF and IUI treatment cycles increase across Europe, along with stable pregnancy rates. *European Society of Human Reproduction and Embryology.* https://www.eshre.eu/ESHRE2024/Media/2024-Press-releases/EIM

Martínez Otero, J. M. (2017). La generalización de los test genéticos y su incidencia en los derechos fundamentales. *Revista Europea de Derechos Fundamentales, 29,* 235–236.

Mediavilla Caro, J. V. (2024). *Errores de diagnóstico prenatal y responsabilidad.* Aranzadi.

Parthasarathy, S. (2025). From the bench to public policy: Enhancing public trust in science. *Social Studies of Science, 55*(2), 295–302.

Pavone, V., & Arias, F. (2012). Beyond the geneticization thesis: The political economy of PGD/PGS in Spain. *Science, Technology & Human Values, 37*(3), 235–261.

Reguera Cabe, M. (2021). Aspectos éticos y legales de los testadores de portadores de mutaciones recesivas en tratamientos de reproducción humana asistida con gametos donados. *Revista de Bioética y Derecho*, n. 51.

Sánchez Patrón, J. M. (2024). La regulación jurídica de la investigación biomédica aplicada al ser humano: entre lo posible y lo incongruente. *Ius et Scientia, 10*(1), 31–51.

Sociedad Española de Fertilidad (SEF). (2007). *Estudio y tratamiento de la pareja estéril.* Adalia.

Sociedad Española de Fertilidad (SEF). (2019). *Cribado genético en donación de gametos.* https://www.sefertilidad.net/docs/newsletter/docs//cribadoGenetico.pdf.

Turrini, M., & Otin-Gavin, J. (2025). Let 'pregnant women choose the destiny for themselves and their child'. How fertility clinic digital platforms frame preimplantation genetic testing (PGT) in Spain. *Sociology of Health & Illness, 47*(1). https://doi.org/10.1111/1467-9566.13876

Urbano, A., et al. (2024). Advantages of carrier screening in gamete donors with a 15-gene panel using next generation sequencing technology. *Medical Genetics and Genomics, 2*, 2.

Part II

Predispositions and Sensibilities

6

Cultivating Genetic Sensibility in Danish Lynch Syndrome Families

Ayo Wahlberg, Laura Louise Heinsen, and Helle Vendel Petersen

Introduction

The question of *when to tell them*—when and how parents should disclose to their children that they may have inherited a genetic predisposition to disease—has become an important topic in discussions about the ethical challenges surrounding familial genetic risk (Derbez, 2018; Stoffel et al., 2008; Stuttgen et al., 2021; Young et al., 2017). For the increasing number of parents with young children who have come to learn that they are carrying a cancer-predisposing genetic (germline) mutation, this question is far from hypothetical. As predictive genetic testing for

A. Wahlberg
Department of Anthropology, University of Copenhagen, Copenhagen, Denmark
e-mail: ayo.wahlberg@anthro.ku.dk

L. L. Heinsen (✉) • H. V. Petersen
Department of People and Technology, Roskilde University, Roskilde, Denmark
e-mail: lauralh@ruc.dk; helven@ruc.dk

V. Argudo-Portal et al. (eds.), *Mundane Genomics*,
https://doi.org/10.1007/978-981-95-4550-6_6

hereditary cancer—as well as a range of other inherited conditions, such as cardiovascular disease—and the subsequent enrolment of "carriers" into preventive medical surveillance or pharmaceutical treatment "for life" are becoming more and more common due to recent advances in medical technology (cf. Dumit, 2012), inherited cancer risk has, in many ways, become part of the fabric of the ordinary. In present-day Denmark, around 500 families have been identified as having Lynch syndrome through a national registry established in the beginning of the 1990s (Myrhøj et al., 1994). Lynch syndrome, the most prevalent among the hereditary, non-polyposis colorectal cancer (HNPCC) syndromes, significantly elevates a person's lifetime risk of a range of extracolonic tumours, especially colorectal cancer. Due to the systematic identification and genetic testing of at-risk individuals, people living with Lynch syndrome in Denmark today lead what we have elsewhere called "surveillance lives", enrolled in lifelong medical surveillance programmes in the form of biannual colonoscopies that aim to catch and remove precancerous polyps to keep disease at bay (Heinsen et al., 2022).

Much literature on the intra-familial disclosure of genetic risk has been principally occupied with identifying "the right time", or the right age, to communicate potentially life-altering information to a child (Forrest Keenan et al., 2009; Bradbury et al. 2009; Holt, 2006; Skirton, 1997). Indeed, the news that a genetic predisposition runs in the family invokes biological kinship ties and biogenetic relatedness in novel ways, as biological children, siblings, and other kin are at risk of carrying the same genetic predisposition until genetic testing has confirmed otherwise (Atkinson et al., 2013). As a consequence, Novas and Rose argue:

> The genetically at risk individual must engage with a communicative problem space.
>
> When should they tell siblings or children that are also at risk that they have decided to undergo predictive genetic testing? When should they tell children or other family members they face the prospect of inheriting a severe [...] disorder? As genetic information is familial, it has the potential to affect our relations with others. Hence the governance of one's own risky genes intersects with the governance of one's communicative relations with others. (Novas & Rose, 2000, p. 505)

Because of the extended impact of genetic knowledge for wider families, identified at-risk individuals are generally encouraged by health professionals to disclose genetic risk to relatives (Baroutsou et al., 2021). From a bioethical and public health perspective, the familial responsibility of informing relatives about genetic risk is couched in terms of expanding individual decision-making and choice, as it allows kin the opportunity to ascertain their own risk and to determine if preventive measures, such as preventive medicine, prophylactic surgery, or surveillance, are relevant (Petersen, 1998; Svendsen & Koch, 2006; Seppälä et al., 2021). At the same time, social scientists have documented numerous ethical challenges surrounding the disclosure of genetic risk information between family members, showing how the disclosure of risk information can be a complex and emotionally charged process (Featherstone et al., 2006; Forrest et al., 2003; Gallo et al., 2005; Gislinge et al., 2024), as well as how a "right not to know" (cf. Andorno, 2004; Beauchamp & Childress, 2001) has come under increasing pressure as expanding possibilities for genetic testing, prevention, and surveillance proliferate (Konrad, 2003; Svendsen, 2005; Hallowell, 1999; Klitzman et al., 2007a; Frey et al., 2022). While such dilemmas are relevant for all family members when a predisposing mutation is detected through testing, in contexts where children's disease risk is not yet known and genetic testing of children for late onset conditions not allowed or recommended until they have reached a certain age (an 18-year-old may consent to being tested although 25 is the age where surveillance is offered to individuals with Lynch syndrome), the question of when to disclose to children that they may be "at elevated risk" for cancer can be even more challenging, as it potentially entails withholding information for a lengthy period (Sie et al., 2013). This can be particularly complicated for parents living with Lynch syndrome who, in preparation for their biannual preventive colonoscopies, must undertake cleansing of the bowel at home and abstain from certain food and drinks, which can be difficult to hide from children, even more so as they grow older. While attention to the timing of genetic risk communication remains influential, it may overlook the situated and relational ways in which families actually live with and make sense of genomic knowledge over time.

In this chapter, we present findings from a qualitative study exploring the experiences of families living with Lynch syndrome, focusing on how parents reflected on how to share, or had shared, genetic risk knowledge. We argue that what is at stake for parents is the cultivation of what we term *genetic sensibility*, meaning the work of "dosing", filtering, and moderating genetic risk knowledge in order to guide children to be responsive towards what it means and takes to be genetically at risk and leading a surveillance life in the near or distant future. To unfold our argument, we present four disclosure stories that exemplify the differences in how parents approach this genetic sensibilization. Rather than adding to the literature on communicative models or styles (Rolland & Williams, 2005; Werner-Lin et al., 2018) or suggesting that we can ever identify a right time or appropriate approach to disclosure in any universal or uniform sense (Derbez, 2018; van der Meer et al., 2006), we expand on the notion of "genetic responsibility" (Kenen, 1994; Lippman, 1998; Hallowell, 1999; Novas & Rose, 2000) by showing how parents feel responsible for not only sharing (or withholding) information until certain moments but also for securing the well-being of their children by creating a balanced genetic awareness. Indeed, while the concept of genetic responsibility aptly describes the social impact of genomic knowledge when genetic carrier testing was new and emerging, the notion of genetic sensibility captures how genetic risk knowledge has moved into "the mundane" of social life, not least considering that in the three decades that have passed since genetic testing became feasible on a larger scale in the 1990s (and since the first wave of social studies of genetic counselling), preventive measures for those identified as genetically predisposed to colon cancer, as well as breast cancer, have become routinized in Denmark and beyond (albeit with limited reach in countries where medical genetics and healthcare more broadly are under-resourced).

In important ways, we might say that the mutated genes that predispose to colon (and breast) cancer have become "enculturated" in the sense proposed by Duana Fullwiley in her ethnographic study of how a "Senegalese" type of sickle cell anaemia came to be seen as "less severe" by geneticists in ways that erased the efforts of patients and doctors to devise methods of living as well as possible with the disease in Senegal. Fullwiley argues that "enculturation, as the vital adaptation techniques that social groups fashion for themselves, involves making sense of and living with specific referent objects […] which become enculturated as both patients and biomedical practitioners engage its supposed dictate of fate and subsequent constraints on (or chances for) living a normal life" (Fullwiley, 2011, p. 20). One of the ways in which the mutated genes underlying Lynch syndrome have become enculturated, we argue, is through the efforts of parents to make children responsive to and prepared for a (potential) future of life-long medical surveillance. Furthermore, our data material reflects culturally specific notions and values about the sanctity of childhood and youth as vulnerable life stages, through which parents are obliged to guide and shepherd children. Indeed, contemporary social science research reflects a growing societal concern with childhood vulnerability, portraying children as more susceptible than ever to challenges that may hinder their physical and emotional growth (Faircloth, 2023; Faircloth et al., 2013; Gulløv, 2018). In response, a powerful narrative has emerged around parenting, casting mothers and fathers as central agents in shaping not just their children's well-being, but their entire life trajectories, or what Furedi has described as "parental determinism" (Furedi, 2008). Notwithstanding the power of genetic sensibility as a heuristic for understanding the mundanity of predictive genetics, we propose that parents' specific approaches to sensibilizing their children must be culturally and socially situated, requiring further exploration of how the cultivation of genetic sensibility takes distinct forms across the globe.

We continue our chapter by outlining what Lynch syndrome is and what prevention of inherited elevated risk of colorectal cancer entails in Denmark, to set the issue of intra-familial risk communication into context. We then go on to discuss the body of literature that has focused on intra-family communication of genetic predisposition, followed by our methods and finally our analysis. Our findings are structured around four "disclosure stories", which we conceptualize as modalities of informing children and/or learning about genetic risk. We outline four disclosure routes that we identified in our engagements with families concerning how information about genetic risk was (actually or planned to be) given to children and/or how children came to find out about Lynch syndrome (cf. Etchegary, 2006; Forrest Keenan et al., 2009). Rather than being either-or forms of (non-)disclosure we understand disclosure routes as configured along a continuum of more-or-less intentional communication of risk, depending on family circumstances and dynamics. Furthermore, families could adjust their forms of disclosure as their children grew older, some being told in one way and others in another. Thus, the disclosure routes we discuss should be understood as representative of the diversity of ways of communicating familial genetic risk, which are highly fluid social and relational communication processes intersecting with everyday life and sociality. Disclosure seldom happens as a single transaction but, indeed, is a continual, sometimes unplanned and unpredictable process (Rowland & Metcalfe, 2013). Hence, we show how parents' disclosure routes amount to what might be seen as a form of "genetic accounting" (cf. Franklin & Roberts, 2006), involving the weighing of pros and cons of when and how to tell. We likewise show how young adults reflected on the ways in which they learned—whether directly or inadvertently—and absorbed that they had possibly inherited an elevated risk for colorectal cancer.

Background

Lynch syndrome, also known as hereditary non-polyposis colorectal cancer (HNPCC), is characterized by autosomal dominant inheritance, which implies a 50% risk of passing on the altered gene to each child. The syndrome is caused by germline mutations in one of the mismatch repair genes (MLH1, MSH2, MSH6, and PMS2), predisposing individuals to early onset colorectal cancer, as well as an increased risk of endometrial, ovarian, urothelial, and skin cancer with varying degrees of expression between the specific genes (Lynch & de la Chapelle, 2003). It is estimated that the lifetime risk of colorectal cancer in Lynch syndrome ranges from 20 to 90%, depending on the gene, compared to a lifetime risk of 5% for sporadic colorectal cancer in the general population. The mean age of onset of hereditary colorectal cancer is 45 years, compared to 65 years for sporadic cancer, but colorectal cancer caused by Lynch syndrome occasionally strikes as early as in a person's twenties. In Denmark, a national HNPCC register plays a central role in predictive cancer genetics and preventive cancer medicine. Initiated through clinical and research efforts at Hvidovre Hospital at the beginning of the 1990s, the register aims to reduce cancer incidence and mortality through prevention and early detection, collecting and organizing genetic, clinical, and familial information on individuals diagnosed with or at risk of Lynch syndrome. Once a pathogenic variant is identified in a family, the register facilitates cascade screening and the enrolment of mutation carriers into lifelong surveillance programs, primarily involving biannual colonoscopies (Lindberg et al., 2017). The register operates as a centralized, nationwide database, integrating genetic counselling with clinical care. As such, over the past three decades, it has played a key role in institutionalizing preventive genetics and shaping how families live with inherited cancer risk in Denmark.

Disclosing Genetic Risk—From When to How

Disclosure of genetic carrier status has been a central question in the field of genetic counselling from its inception, as juridical notions of a "right (not) to know", Hippocratic oaths of doing no harm, and ideas associated with just what genetic responsibility might mean can come into tension, not least in situations of frayed or non-existent family relations. The majority of studies on disclosure when it comes to inherited elevated risk have focused on family disclosure and genetic responsibility in regard to wider extended families (Svendsen, 2005; Forrest et al., 2003; Stoffel et al., 2008; Arribas-Ayllon et al., 2008; Baroutsou et al., 2021; Frey et al., 2022). These studies have shown that a (albeit contested) moral duty to disclose this knowledge to kin has taken form which has put the right not to know under pressure, especially since hereditary cancers such as breast and colon cancer have become preventable through mastectomies and colonoscopies, or detected earlier thereby improving treatment possibilities and outcomes (Svendsen & Koch, 2006). More recently, scholars have begun focusing on the particular difficulties parents face concerning disclosure of genetic risk to their offspring (Aktan-Collan et al., 2011; Forrest et al., 2003; Forrest Keenan et al., 2009; Holt, 2006; McConkie-Rosell et al., 2009; Metcalfe et al., 2008; Metcalfe et al., 2011; Rowland & Metcalfe, 2013; Werner-Lin et al., 2018; Young et al., 2017).

Some of these studies suggest that being informed earlier helps offspring cope and deal with their risk better than being told later (Forrest Keenan et al., 2009; Holt, 2006; Skirton, 1997). Conversely, in their interview study of families disclosing genetic information about Huntington's disease, Klitzman and colleagues argue that withholding risk information can cause harm and stir up feelings of resentment, anger, and disappointment, showing how "parents may justify non-disclosure to children based on desires to avoid burdening these offspring [while] partial and incomplete information may also be given piecemeal over several years, generating misunderstandings [and] creating tensions" (Klitzman et al., 2007b, p. 1841). At the same time, geneticists and ethicists have debated whether there is an optimal age for disclosure, with disagreements in the literature as to when this might be (Forrest Keenan et al., 2009, p. 1893; Holt, 2006; Klitzman et al., 2007b; Skirton, 1997; Van der Meer et al., 2006). It is not surprising then, that in their meta-analysis of communicated inherited genetic risk between parent and child, Rowland and Metcalfe conclude, "disclosure of genetic risk information within families is a highly complex and affective process often resulting in delayed disclosure [which] can lead to increased family tensions generated by misunderstanding, blame and secrecy" (2013, p. 870).

Focusing more specifically on how Finnish parents with Lynch syndrome communicate with their offspring, Aktan-Collan and colleagues have explored the disclosure dilemmas parents face, showing how they often find themselves stuck between a rock and hard place, "on one hand, facing the desire to protect children from adverse feelings, and on the other hand, being aware of the importance of undergoing testing and cancer surveillance" (Aktan-Collan et al., 2011, p. 48). Amongst the 248 parents who tested positive for a Lynch syndrome mutation, they found that 83% had communicated their genetic test results to at least one of their children. However, only 12 individual parents had shared this information with children under the age of 18. This relatively low number reflects broader concerns about minors' capacity to understand the implications of genetic testing. As the authors note, "minors are considered unable to fully understand implications of genetic testing until the development of their ability to conduct abstract thoughts" (Aktan-Collan et al., 2011, pp. 46, 48). Nonetheless, given that "research shows that children and young people want their parents to engage in open and honest discussions about [their] genetic condition", scholars have argued that early, age-appropriate disclosure can better prepare children for the future, contribute to effective coping strategies, promote enhanced adaptation and emotional well-being, and, not least, reduce parental anxieties concerning disclosure from an unwitting source (Rowland & Metcalfe, 2013, p. 870).

In what follows, we shift the focus away from trying to identify what the most appropriate time to disclose might be, towards exploring what families living with Lynch syndrome *do* when it comes to disclosing genetic risk to children (cf. Geelen et al., 2011). We show that disclosure hinges on both the weighing of burdens and benefits of genetic knowledge for the individual child/young adult *and* on culturally specific ideas, norms, and values about childhood and youth as not only pervasively vulnerable life stages but also as a time of innocence that should not be unreasonably disturbed.

The Study

The chapter is based on a shared interdisciplinary research project between the HNPCC register at Hvidovre Hospital and the Department of Anthropology at the University of Copenhagen. The data on which our analysis of family communication about genetic risk is based comes from over ten years of qualitative engagement with families with Lynch syndrome. Interviews with 12 people with Lynch syndrome were carried out by the third author between 2009 and 2011 followed by a further 6 interviews with individuals with Lynch syndrome conducted between 2017 and 2019 by the first author. Moreover, all authors conducted three focus group interviews with families (parents and their adolescents) and one interview with a parent living with Lynch syndrome and one young adult during 2019. To our knowledge, all individuals came from families with mother, father, and siblings, and the study is thus limited to heteronormative family forms. And finally, information evenings for families with Lynch syndrome and other variants of inherited elevated risk of colorectal cancer organized by the Danish HNPCC register were attended and observed, and interviews with attending families were conducted. All interviews were transcribed verbatim, and we carried out thematic analyses of all of them with a focus on the disclosure stories of the families we spoke to.

Research took place in accordance with Hvidovre Hospital's ethical requirements for qualitative research, and those of the University of Copenhagen, governed by Danish law (LBK nr 1338). Recruitment of interviewees and storage of observation notes and interviews were in compliance with data protection requirements and all participants provided their written or oral consent to participate in interviews following clearly explained information about the purpose of research. Given our many years of collaborative work with the HNPCC register, we ensured that all participants were directed to relevant healthcare professionals in case they had any follow-up queries about Lynch syndrome.

Disclosure Stories

For a person at elevated risk of colorectal cancer, the pathways to finding out that Lynch syndrome "runs in the family" are diverse and manifold. The individuals and families we met represented a continuum from newly diagnosed who had not yet had a chance to tell children about their new-found "at risk" status to those who were contemplating how and when to tell as well as parents who had known for years and had already "had the talk" with all or some of their children. The types of disclosure, or what we call disclosure stories, that we present in the following are thus highly contingent upon the locus in a person's own Lynch syndrome "career", their life stage, and the life events or "vital conjunctures" (Johnson-Hanks, 2002) that have already occurred or are likely to in the future (see also Etchegary & Fowler, 2008). Amongst those of our informants who had children, all believed that children had a right to be informed about the genetic predisposition they might be carrying, but, as we learned, their approaches to informing them were different. For some, the major issue concerned timing of disclosure, not only in terms of finding the right age, but finding the right situation, while for others the question of "how to tell" was a cause of consternation. Hence, deciding how and when to tell them depended on a number of factors including age, whether or not parents felt their child was "ready", as well as how telling them and talking about it might invoke feelings of anxiety, tensions, and guilt in the parent carrying the "defective" gene.

Catalysts: Family Conference in the Wake of a Positive Test Result

When Poul, an elderly man in his seventies, tested positive for Lynch syndrome at the age of 60, he immediately summoned his and his wife's three girls, the youngest being 15 at the time and the oldest 20, to share the news. The genetic predisposition that ran in his maternal lineage had already caused several deaths in his family; however, the thought that it could be hereditary had not hit home until his mother's sisters were tested, snowballing himself and his brothers into genetic testing. Of the three brothers, Poul was the only one with the faulty mismatch repair gene. Poul and his wife took it as self-evident that they would inform their daughters of the hereditary risk as well as the possibility of being tested themselves; as he reasoned, "I was never in doubt that if it turned out to be hereditary, I would be open about it. You know, so that it could be shared. I really had no doubt about that. But I guess that's also just part of who I am—I don't go around hiding things like that. There's no reason to. I don't think keeping it to myself would make me a better man". Poul's wife supported his openness, observing, "We talked a lot about it. Also, we talked about that just because you have the gene, it's not given you will get cancer". Here, the event of genetic counselling and testing that culminated in a positive test result catalysed the sharing of information with his daughters, which was supported by his wife. For Poul, having girls well on their way into adulthood meant that the timing of disclosure was not the issue for him; rather, informing his daughters as quickly as possible gave them the opportunity to ascertain their own risk status and prepare for a potential surveillance life akin to his own. As the quote illustrates, there are things to be done to prevent cancer when given a diagnosis like Lynch syndrome, a point to which we return later. It turned out that Poul's middle daughter had also inherited the gene and surveillance life thus became a joint "fate" for the two.

Planned: Finding "the Right Moment"

> It's about the right moment when the channels are open, you know. It's not something you just come waltzing in, and telling over dinner. I sometimes have these deep conversations with my oldest son, about the universe, where he's, like, he wants to lean into me. It would be at such a time that I would tell [him].
>
> (Tine, mother with Lynch syndrome)

As a mother of two sons aged 16 and 19, Tine had told them that she had a gene defect as soon as she was diagnosed. However, unlike Poul, she had not yet informed her sons that the "defect gene" could have been passed on to them as well (and according to Tine's assessment, the sons had not figured it out). Tine had lost both her parents to cancer and battled duodenal cancer herself, which she was told by doctors could be related to the genetic predisposition. With a considerable familial cancer history, Tine found it difficult to disclose the risk information to her children. She reasoned that telling them would demand that she compromised her own coping strategy of keeping her risk status at an emotional distance. She had somehow taught herself not to think too much about it. "You can try and push it away and then you get a bit distanced and rational", she told us. Later she elaborated that she would tell her sons together with her ex-husband, the father of the boys, to tackle the issue as a family and get support from him in making sure "it isn't talked about in too distanced a manner" as "I want them to feel secure and loved and protected by us, which doesn't harmonize with being as […] distanced, rational and detached [as I am]". So, Tine struggled with both finding the right moment to disclose, when she felt "the channels are open", and how to disclose in a way that would make her sons feel safe.

For others, the burden of informing centred only around finding the right time, especially in families who self-identified as having an open and honest parenting style. Henriette, a mother of four sons who prior to the interview had informed her two oldest teenage sons, recalled when asked what considerations she had had when telling:

> It wasn't like a thing where we had to consider how to tell it. Because we're a family where we can talk about everything, and we've experienced horrible things before which we were open about. So, it wasn't like, "Oh my God, how are we gonna talk about this thing?" It was more like, "Well, we're gonna have the talk at some point and they should know what's relevant, but no need to add a bunch of information that will just make them nervous or confuse them". So, it was just about having it [the talk] when we felt it was the right time.

Past, present, and future seemed to fold into one another in the complex process of weighing when to tell their children, for those parents who searched for the "right" moment. As they evaluated their children's current maturity and emotional readiness, they simultaneously revisited memories of illness and loss within the family, while also anticipating not just a singular future moment but a range of possible futures—near and distant—in which disclosure might be revisited, misinterpreted, or felt as harmful. The potential for "damage" was not limited to some distant horizon, but was deeply felt in the presence of disclosure itself, where parents feared overwhelming or burdening their child in a moment that could ripple forward in unforeseen ways. This was evident during a conversation with John and his wife Ulla. John had recently been informed that he had Lynch syndrome after being diagnosed with prostate cancer, for which he had undergone successful treatment. John and Ulla's two children, both teenagers, had not been informed yet, but when asked when John and Ulla would tell them, the following conversation took place:

John: Our kids are very different. Our son is the one we need to pay the most attention to, because he worries a lot.

Ulla: He worries about many things.

John: Our daughter is more robust. So, he should probably be around 18 or 20 before we tell them about the risk. Because before that, there are just too many things for him to tackle, you know teenage things.

Ulla: He could also be like 17, it depends.

John: Yeah, about that age, but he doesn't need more worries right now, so we'll wait until, yeah 17 to 18, something like that.

Ulla: He's a bit fragile.

John: Yeah, but it's not easy, because sometimes he's really strong, but then you don't know how they'll react. It would be a shame putting a bunch of worries in his head, because it's too soon. There's nothing to be done now anyway.

Ulla: No, there isn't. Not until 25 years of age.

John: But as Ulla says, you shouldn't keep them out of the loop either.

Ulla: No, because our son is the type that if he's told when he's 25, he would be furious.

John: Yes, that's too late.

Ulla: He would understand us if we told him we chose to wait until you turned 18, or whatever we think, but of course we shouldn't wait until he's 25.

Writing in the context of assisted reproduction, sociologists Sarah Franklin and Celia Roberts have termed weighing the costs of continuing or ending treatment in the wake of unsuccessful preimplantation genetic diagnosis treatment "reproductive accounting" (Franklin & Roberts, 2006, p. 172). Likewise, it is clear from this conversation that finding the right time to disclose genetic risk involves a kind of genetic accounting, a balancing act of weighing the damage of imparting against the damage of withholding information, and the potential relational friction that choosing wrongly could trigger. Considerations about when to tell are thus rooted in moral reasoning about good parenting as well as about good, safe childhood and youth. Norbert Elias (1998) has described how a general democratization and emotionalization of relations between parents and their children characterize contemporary European parenting cultures. This transformation towards egalitarian parenting values has, as Eva Gulløv notes, reconfigured the expectations of the role as parent from "authoritarian and instructive [...] to listening, including and supporting of the child's development and emotional life" (Gulløv, 2018, p. 13; see also Gulløv, 2016). This may, as the exchange between John and Ulla illustrates, put pressure on parents to make social judgements about, for instance, how children will respond to and deal (differently) with genetic at-risk knowledge.

Accidental: A Case of Talking Out of Turn

Asbjørn, aged 16, told us with a grin on his face that the news that his mother was a Lynch syndrome carrier was "blurted out" accidentally by his grandmother during one of his regular visits to her house:

> I was told last year. I was told because my mom had to go to all those examinations, and she had to do a bunch of stuff beforehand. I mean, she ate some really weird things... And then I was at my granny's and we were chatting about all that weird food and then she just said, "Well that's because of those examinations". And then she just told me, "Well, your parents' probably haven't told you". And I was like NO, they hadn't!

Numerous studies in the social sciences have pointed to the social, emotional, and existential impact of being diagnosed with a serious disease, that it can amount to a form of "biographical disruption" (Bury, 1982) in life and/or bring about a "loss of self" (Charmaz, 1983). People who experience being diagnosed with a serious disease are often able to recount the exact moment when given the diagnosis. Likewise, in Monica Konrad's (2003) ethnography on families living with the risk of the hereditary, neurologically degenerative Huntington disease, the moment of disclosure is also recollected vividly, with one of her interviewees explaining that "life does completely change and your whole family's life completely changes and it's this gradual, gradual change, you know. You think you're coping fine. And you're not. Nobody is coping. None of us are coping, you know" (Konrad, 2003, p. 40). This can also be the case amongst people living with Lynch syndrome, as we have discussed in a previous study where one of our interviewees recalled, "I did feel like I had been hit by a freight train when I found out that I was carrying the gene. In that moment, it was as if my world fell apart" (see Heinsen et al., 2022, p. 35). However, many of the young adults we met who learned they were *possibly* (50% chance) carrying a cancer-predisposing gene from their parents could hardly recall when they were told and how. In fact, at times it did not even come as a surprise, because some had overheard various conversations through the years without necessarily thinking much of them at the time. As Jonathan, 20, said, "I don't remember the exact situation. I don't remember how old I was, but I do remember I wasn't that surprised at all. I remember overhearing a conversation a while ago, and I didn't know what it was at that time, but when they told me it was like, then the penny dropped, then it all made sense".

In these ways, overhearing can be a "natural" route of disclosure, akin to the ways that children may come to learn of many different matters that are considered adult in nature by parents. And, as Jonathan points out, the act of overhearing itself might in some ways "just" be the final piece that allows the "penny to drop".

Continual Process

During a focus group interview with young adults who had tested positive for Lynch syndrome, Andreas, aged 25, told us, "It's kind of like something I grew up with". Karla, aged 27, chipped in, "My knowledge of risk came gradually. The older I got, the more I understood that something in my family was like really crap". This pattern of coming to understand risk knowledge as a continual process of accumulating insightfulness was mostly recounted by young adults who had grown up in families where a parent had known about his or her risk for a long time and where surveillance colonoscopies had been a part of their life. Being a bystander to the preparatory work a parent had to do prior to a colonoscopy appointment had provided clues to something not quite ordinary (as Asbjørn had also mentioned), allowing them to put the pieces together bit by bit as the years went by. As Jonas said, "I remember my dad went to those check-ups. So I remember it from when I was very young that he went to those examinations. And he told us a bit about what it was and why he had to go. But it wasn't until after I had turned 18 that they asked me if I wanted to go for genetic testing. I said yes, got tested, got the answer".

Andreas had recently had a baby daughter and reckoned that passing on the information about Lynch syndrome to her would be a "continual process", telling her "quietly", responding to her questions as she grew older, thereby "turning it into something natural". In these cases, how children come to learn about risk can be seen as a gradual and careful process of genetic sensitization. As Sofie, a 22-year-old medical student who had inherited the "faulty gene" from her father told us:

> I don't remember ... I was quite old when I was told that there was a risk we had the gene. It's not that I've always known but I don't remember when they told me. I guess my father wasn't that chit chatty so we were told when he started going for those check-ups. It wasn't dramatic at all, because we were told that you can do something about it and you get to go to these surveillance programmes.

Thus, disclosure can be experienced almost as a non-event, as, upon reflection, it is recollected as a gradual process of learning about an inherited risk running in the family. At the same time, both the parents and young adults we interviewed stressed that intra-family communication around Lynch syndrome is just as much about communicating risks as available preventive measures. For the young adults we met, the fact that their (potentially) genetically elevated risk and the cancer disease it is likely to trigger can be prevented makes it somehow more manageable. This also goes for the parents who had to tackle how and when to tell. As one mother said when asked how her children reacted to the news that Lynch syndrome runs in the family, "They were cool about it. They also underlined, you know, we were told that if you go to your surveillance colonoscopies, then your risk of dying from colon cancer is lower than amongst the rest of the population, you could say".

Discussion

In many of the social and ethical studies of disclosure dilemmas, there has been an emphasis on understanding how "genetic responsibility" and "genetic citizenship" generate new senses of obligation and duty as well as forms of care within wider families. In Hallowell's study of the genetic counselling of women who carry a BRCA gene, for example, she argues that "by labelling individuals as 'at-risk', and presenting genetic risks as manageable, genetic counselling implicitly places individuals under an obligation to attempt to modify these risks [with …] responsibility to their kin to establish the magnitude of their risk, and the risks to other family members, and to act upon this information by engaging in some form of risk management" (Hallowell, 1999, p. 615). In these discussions, the central ethical concern of protecting people's "right (not) to know" is what is at stake, giving rise to moral quandaries regarding how to reach out in potentially wide kinship webs, and to whom. As we have seen in this chapter, when it comes to intra-familial disclosure of the risk of Lynch syndrome between parents and children, the "right not to know" is not considered an option, it is "only" a matter of figuring out how best to communicate the possibility of having inherited an elevated risk. As a result, what we have shown is how intra-familial communication of genetic risk can be seen as a range of more or less explicit strategies and efforts on the part of the parents to cultivate a kind of *genetic sensibility* in their families.

In Denmark, parenting is shaped by culturally embedded ideals that emphasize the child's gradual development into an emotionally balanced, self-reflective, and socially responsible individual. Danish parents are expected to guide their children with care, not only by protecting them from harm but also by preparing them to navigate the complexities of life with emotional maturity. This requires a delicate balancing act: shielding the child when necessary, while also fostering autonomy and awareness. As Gulløv (2016) argues, emotional management has become central to contemporary Danish parenting. Such expectations are tied to a broader cultural understanding of childhood and youth as life stages where timing, tone, and emotional calibration matter. Likewise, Gullestad (1997) has highlighted how parent-child relations are framed by social norms that demand sensitivity to the boundaries between how childhood and adulthood are managed. Within this framework, parenting is not only about meeting physical or practical needs but about shaping the child's inner life—especially their capacity to handle difficult or abstract knowledge, such as genetic predispositions.

This cultural backdrop might help explain why Danish parents who carry a hereditary condition like Lynch syndrome often approach the disclosure of genetic risk with caution and care. Their concern is not only *what* to tell their children, but *when* and *how* to do so in a way that is attuned to the child's emotional readiness, as we saw in the cases of both Tine and the couple, John and Ulla. Cultivating what we term "genetic sensibility"—preparing children to live with the possibility of being genetically at risk without overwhelming them—by assembling family conferences to discuss genetic risk collectively, or "dosing" children with information in appropriate amounts as they grow older, distinguishing between children on the basis of when they were respectively seen to be robust enough to deal with risk knowledge—or having to react following accidental disclosure when a child overheard a conversation or a family member spoke out of turn—is deeply rooted in these cultural norms. It reflects a parenting ethos where knowledge is not simply transferred, but carefully mediated, filtered, and timed. Yet, there is no one size that fits all, as the timings when parent carriers come to know of their own genetic status can vary greatly, which shapes the context of informing children. Family conferences made sense to Poul, who had adult children, not only because of Poul's girls' age, but also because of the urgency of getting them tested and enrolled in surveillance. While the surveillance life that living with Lynch syndrome entails has not become completely normalized for all, family experiences which accumulate over time come to be passed on, as we saw with Andreas and his reflections on how he would disclose the possibility that his newly born daughter had inherited an elevated risk for colon cancer. Genetic sensibility is something that is cultivated more or less consciously over time within families.

Conclusion

In this chapter, we have argued that when it comes to communicating hereditary risk to children there is a need to shift away from understandings of disclosure as a singular, demarcated, and critically defining moment (cf. Forrest Keenan et al., 2009) to an understanding that intra-familial communication of risk information is a distributed, fluid, and filtered process that amounts to what we see as a part of families' attempts to cultivate genetic sensibility, not least as a way to counteract or fend off fatalism. We suggest that our findings can helpfully be introduced into genetic counselling sessions as a standard part of the issues and matters to be discussed with parents of young children or young adults following a positive genetic test. Such parents should be given advice that reassures them that there is no single form of intra-familial disclosure, for example, at a certain age that is right for everyone. Rather the "right time" can more helpfully be thought of as putting together the "right approach" for cultivating genetic sensibility within the family when Lynch syndrome becomes inescapably a part of their new "surveillance life".

At the same time, it is important to remember that the enculturation of Lynch syndrome through the cultivation of genetic sensibility that we have described in this chapter is very particular to a welfare state like Denmark that provides medical surveillance as part of a taxpayer-financed healthcare system. We have visited and interviewed families who have actively chosen to lead surveillance lives, but there are of course those families who, for different reasons, decide not to be genetically tested, or when learning of their "at risk" status, opt not to take part in medical surveillance programmes. More research is needed to understand their perspectives as well.

Acknowledgements Research for this chapter was made possible by a grant from the European Research Council (ERC) under the European Union's Horizon 2020 research and innovation programme for the VITAL project (grant agreement No ERC-2014-STG-639275) and by the grant "Young Adults in Families with inherited Colorectal Cancer: A Double Helix" funded by the Danish Cancer Society's *Knæk Cancer* funds (grant agreement R210-A12878).

References

Aktan-Collan, K. I., Kääriäinen, H. A., Kolttola, E. M., Pylvänäinen, K., Järvinen, H. J., Haukkala, A. H., & Mecklin, J.-P. (2011). Sharing genetic risk with next generation: Mutation-positive parents' communication with their offspring in Lynch syndrome. *Familial Cancer, 10*(1), 43–50. https://doi.org/10.1007/s10689-010-9386-x

Andorno, R. (2004). The right not to know: An autonomy based approach. *Journal of Medical Ethics, 30*(5), 435–440. https://doi.org/10.1136/jme.2002.001578

Arribas-Ayllon, M., Sarangi, S., & Clarke, A. (2008). The micropolitics of responsibility vis-à-vis autonomy: Parental accounts of childhood genetic testing and (non)disclosure. *Sociology of Health & Illness, 30*(2), 255–271. https://doi.org/10.1111/j.1467-9566.2007.01037.x

Atkinson, P., Featherstone, K., & Gregory, M. (2013). Kinscapes, timescapes and genescapes: Families living with genetic risk. *Sociology of Health & Illness, 35*(8), 1227–1241. https://doi.org/10.1111/1467-9566.12034

Baroutsou, V., Underhill-Blazey, M. L., Appenzeller-Herzog, C., & Katapodi, M. C. (2021). Interventions facilitating family communication of genetic testing results and cascade screening in hereditary breast/ovarian cancer or lynch syndrome: A systematic review and meta-analysis. *Cancers, 13*(4), Article 4. https://doi.org/10.3390/cancers13040925

Beauchamp, T. L., & Childress, J. F. (2001). *Principles of biomedical ethics* (5th ed.). Oxford University Press.

Bradbury, A. R., Patrick-Miller, L., Pawlowski, K., Ibe, C. N., Cummings, S. A., Hlubocky, F., Olopade, O. I., & Daugherty, C. K. (2009). Learning of your parent's BRCA mutation during adolescence or early adulthood: A study of offspring experiences. *Psycho-Oncology, 18*(2), 200–208. https://doi.org/10.1002/pon.1384

Bury, M. (1982). Chronic illness as biographical disruption. *Sociology of Health & Illness, 4*(2), 167–182. https://doi.org/10.1111/1467-9566.ep11339939

Charmaz, K. (1983). Loss of self: A fundamental form of suffering in the chronically ill. *Sociology of Health & Illness, 5*(2), 168–195. https://doi.org/10.1111/1467-9566.ep10491512

Derbez, B. (2018). Is there a "right time" for bad news? Kairos in familial communication on hereditary breast and ovarian cancer risk. *Social Science & Medicine, 202*, 13–19. https://doi.org/10.1016/j.socscimed.2018.02.022

Dumit, J. (2012). *Drugs for life: How pharmaceutical companies define our health*. Duke University Press. https://doi.org/10.1215/9780822393481

Elias, N. (1998). The civilizing of parents. In J. Goudsblom & S. Mennell (Eds.), *The Norbert Elias reader*. Blackwell.

Etchegary, H. (2006). Discovering the family history of Huntington disease (HD). *Journal of Genetic Counseling, 15*(2), 105–117. https://doi.org/10.1007/s10897-006-9018-7

Etchegary, H., & Fowler, K. (2008). They had the right to know. Genetic risk and perceptions of responsibility. *Psychology & Health, 23*(6), 707–727. https://doi.org/10.1080/14768320701235249

Faircloth, C. (December, 2023). Intensive parenting and the expansion of parenting. In Lee, E., Bristow, J., Faircloth, C., & Macvarish, J. (eds), *Parenting Culture Studies* (pp. 33–67). Cham: Springer International Publishing. https://doi.org/10.1007/978-3-031-44156-1_2

Faircloth, C., Hoffman, D. M., & Layne, L. L. (2013). *Parenting in global perspective*. London: Routledge. https://doi.org/10.4324/9780203103906

Featherstone, K., Atkinson, P., Bharadwaj, A., & Clarke, A. (2006). *Risky relations: Family, kinship and the new genetics*. Routledge.

Forrest, K., Simpson, S., Wilson, B., Van Teijlingen, E., McKee, L., Haites, N., & Matthews, E. (2003). To tell or not to tell: Barriers and facilitators in family communication about genetic risk. *Clinical Genetics, 64*(4), 317–326. https://doi.org/10.1034/j.1399-0004.2003.00142.x

Forrest Keenan, K., van Teijlingen, E., McKee, L., Miedzybrodzka, Z., & Simpson, S. A. (2009). How young people find out about their family history of Huntington's disease. *Social Science & Medicine, 68*(10), 1892–1900. https://doi.org/10.1016/j.socscimed.2009.02.049

Franklin, S., & Roberts, C. (2006). Born and made: An ethnography of preimplantation genetic diagnosis. In *Born and Made*. Princeton University Press.

Frey, M. K., Ahsan, M. D., Bergeron, H., Lin, J., Li, X., Fowlkes, R. K., Narayan, P., Nitecki, R., Rauh-Hain, J. A., Moss, H. A., Baltich Nelson, B., Thomas, C., Christos, P. J., Hamilton, J. G., Chapman-Davis, E., Cantillo, E., Holcomb, K., Kurian, A. W., Lipkin, S., et al. (2022). Cascade testing for hereditary cancer syndromes: Should we move toward direct relative contact? A systematic review and meta-analysis. *Journal of Clinical Oncology, 40*(35), 4129–4143. https://doi.org/10.1200/JCO.22.00303

Fullwiley, D. (2011). *The enculturated gene: Sickle cell health politics and biological difference in West Africa*. Princeton University Press. https://doi.org/10.1515/9781400840410

Furedi, F. (2008). *Paranoid Parenting. Why Ignoring the Experts may be the Best for Your Child*. Continuum, London and New York.

Gallo, A. M., Angst, D., Knafl, K. A., Hadley, E., & Smith, C. (2005). Parents sharing information with their children about genetic conditions. *Journal of Pediatric Health Care, 19*(5), 267–275. https://doi.org/10.1016/j.pedhc.2005.05.008

Geelen, E., Van Hoyweghen, I., & Horstman, K. (2011). Making genetics not so important: Family work in dealing with familial hypertrophic cardiomyopathy: Genetics, healthcare, family and kinship in a global perspective: Situated processes of co-construction. *Social Science & Medicine (1982), 72*(11), 1752–1759. https://doi.org/10.1016/j.socscimed.2010.06.012

Gislinge, J. I. P., Byrjalsen, A., Naver, K. V., Clausen, H. V., Ravn, P., Petersen, K. R., et al. (2024). Living a cancer surveillance life: A meta-ethnographic synthesis of everyday experiences and ambivalences for women living with hereditary risk of breast and/or ovarian cancer. *Psycho-Oncology, 33*(12), e70054.

Gullestad, M. (1997). A passion for boundaries: Reflections on connections between the everyday lives of children and discourses on the nation in contemporary Norway. *Childhood, 4*(1), 19–42. https://doi.org/10.1177/0907568297004001003

Gulløv, E. (2016). Følelseshåndtering i et socialhistorisk perspektiv. *Dansk Sociologi, 27*(3–4), 8–27.

Gulløv, E. (2018). Diskussionen om opdragelse: om ansvar, autoritet og balancer. *Forskning i pædagogers profession og uddannelse, 2*(2), 15–15.

Hallowell, N. (1999). Doing the right thing: Genetic risk and responsibility. *Sociology of Health & Illness, 21*(5), 597–621. https://doi.org/10.1111/1467-9566.00175

Heinsen, L. L., Wahlberg, A., & Petersen, H. V. (2022). Surveillance life and the shaping of 'genetically at risk' chronicities in Denmark. *Anthropology & Medicine, 29*(1), 29–44. https://doi.org/10.1080/13648470.2021.1893654

Holt, K. (2006). What do we tell the children? Contrasting the disclosure choices of two HD families regarding risk status and predictive genetic testing. *Journal of Genetic Counseling, 15*(4), 253–265. https://doi.org/10.1007/s10897-006-9021-z

Johnson-Hanks, J. (2002). On the limits of life stages in ethnography: Toward a theory of vital conjunctures. *American Anthropologist, 104*(3), 865–880.

Kenen, R. (1994). The human genome project: Creator of the potentially sick, potentially vulnerable and potentially stigmatized? In I. Robinson (Ed.), *Life and death under high technology medicine* (pp. 49–64). Manchester University Press.

Klitzman, R., Thorne, D., Williamson, J., Chung, W., & Marder, K. (2007a). Decision-making about reproductive choices among individuals at-risk for Huntington's disease. *Journal of Genetic Counseling, 16*(3), 347–362. https://doi.org/10.1007/s10897-006-9080-1

Klitzman, R., Thorne, D., Williamson, J., Chung, W., & Marder, K. (2007b). Disclosures of Huntington disease risk within families: Patterns of decision-making and implications. *American Journal of Medical Genetics Part A, 143A*(16), 1835–1849. https://doi.org/10.1002/ajmg.a.31864

Konrad, M. (2003). Predictive genetic testing and the making of the pre-symptomatic person: Prognostic moralities amongst Huntington's-affected families. *Anthropology & Medicine, 10*(1), 23–49. https://doi.org/10.1080/13648470301269

Lindberg, L. J., Ladelund, S., Frederiksen, B. L., Smith-Hansen, L., & Bernstein, I. (2017). Outcome of 24 years national surveillance in different hereditary colorectal cancer subgroups leading to more individualised surveillance. *Journal of Medical Genetics, 54*(5), 297–304.

Lippman, A. (1998). The politics of health: Geneticization versus health promotion. In S. Sherwin (Ed.), *The politics of women's health: Exploring agency and autonomy* (pp. 64–82). Temple University Press.

Lynch, H. T., & de la Chapelle, A. (2003). Hereditary colorectal cancer. *New England Journal of Medicine, 348*(10), 919–932. https://doi.org/10.1056/NEJMra012242

McConkie-Rosell, A., Heise, E. M., & Spiridigliozzi, G. A. (2009). Genetic risk communication: Experiences of adolescent girls and young women from families with fragile X syndrome. *Journal of Genetic Counseling, 18*(4), 313–325. https://doi.org/10.1007/s10897-009-9215-2

Metcalfe, A., Coad, J., Plumridge, G. M., Gill, P., & Farndon, P. (2008). Family communication between children and their parents about inherited genetic conditions: A meta-synthesis of the research. *European Journal of Human Genetics, 16*(10), Article 10. https://doi.org/10.1038/ejhg.2008.84

Metcalfe, A., Plumridge, G., Coad, J., Shanks, A., & Gill, P. (2011). Parents' and children's communication about genetic risk: A qualitative study, learning from families' experiences. *European Journal of Human Genetics, 19*(6), Article 6. https://doi.org/10.1038/ejhg.2010.258

Myrhøj, T., Bernstein, I., Bisgaard, M. L., Svendsen, L. B., Søndergaard, J. O., Mohr, J., Dahl, S., & Bülow, S. (1994). The establishment of an HNPCC register. *Anticancer Research, 14*(4 B), 1647–1650.

Novas, C., & Rose, N. (2000). Genetic risk and the birth of the somatic individual. *Economy and Society, 29*(4), 485–513. https://doi.org/10.1080/03085140050174750

Petersen, A. (1998). The new genetics and the politics of public health. *Critical Public Health, 8*(1), 59–71. https://doi.org/10.1080/09581599808409212

Rolland, J. S., & Williams, J. K. (2005). Toward a biopsychosocial model for 21st-century genetics. *Family Process, 44*(1), 3–24. https://doi.org/10.1111/j.1545-5300.2005.00039.x

Rowland, E., & Metcalfe, A. (2013). Communicating inherited genetic risk between parent and child: A meta-thematic synthesis. *International Journal of Nursing Studies, 50*(6), 870–880. https://doi.org/10.1016/j.ijnurstu.2012.09.002

Seppälä, T. T., Latchford, A., Negoi, I., Sampaio Soares, A., Jimenez-Rodriguez, R., Sánchez-Guillén, L., Evans, D. G., Ryan, N., Crosbie, E. J., Dominguez-Valentin, M., Burn, J., Kloor, M., von Knebel Doeberitz, M., van Duijnhoven, F. J. B., Quirke, P., Sampson, J. R., Møller, P., Möslein, G., & on behalf of the European Hereditary Tumour Group (EHTG) and European Society of Coloproctology (ESCP). (2021). European guidelines from the EHTG and ESCP for Lynch syndrome: An updated third edition of the Mallorca guidelines based on gene and gender. *British Journal of Surgery, 108*(5), 484–498. https://doi.org/10.1002/bjs.11902

Sie, A. S., Prins, J. B., Spruijt, L., Kets, C. M., & Hoogerbrugge, N. (2013). Can we test for hereditary cancer at 18 years when we start surveillance at 25? Patient reported outcomes. *Familial Cancer, 12*(4), 675–682. https://doi.org/10.1007/s10689-013-9644-9

Skirton, H. (1997). Telling the children. In A. Clarke (Ed.), *The genetic testing of children*. Garland Science.

Stoffel, E. M., Ford, B., Mercado, R. C., Punglia, D., Kohlmann, W., Conrad, P., Blanco, A., Shannon, K. M., Powell, M., Gruber, S. B., Terdiman, J., Chung, D. C., & Syngal, S. (2008). Sharing genetic test results in Lynch syndrome: Communication with close and distant relatives. *Clinical Gastroenterology and Hepatology, 6*(3), 333–338. https://doi.org/10.1016/j.cgh.2007.12.014

Stuttgen, K., McCague, A., Bollinger, J., Dvoskin, R., & Mathews, D. (2021). Whether, when, and how to communicate genetic risk to minors: I wanted more information but I think they were scared I couldn't handle it. *Journal of Genetic Counseling, 30*(1), 237–245. https://doi.org/10.1002/jgc4.1314

Svendsen, M. N. (2005). Pursuing knowledge about a genetic risk of cancer. In R. Jenkins, H. Jessen, & V. Steffen (Eds.), *Managing uncertainty. Ethnographic studies of illness, risk and the struggle for control* (pp. 93–122). Museum Tusculanum Press.

Svendsen, M. N., & Koch, L. (2006). Genetics and prevention: A policy in the making. *New Genetics & Society, 25*(1), 51–68. https://doi.org/10.1080/14636770600603485

Van der Meer, L., Timman, R., Trijsburg, W., Duisterhof, M., Erdman, R., Van Elderen, T., & Tibben, A. (2006). Attachment in families with Huntington's disease: A paradigm in clinical genetics. *Patient Education and Counseling, 63*(1), 246–254. https://doi.org/10.1016/j.pec.2005.11.019

Werner-Lin, A., Merrill, S. L., & Brandt, A. C. (2018). Talking with children about adult-onset hereditary cancer risk: A developmental approach for parents. *Journal of Genetic Counseling, 27*(3), 533–548. https://doi.org/10.1007/s10897-017-0191-7

Young, A. L., Butow, P. N., Vetsch, J., Quinn, V. F., Patenaude, A. F., Tucker, K. M., & Wakefield, C. E. (2017). Family communication, risk perception and cancer knowledge of young adults from BRCA1/2 families: A systematic review. *Journal of Genetic Counseling, 26*(6), 1179–1196. https://doi.org/10.1007/s10897-017-0125-4

7

Cancer Risk Assessment Scores: Personalized Risk and Mundane Genomics

Lisa M. Hoffman and Josh Tenenberg

In a recent US-based *People Magazine* cover story, Olivia Munn, a 43-year-old American actress, told of her breast cancer diagnosis and double mastectomy. In the article, Munn explained that her ob/gyn doctor "decided to calculate" a Breast Cancer Risk Assessment Score for her. "The fact that she did saved my life". Munn explained that her doctor "looked at factors like my age, familial breast cancer history, and the fact that I had my first child after the age of 30. She discovered my lifetime risk was at 37%. Because of that score I was sent to get an MRI, which led to an ultrasound, which then led to a biopsy. The biopsy showed I had Luminal B cancer in both breasts. Luminal B is an aggressive, fast-moving

L. M. Hoffman (✉)
School of Urban Studies, University of Washington Tacoma,
Tacoma, WA, USA
e-mail: hoffmanl@uw.edu

J. Tenenberg
School of Engineering & Technology, University of Washington Tacoma,
Tacoma, WA, USA
e-mail: jtenenbg@uw.edu

V. Argudo-Portal et al. (eds.), *Mundane Genomics*,
https://doi.org/10.1007/978-981-95-4550-6_7

cancer". Munn "urged others to be proactive about their health and ask their doctors to calculate their 'Breast Cancer Risk Assessment Score' and to get annual mammograms and breast MRIs starting at age 30 if the [risk] number is greater than 20%" (Etienne, 2024). To encourage this, the online version included a link to the Susan G. Komen website that explains the risk assessment tool they use to estimate "a woman's risk of developing breast cancer in the next 5 years and within her lifetime (up to age 90)".[1]

These risk assessment scores are both complex and precise, incorporating multiple factors as well as producing a single number, such as 37%. They are future-oriented entanglements that produce knowledge about our risks, about us, and about how we should be in the world. For individuals like Munn who are identified as having an elevated risk, their entanglements are high stakes—life and death—forcing choices that are difficult as well as clear. As we review in this chapter, risk assessment scores are moving beyond the lower-tech "Gail Model" to include a dizzying variety of cutting-edge technologies, including advances in genetic sequencing, epigenomics, microbiomics, machine learning, and what is called precision or personalized medicine (Prainsack, 2017, 2023). The confluence of these practices produces sociotechnical imaginaries (Jasanoff, 2015a, 2015b) in which people are subject to new categorizations and understandings of risk, speculations about the future, and transformations in health care as well as self-care. We argue that these "scores" are becoming more individualized as well as mundane, organizing risk in more personalized ways through the production of "your" score, based on your risk factors, your genome. We also suggest that a new configuration of biopolitics is emerging in which screenings of one's blood or saliva, along with analysis of other risk factors, is proliferating riskiness through these calculation practices (see Rose, 2001; Rabinow & Rose, 2006; Novas & Rose, 2000). The incorporation of the scores into people's everyday lives shapes temporalities as well as subjectivities, emphasizes responsibility, incorporates ideals of rational decision-making, and, for many, creates "a highly potent zone of anxiety" (Lock, 2011, p. 223).

[1] The original online article provided the Susan G. Komen link while the article accessed on 6/2/2025 provided a link to the National Cancer Institute page about the Gail Model.

More specifically, this chapter considers how risk—itself a "technology for predicting and managing the future" (Alaszewski, 2023, p. 324)—is assessed and calculated. We begin by discussing "lifetime risk", a common concept in risk scoring, and then turn to the implications of the rise of risk factors and surveillance medicine (Armstrong, 1995). We then turn to the complexity of current risk calculation tools, the use of vital statistics, genome-wide association studies, and existing databases. In the final section, we address questions of the increasing mundane-ness of genomic risk, with particular attention to the way these practices individualize and personalize surveillance medicine. Disease risk scoring and genetic screenings are being *normalized and made commonplace*, appearing in *People Magazine*, on myriad websites, in doctor's offices, through health advocacy organizations, and in direct-to-consumer (DTC) genomic risk assessment companies. The addition of genomic and other "multi-omic" information also produces a more personalized risk assessment that is derived from multiple forms of one's own data. "Responsible" citizen subjects are encouraged to engage in these assessments, making risk "visible" in new ways (Turrini & Bourgain, 2021).

Anticipation and Lifetime Risk

Health risk scores are anticipatory, specific, and yet uncertain, like the hedging of one's bets over time, knowing that loss is as possible as gain. Participating in these risk assessments, even for a famous actress, transforms what had no value, what was not known, into actionable information that is understood as valuable to a wide (enough) audience. The "cancer risk assessment score" is, then, a personalized understanding and experience of risk and the possibility of illness, a number that emerges through a "series of sociotechnical processes" and produces new forms of value (Verran, n.d.). It is this nexus of anticipation, future-crafting (Kerr et al., 2021, p. 9), the possibility of cancer in particular, and what is called *personalized* medicine that is of interest to us, a specific form of sociotechnical imaginary (Hood & Friend, 2011; Heinsen et al., 2022; Jensen et al., 2023; Prainsack, 2017, 2023). Jasanoff describes sociotechnical imaginaries as "collective, durable, capable of being performed; yet they are also temporally situated and culturally particular", emphasizing that science and technology are *social* processes, materializing in configurations of power (2015a, p. 27).

Yet, in human practice, anticipating one's future not only forms opinions, but also conjures images, predictions, possibilities, and potential actions. It is a subject-making process as well. In this sense, it embodies what Adams et al. (2009) refer to as an anticipatory episteme. Anticipation is "an affective state … not just a reaction, but a way of actively orienting oneself temporally" (Adams et al., 2009, p. 247; see also Novas & Rose, 2000). So, while cancer risk assessment scores are statistically based computations, they are also "lived condition[s] or orientation[s]" that encompass both hope and fear, possibility and preparedness, temporal politics as well as a will to anticipate (Adams et al., 2009, p. 249). Indeed, "preventing preemptively" and "intercepting" disease before it occurs, hallmarks of precision medicine (and post-9/11 foreign policy, see Cooper, 2006), are complex processes of anticipating, involving patients, their loved ones, doctors, healthcare systems, screening technologies, testing procedures, insurers, and corporations—animating a desired future for all (see Rose, 2001; Rabinow & Rose, 2006; Novas & Rose, 2000). On receiving an elevated lifetime risk score, individuals are encouraged to "become an active partner in the drive for health, accepting their responsibility for securing their own well-being" (Rose, 2001, p. 6), shaping a mutated biopolitics that regards health scoring and risk assessment as "normal" and mundane, turning it from something that was "previously exceptional, remarkable" into that which is "routinely thinkable, perhaps even expected" (Rose, 2001, p. 16).

These scores, and the imperative to prevent something from happening, are "thinkable" and "reasonable" as a part of personalized and genomic medicine, a shared imaginary, a "postgenomic possibility" (Sunder Rajan, 2005) in our "postgenomic condition" (Reardon, 2017). Not only are they understood as "thinkable" and accepted as ways to understand one's health, but they also are "proliferating and globalizing" (Rose, 2001, p. 7). Such practices fit within the liberal political tradition of responsibility and security as well. Indeed, managing one's risk is a key aspect of liberal thinking and acting, designed for "rational" decision-making and action now for a secure future. Pat O'Malley notes that within liberalism, both calculability *and* incalculability are needed for individuals to be "free" to "chart a course", to engage in "life planning" as well as to "make rational decisions and take responsibility" for their future and their health (O'Malley, 2004, p. 173; Rabinow & Rose, 2006, p. 209; also Alaszewski, 2023; Lemke, 2007). Being "free" to manage ourselves through risk and uncertainty "create[s] demands for new risk-diagnostic tests, routines, equipment" such as these scores (O'Malley, 2004, p. 177). As we discuss below, these techniques of governing have been shifting along with technological and medical ways of doing things, resonating with our own experiences of cancer risk.

We come to these issues not only through intellectual curiosity, but also through personal experience. Josh Teneberg's father was diagnosed with pancreatic cancer in December 1988. As a consequence, Josh immersed himself in medical research on the disease, desperately seeking the latest treatment to prolong the very short remaining life span that the literature indicated his father had. His father died in August 1989, less than nine months after diagnosis. While in many ways we can understand Josh's life as enmeshed in what Ulrick Beck (1992) termed a "risk society", in 1989 not a word was said to Josh or his sisters about how their father's pancreatic cancer might have a bearing on their own chances of getting it, that is, on their own futures. Genetic and hereditary connection and speculation were not the organizing logics in the hospital and doctors' offices where his father received treatment; the future had not yet "colonis[ed] the present" for his family (O'Malley, 2004, p. 178). Lisa Hoffman, on the other hand, was identified as "high risk" for pancreatic cancer and given terrifying statistics about lifetime risk after her mother died from this disease in 2015, only three months after diagnosis, and then her brother passed from it in 2018. She learned from further research that having multiple family members with pancreatic cancer increased one's lifetime risk sevenfold, which increased significantly if a family member had "young-onset" of the disease, like her brother (Brune et al., 2010; see also Klein, 2021; Mizrahi et al., 2020; Ilic & Ilic, 2016).

Through conversations together—between a computer scientist (Josh Tenenberg) and cultural anthropologist (Lisa Hoffman)—we recognized that Josh had never considered his *own* lifetime risk for pancreatic cancer until discussions with Lisa, nor had he internalized the personalized responsibility to avoid and manage risk (O'Malley, 2004, Chapter 9). The sharing of personal connections to pancreatic cancer that occurred at two distinct historical moments raised questions for us about how risk was assessed, predicted, communicated, and deployed as a technique of self-making and being-made. As Lisa's own experiences with a high-risk designation propelled her into an anthropological study of risk, cancer, genetics, and precision medicine, Josh also faced the possibility that these reflections on his own experience in 1989 meant he too was being drawn into and governed by these regimes of knowing and "solidarities of fate" (O'Malley, 2004, p. 181). We found ourselves sharing a similar yet distinct biosociality that we were not sure we wanted to embrace—and yet we wanted *to live*, a biosociality of vitality (Rabinow & Rose, 2006, p. 211). As we considered our shared condition, we also considered alternative modes of being in the cancer "risk" world (see Krupar, 2012; Cooper, 2006).

What is particularly interesting is that the use of "lifetime risk" as a way to speculate about future well-being, disease prevention, and early detection has had a relatively short lifetime. Figure 7.1 shows the citation counts for a search on "lifetime risk" (in quotes) across the years 1980 to

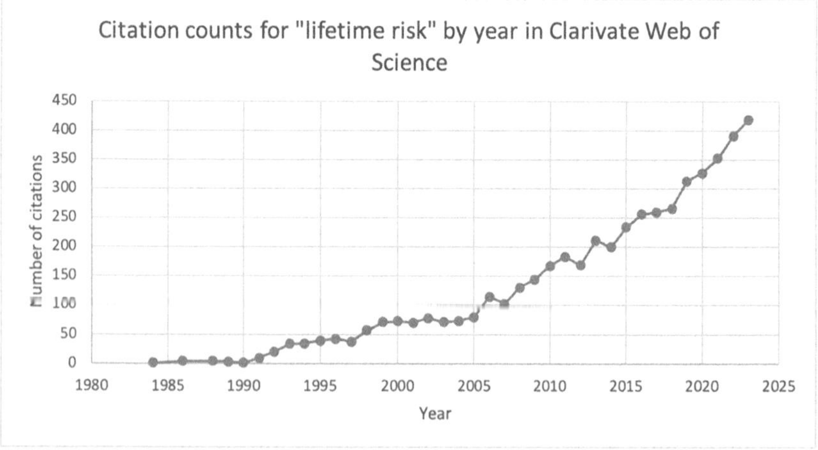

Fig. 7.1 Citation counts from a Web of Science search on "lifetime risk" (in quotes), April 2024

2023 in the Clarivate Web of Science database (see also Armstrong, 2012). Prior to 1984, "lifetime risk" was not a part of medicine's lexicon, and its current meaning must therefore depend upon the discursive practices of those academic, research, and clinical communities where it is now used; from these domains, the term has spread into the common vernacular, so that one can encounter it in such places as *People Magazine*. And it is in this magazine that Olivia Munn takes on the "duty" to share risk knowledge with others very seriously (see Lemke, 2007). Seeking out cancer risk assessment scores that sort out one's "lifetime risk" is, then, a relatively recent mutation of health assessment and speculation, albeit one that has become commonplace, accepted, expected. The mundaneness of such postgenomic thinking inserts a "vital politics" into everyday living (Rose, 2001). Indeed, as Jasanoff (2015a, p. 9) emphasizes, there are "myriad ways in which scientific and technological visions enter into the assemblages of materiality, meaning, and morality that constitute robust forms of social life". Lifetime risk assessment scores are one technique by which sociotechnical imaginaries shape what we mean by un/healthy lives.

Risk and Screening: Mutating Surveillance Practices

The availability of these risk assessment scores builds on what David Armstrong (1995) called "Surveillance Medicine". In Surveillance Medicine, which emerged early in the twentieth century, the target of medicine expanded from those who were already ill to the *entire population*. Achieving this universalization required blurring the border between illness and health, achieved through two distinct but intertwined practices, one epidemiological and the other in relation to health screening. The epidemiological practices involve the collection of statistical data about health, mortality, and illness for populations. The most important of these are vital statistics, public health surveillance, and cohort studies. National registries of vital statistics track births, deaths, and causes of death in a population, and their use in the US goes back to the earliest colonial settlements. Systems of public health surveillance, defined as "ongoing systematic collection, analysis, and interpretation of data on specific health events affecting a population" (e.g., for occupational accidents, adverse drug reactions, and product safety) began to be collected in the latter half of the twentieth century (Perrin et al., 2001, p. 12). One of the most important of these is the National Cancer Institute's Surveillance, Epidemiology, and End-Results (SEER) Program, the "authoritative source for cancer statistics in the United States" (*Surveillance, Epidemiology, and End Results Program*, n.d.). It is from these surveillance and vital health statistics that the *risk* of particular diseases for a population are derived, indeed, risks that even those who are "healthy" may harbour.

But it was only with the advent of the large-scale epidemiological *cohort studies* that began in the mid-twentieth century that the concept of the *risk factor* was developed (see also Hacking, 2006). In a cohort study (with the Framingham Heart Study being one of the most well-known), the researcher followed a group of individuals over time (the "cohort"), characterizing each individual according to a standardized set of categories (such as tobacco use, exercise, age, and/or diet). The cohort data was analysed to determine statistical relationships between these categories and various illnesses and mortality that later arose in this cohort (Capili & Anastasi, 2021). When the subset of the cohort with a particular value for a category measure (such as smoking, or blood pressure above a certain level) had a higher incidence of a disease (such as heart disease or cancer) compared to the rest of the cohort, this category value was defined as a *risk factor* for the disease.

Risk and risk factors, then, obtain their meaning from the underlying statistics from which they are derived. Yet, the statistic is a characteristic of a *population*, not an individual. The way in which these population-based statistics became individualized in Surveillance Medicine was through the *health screen*: "the process of identifying individuals who may be at higher risk of a disease or condition amongst large populations of healthy people" (Evidence and recommendations: NHS population screening, 2015). The term derives from the first clinical uses of X-rays at the start of the twentieth century for the early detection of tuberculosis, where "the radiographic image was viewed on a screen", as well as from public health use of mesh screens on the windows of buildings to keep out disease-carrying flies and mosquitoes (Armstrong, 2012, p. 18). These two etymological roots, from the clinic and from public health, highlight the two "registers of professional uses" where screens had widespread uptake (Turrini & Bourgain, 2021, p. 1630).

Initially, the health screen was used in the clinic for the detection of signs of illness before symptoms emerged, with an intent "to identify 'latent' disease" (Armstrong, 2012, p. 19). Yet, with the recent increase in the number of risk factors assessed for diseases, screening tests provide a systematic basis for *categorizing the individual* according to whatever risk factors the screens detect. Risk, then, is not only a statistical category, but also becomes "a social condition" (Gillespie, 2012, p. 34; Armstrong & Eborall, 2012). While statistics are calculated from a population, the presence of risk factors as detected through health screens *manifest in the individual* who may, or may not, be at-risk, producing what Novas and Rose have called "somatic individuality" (2000). Thus, as these practices led to the development of "risk profiles, scales and indices", the individual who crossed a threshold and obtained a risk factor became a "risky subject" (Rose, 2001, p. 8).

The location where screens could be concentrated and under the control of the medical profession was in the *general health check*, also known as the *periodic health examination* (PHE). "General health checks involve a contact between a person and a healthcare professional to identify signs, symptoms, or risk factors for disease that were previously unrecognized. They are combinations of screening tests" (Krogsbøll et al., 2012, p. 1). The examination indeed combines "hierarchical surveillance and normalizing judgement ... disciplinary functions of distribution and classification", leading to "the fabrication of cellular, organic, genetic and combinatory individuality" (Foucault, 1995, p. 192, in Krupar, 2012, p. 52). Starting from early in the twentieth century, periodic health examinations became a common characteristic of the doctor-patient encounter in the US, UK, and many countries worldwide (Holland, 2009). As a result, "[h]ealth examinations helped to foster the belief that Americans needed more medical care and health supervision, for they almost uniformly showed that very few people were healthy and normal" (Starr, 1982, p. 193).

Once risk was established as a concept materially rooted in the categorization, measurement, and statistical analysis of large cohorts of individuals in a population, and individualized through the health screen, illness could no longer be defined strictly by its signs and symptoms, and treatment was no longer simply about curing the ill. Rather, "illness was redefined by treatment as risk and health as risk reduction" (Dumit, 2012, p. 116; also Aronowitz, 2009, 2015; Gillespie, 2012). Dumit explained this meant "we are dedicated to measuring our health by understanding our risks and taking practical steps to reduce them" (2012, p. 197), laying the groundwork for the contemporary expansion of risk assessment scoring. The practices of "chronic treatment for risk reduction" are what Dumit (2012, p. 14) calls *mass health*, and as a part of surveillance medicine produce "a different form of identity as its monitoring gaze sweeps across innovative spaces of illness potential … Its calculability is given in the never-ending computation of multiple and interrelated risks", and, Armstrong concludes, "Its subject and object is the 'risky self'" (1995, p. 403). The risky subject, now quite commonplace and self-perpetuating, is a subject that must "optimize" their "overall 'well-being' " (Rose, 2001, p. 17) by recognizing themselves as at-risk, genetically at-risk, and taking responsibility; indeed, what Monica Konrad has termed "the making of the 'pre-symptomatic' person" (2003, p. 24; also Hacking, 2006). Nikolas Rose refers to this as a shift in the "focus of regulatory strategies from group risk to individual susceptibility" (the advent of somatic individuality), which moves past being a part of a risk group to the "identification" of "particular individuals" who are considered to have an elevated risk (2001, p. 11). Additionally, screening for "latent disease" produces a "temporal dimension of illness" that requires screening of younger people for diseases with onset in older adults (Armstrong, 2012, pp. 19, 20), resonating with what more recently have been termed "genetically at risk chronicities" (Heinsen et al., 2022).

Through the transition to surveillance medicine and mass health, prevention has become a central concern of health. No longer should an individual only seek medical care when they experience feelings of unwellness; rather, they should now seek *preventative* medical care, using the periodic health examination and screenings to determine whether any silent killers might be stalking them. The addition of genetic testing to the types of screenings that are available has helped to fuel a more personalized risk, a topic we address in the next section. Genomic medicine focuses even more on a personalized understanding of risk, also addressed below. The temporalities of illness seem to condense with these practices as well, such that the asymptomatic individual is assessed and tested and then expected to "modify their lifestyle responsibly in relation to their genetic risk" (Novas & Rose, 2000, p. 495) with the aim, in precision/personalized medicine, of "disease interception" (Prainsack, 2023, p. 445, also NCI, n.d.-b).

Genetics, Cancer, and the Calculation of Risk

As the price of genetic testing has decreased and more medical professionals are receiving training, it has become more common. Not only is it occurring in reproductive, prenatal space, but also as a part of the periodic health examination. In addition, direct-to-consumer genetic (DTC) testing is available for those who are willing and able to pay out of pocket, and computational tools are available online for those who input non-genetic, family history, and lifestyle information. As these risk calculators use artificial intelligence and machine computation (i.e., new ways of knowing risk) and become increasingly common, reconfigured forms of value and "moral trouble" are emerging (Verran, 2010). Drawing on semiotics, Helen Verran pushes us to think about how numbers may be symbolic as well as iconic, which also references the ways that numbers may transition "from representation to constitution" (2010, p. 177). In this sense, the "scores" are not simply a symbolic "*representation*" of risk, but also may be understood to be "*constitutive* of the entity itself. They work as icons" (Verran, 2010, p. 176, emphasis in original). These numbers help people "know" their risk, while also materializing their disease risk through the risk calculation itself, such that it becomes "visible when somebody recognizes it as relevant" (Turrini & Bourgain, 2021, p. 1630). Each risk factor, each biomarker, each individual number used in the calculations "are collected together as a many" (Verran, 2010, p. 176, one/many), shaping, making visible, constituting disease risk itself. As a technology of governing, risk assessment includes outcomes, categorizing via probability, an orientation towards the adverse, and time (Heyman et al., 2012). Yet, this “collection of many” never distils into a single number, a single representation of *one*'s risk of any particular disease. Instead, it disperses into an increasingly large number of risk scores for the same individual for the same disease that can be calculated in a myriad of ways; this sounds like moral trouble. Indeed, the desire to identify an individual's susceptibility “offer[s] no clear-cut algorithm for the decisions of doctors or their actual or potential patients” (Rose, 2001, p. 12).

For example, the Gail Model calculator on the National Cancer Institute (NCI, also NCI, n.d.-a) website gives a five-year risk of developing breast cancer of 0.6% for a 40-year-old, childless, white woman with no prior history of breast cancer, no benign breast biopsy, first menses at age 14, and no knowledge of BRCA variants or of first-degree relatives with breast cancer. The “know your chances” calculator, in a different corner of the NCI's vast website, indicates the same person's chance of dying (checkboxes only for female, white, age 40, and never having smoked) from breast cancer in the next 10 years as 0.1256%. On the “Cancer Query System” of the Surveillance, Epidemiology, and End Results (SEER) Program, yet another part of the NCI's domain, the chance of the same person dying of breast cancer in their remaining lifetime is 2.361% (checked boxes for “U.S. Mortality 2000–2022” database, 2020–2022, White non-Hispanic), while the SEER*Explorer of the NCI-SEER website indicates that in the most recent year for which there is data (2021), 243.4071 White non-Hispanic women in the age range 40–64 were diagnosed with breast cancer out of every 100,000 women in the US, or 0.2434%. And if one instead chooses the age range “< 50” instead of “40–64” on this same website, the breast cancer incidence drops to 0.0528 for 2021. So considering just a few demographic categories into which one individual can be placed, there is no simple formula that yields a single risk score: these numbers are multiple, ambiguous, and uncertain, as well as precise.

Yet, if one uses the information about each of the 31 additional risk factors for breast cancer available on the Susan G. Komen website, recommended in the Olivia Munn article, one can learn that one's risks of getting or dying from breast cancer are not any of the numbers obtained from the different risk calculators on the sprawling and multi-divisional

NCI website. Rather, risk of breast cancer corresponds to one's Ashkenazi Jewish status, breast density, consumption of carotenoids, and height, among many other risk factors.[2] With the intensification of what comes into biomedical surveillance as relevant factors, there is a "reterritorializ[ation] and expan[sion] of the domains and sites" that become salient to risk assessment (Adams et al., 2009, p. 253). While the amount that each of these separate risk factors contributes to one's risks is never stated in a precise way, almost all are indicated as having a "small or modest" impact, even if it is never clear whether several small or modest impacts might together amount to a large and deeply distressing impact.

These differences in risk score are not only due to the different diseases whose risk one might inquire into, or different choices required of the savvy risk-score seeker for input into a risk model, or that some statistics are in terms of dying while others are in terms of getting cancer. Nor is the main reason for risk score differences due to the proliferation of predictive models for a large variety of diseases, as important as these are for accounting for score differences. "For breast cancer prognostication, around 60 models currently exist, of which the majority have the same set of predictors. Likewise, a similar set of predictors can be found in over 300 models predicting cardiovascular disease risk, in over 40 models predicting prognosis for esophageal or gastric cancer, and so on" (Binuya et al., 2022, p. 3). As much as anything, the differences in the risk scores that the same individual can obtain for getting or dying from one or another affliction is that the underlying databases from which each of the disparate risk-score websites and models draw has data on subsets of the risk factors that have been identified for various diseases. For instance,

[2] Examples for breast cancer are overwhelmingly of women, drawing on a sociocultural understanding that women should be tested rather than men. This is borne out in the statistic that in the US, women are ten times more likely to get tested for breast cancer than men (Pritchard, 2019, p. 28). Medical oncologist Colin Pritchard argues that the "Hereditary Breast and Ovarian Cancer" syndrome should be renamed the King syndrome "if we're going to educate the public about the importance of these genes for cancer risk for everyone, not just women" (quoted in Brown, 2024). Pritchard is advocating the "rebranding" of BRCA 1 and 2 mutations as the King syndrome, named for the Mary-Claire King who discovered the mutation in 1990. This also follows the convention of naming pathogenic variants after the scientists who discovered them, such as Lynch syndrome and Li-Fraumeni syndrome.

NCI-SEER collects "cancer incidence, prevalence, and survival data from population-based cancer registries covering approximately 34.6 percent of the U.S. population. The SEER registries routinely collect data on patient demographics, primary tumor site, tumor morphology, stage at diagnosis, and first course of treatment" (*Accessing SEER Data*, n.d.). What is not recorded, however, is information on consumption of alcohol, height, weight, age of first menses, let alone Ashkenazi Jewish status, breast density, consumption of carotenoids, or virtually any of the other risk factors identified in the medical literature for all of the cancers under SEER's purview. This means that each database, each disease, each way to calculate risk, and each identified risk factor leads to a different risk score for the same individual. So "my risk" is not a single number, but a *multiplicity* of risk scores for having or dying from any particular disease over any given period of time, such as a year or a lifetime. Nevertheless, these numbers are meaningful, and in Olivia Munn's case, likely saved her life, underscoring the stakes at play with risk assessment scores.

Moreover, if this multiplicity of scores results in uncertainty about one's risk, recent genomic research dramatically increases uncertainty, since these risk scores take account of the complexity and particularity of an individual's genetic makeup. These individualized genetic risk factors are called single-nucleotide polymorphisms (SNPs), indicating "a genomic variant at a single base position in the DNA" (Gunter, n.d.). SNPs are detected not from large population disease incidence and mortality databases such as those of the NCI-SEER programme, or from cohort studies, but from *genome-wide association studies* (GWAS).

> The basic procedure of GWAS is to scan for genetic variants that are statistically significant when comparing two groups, typically a group with and without a specific disease. Risk alleles, or single nucleotide polymorphisms (SNPs), are said to be associated with a specific disease if its appearance among a group with a given disease is statistically significant compared to a group without the disease. (Green & Vogt, 2016, p. 114)

The identified SNPs are not infrequent and few, and far more than the 13 genetic variants identified as increasing risk for breast cancer on the Susan G. Komen site (*Factors Linked to Breast Cancer Risk*, 2024). In fact, "[t]here are currently hundreds of variants associated with breast cancer susceptibility", and "thousands of SNPs associated with diseases such as various cancers, coronary artery disease (CAD) and psychiatric disorders" have been found (Yanes et al., 2020, pp. R168, R166), with every indication that thousands (or tens of thousands) of additional ones will be identified. The continuous expansion of factors, mutations, behaviours, and environments that could be entangled in the scores fuels ongoing speculation and the requirement to responsibly "optimize" one's future mode of being (Rose, 2007). Moreover, as individualized risk assessments are the outcome of these biomedical and calculative procedures, emergent technologies of subject formation take shape.

While a few individual SNPs have been identified that result in substantial increases in risk of a disease by a factor of 4 or more (what are called *monogenic* risk factors), such as the BRCA1 and BRCA2 pathogenic variants for breast and ovarian cancer (Kuchenbaecker et al., 2017), most of the increases to risk associated with these genetic variants are relatively small. "A large portion of genetic risk for complex diseases stems from the interactive effect of multiple low- or modest-risk single-nucleotide polymorphisms (SNPs)" (Yanes et al., 2020, p. R165). As a result, methods for predicting risk have been developed that take into account the small and multiple SNPs that any individual will have, resulting in an individual's *polygenic risk score* (PRS). The US-based National Human Genome Research Institute explains that "Genetic tests look at the small differences between individuals' genomes", while "polygenic risk scores are tools for assessing many genomic variants across the genome to determine a person's risk for disease" (Rogers, 2025). Significantly, these PRSs "are calculated as weighted counts of thousands of risk variants that they carry, where the risk variants and their weights have been identified in genome-wide association studies" (Wray et al., 2021, p. 101). Genetic information coming from GWAS and the identification of an increasing number of risk alleles is leading some medical researchers to conclude that "each of us have some risk variants in our DNA for all diseases" (Wray et al., 2021, p. 102).

GWASs, however, are generally not predictive for all populations due to the disproportionate number of Europeans in the databases resulting in a "weaker ... predictive performance of a polygenic score model"

(Moorthie et al., 2023, p. 439). Jonathan Kaplan and Stephanie M. Fullerton have examined, for instance, how polygenic risk scores make estimates related to "specific ancestral genetic backgrounds"; these are, however, "not obviously translatable to patients who self-identify with (or are ascribed to) racial and/or ethnic categories rather than genetic ancestries" (2022, p. 2).[3] This is critical to understand as population structure is highly complex, even though many geneticists have turned to genetic explanations and PRS as a way "to address health disparities" (Kaplan & Fullerton, 2022, p. 3; also Raz & Minari, 2023). Moreover, Kaplan and Fullerton argue that the ways in which PRSs calculate disease risk essentially "lowe[r] the statistical standards for counting a marker as trait-associated", which means that "predictive accuracy is increased" but "at the expense of a loss in terms of being able to distinguish spurious from robust associations at any single locus" (2022, p. 3). Thus, because PRS have been developed with data from one population subgroup (predominantly European), they "tend to be much less predictive when applied to other populations" (2022, p. 4). Ultimately, this means that even as such scores proliferate, moving from a "population-level understanding of disease risk" to a number representing *an individual's risk* is a murky endeavour, especially as they cannot account for the environmental and social realities of people's lives. And as with non-genetic risk factors, there is considerable variation in the models and algorithms used to calculate these risk factors, particularly given the number of choices available in the commercial direct-to-consumer genetic test market. Indeed, groups "use different methods to develop PGS [polygenic risk scores]" and, not surprisingly "results can diverge" (Chapman, 2023, p. 448). In this market, companies use their own proprietary algorithms for calculating the PRS, sometimes based on proprietary data and GWASs that the companies have collected from consumers.

[3] They term such identities, which are highly social and embedded in racialized forms of power, "non-genetic population arrangements" (Kaplan & Fullerton, 2022, p. 4).

An Emerging Multi-Omic Paradigm

There is an increasing number of third-party interpretation services (TPIs) available to consumers, where an individual can upload their "raw data" from a genetic test to have an explanatory report provided to them that identifies their risk across a wide range of diseases. This was also made possible by 2014 HIPAA rules in the US that gave patients access to their designated record set (DRS), which could include "uninterpreted" genetic sequence data that patients could then supply to the TPIs (Nelson & Fullerton, 2018). For example, Nucleus Genomics offers both its own genetic testing and interpretation of genetic tests by other DTC genetic-testing companies. "We use cutting-edge DNA analysis to explore a greater breadth of diseases and traits", which include breast cancer, coronary artery disease, age-related macular degeneration, migraine, asthma, ADHD, type-2 diabetes, insomnia, Alzheimer's disease, prostate cancer, alcohol dependence, rheumatoid arthritis, hypertension, schizophrenia, "and many more" (Nucleus Genomics, 2024). This resonates with a study of 23 TPIs by Nelson and Fullerton (2018) that the authors classified into groups of TPIs providing genetic ancestry, those providing genealogy, and those providing health and wellness information (from diet to medication side effects to brain function) to individuals. They found that the TPI tools providing health/wellness information referenced peer-reviewed scientific studies, but the TPI reports varied from database linking to aggregation (via a formula to provide a risk score), contextualization (e.g., a trait score in terms of a given population), and/or recommendation (Nelson & Fullerton, 2018, pp. 775–6). Nevertheless, interviews with PRS tool developers affirmed their reluctance to claim "interpretation" of an individual's data and instead underscored their commitment to *linking* individuals with further information (e.g., databases, scientific literature) (Nelson & Fullerton, 2018, p. 776).

Not only is it the case that "TPI lacks expert oversight and might introduce additional sources of inaccurate or incomplete information" but "many consumers often use multiple TPI platforms to access health-risk information" (Nolan & Ormondroyd, 2023, pp. 4, 6). Given this level of variation, uncertainty, and dynamic change in genetic testing and the calculation and interpretation of risk scores, it is not surprising that the Board of Directors of the American College of Medical Genetics and Genomics asserted that "even with the most sophisticated technologies available, genetic studies on complex traits and disease constitute an inexact science" (ACMG Board of Directors, 2021, p. 2027). Other medical researchers have reached similar conclusions, including the warning that "the scale and complexity of information that may be conveyed from third-party interpretation of raw genetic data is unprecedented and likely to pose significant new challenges for health care providers" (Nelson & Fullerton, 2018, p. 779). At the same time these numbers have the potential to reshape healthcare and disease risk itself as people "play the numbers" (Good, 2001, p. 403); they also are embedded in a multimillion-dollar DTC-TPI marketplace. Thus, while the CEO of Nucleus Genomics waits for "a human genome on every iPhone" so that their product can combine the genetic and non-genetic risk factors to assess disease risk (Beteul, 2021), "conventional GWAS studies" do not combine the genetic with non-genetic interactions that affect risk (National Research Council, 2011, p. 43).

Nevertheless, precision medicine has the explicit goal of encompassing a multitude of "omics" to account for disease incidence and mortality disparities, including "proteomics [the proteins expressed by an individual's genomes], epigenomics [chemical compounds that attach to a person's DNA to modify its function] … metabolomics [small molecules in living tissue]" (Neagu et al., 2024, p. 19), " 'phenomics' technologies that enable monitoring and self-monitoring of phenotypic biomarkers …'exposomics' that provide inputs on environmental exposure" (Green & Vogt, 2016, p. 106), and microbiomics, which are "the microbiomes present in our gut and other epithelial surfaces [that] can have a major role in the wellness–disease balance" (Hood & Friend, 2011, p. 104). The emerging field of nutrigenomics is an excellent example in which an individual receives "dietary and lifestyle advice based on single nucleotide polymorphism genetic analysis", which is considered functional (not structural) genomics (Harvey, 2009, 2010, p. 368). Alison Harvey argues that nutrigenomics aims to "empower" people to optimize their wellness, while also focusing on *individuals* themselves. She explains that "clinical genetic medicine" understands "the gene as an element linking family members together in risk", but that "nutrigenomics considers the genome as placing the *individual* uniquely at risk" (Harvey, 2009, p. 126, emphasis added).

Research has documented that DTC genetic testing companies also "valorize" the individual and their "autonomy" to manage their health, producing an "individualized biopolitics" in which "customers" ignore complex factors to focus on genetics, "even endow[ing] their results with a diagnosis value" (Ducournau & Beaudevin, 2011, pp. 5, 7). Others have argued that identities become wrapped up in genetic mutations, producing a "genetic essentialism" in which the self is known through the genetic condition (Nordgren & Juengst, 2009). Yet, attempts to individualize risk through considering increasing amounts of data in relation to an increasing number of risk factors do not lead to a single number that represents "my chances" of getting or dying from any particular disease. Instead, they lead to a collection of different risk scores, and an almost endless number of contributing "risk factors". Nevertheless, fields like nutrigenomics reframe the genome into a more personalized expression of risk. Such framings work through the idea that we can now understand how you are "uniquely" at risk.

Risk then is not just a matter of one's age, gender, smoking, ancestry, or genes, but has expanded to include virtually all of the material and biological constituents of an individual, along with the interactions of the individual and the environment through time. Risk in the multi-omic paradigm of precision medicine encompasses and is calculated from "a gradient of layers spanning the heritable foundations of disease to more direct measures of environmental exposures on biological processes" (Mohammadi-Shemirani et al., 2023, p. 56). And in this world of precision and personalized medicine "each patient will be surrounded by a 'virtual cloud' of billions of data points that will uniquely define their past medical history and current health" (Hood & Friend, 2011, p. 185). From this data, one's risk status for any and all diseases can be measured and monitored on an ongoing basis, since "non-genetic risk factors, including epigenetic factors ... can change over time and interactions with different environmental factors and conditions can produce different outcomes" (Jansen et al., 2024, p. 5). This is a highly varied exercise that not only produces immense quantities of calculations and data, but also "render[s] the future as probabilistic" and open "to hope" as well as "technical intervention" (Rabinow & Rose, 2006, p. 208). In the final section, we address our argument that through the increasing mundaneness of these probabilistic and statistical practices, the notion of individualized and personalized medicine is shifting.

Personalized Risk and Mundane Genomics

In the late twentieth and early twenty-first centuries, genetic medicine has opened up new, expanded territory in surveillance medicine and has developed new screening techniques through which "to see" risks that were previously invisible (see also Turrini & Bourgain, 2021). With the expansion of genetic and disease-risk testing, personalized risk calculations move across doctors' offices, direct-to-consumer products, and popular media, making the governance of life through genomic knowledge increasingly common. Risk assessment calculations not only muddle distinctions between health and illness, but they also place an individual in multiple computational axes of genetic, non-genetic, lifestyle, and environmental risk factors. Chris Gillespie has in fact argued that "instead of a biological entity, being at risk is an epidemiological or probabilistic entity, one that reflects a person's place within a numerical distribution" (2012, p. 42). Not only does each gene, SNP, allele, and its possible variants serve as risky formations, but subsets of these genes are combined with one another and with non-genetic risk factors to provide a virtually infinite set of risk categories that an individual must navigate. At the same time, a plethora of methods for calculating risk are circulating, governing daily life with statistics and probabilities.

Receiving a high lifetime risk score for breast cancer launched Olivia Munn into a sequenced set of medical interventions. She encouraged others to ask for their own breast cancer risk assessment, to get their own risk calculation done, and then to take action. The scores offer a way to know about one's health through machine-based calculations that assess one's life history, genome, multiple-omics, as well as everyday habits to produce a *more personalized* risk assessment. These processes "delineat[e] who are risky subjects" (Krupar, 2012, p. 54), and at the same time produce those very subjects. This subject is expected to "become skilled, prudent and active, an ally of the doctor, a proto-professional" (Novas & Rose, 2000, p. 489). Despite the numerous tools, methods, algorithms, and variety of risk assessment scores a person may receive, these scores *do* things; in other words, they have very real, material consequences.

The subject emerging through these processes learns to take action now in order to "maximise their 'vital capital' (their health)" (Harvey, 2010, p. 365). The obligation to prevent (and detect early) future illness becomes paramount, with an individualized surveillance program based on the risk assessment. Scholars have described this in terms of cancer as the "crafting" one's future (Kerr et al., 2021), not only in terms of individual risk factors but also in relation to scores and "raw" genetic data. In fact, those who are "not-yet-sick" but at-risk for cancer (Svendsen, 2006, p. 140), and those participating in early detection screenings because they are high-risk, are "increasingly" included in cancer statistics (Bennett & Manderson, 2024, p. 390). The *pre*cancerous are considered to be in "Stage 0" and thus "no longer 'cancer free' " (Krupar, 2012, p. 59). This further breaks down the boundary between sickness and health; one is already in Stage 0 when undergoing early detection colonoscopies, MRIs, or endoscopic ultrasounds. Under such a logic, we are all risky subjects, suffering from some latent, invisible disease, underscoring how mundane such conditions have become. Indeed, in Norway, studies "have shown that if one follows authoritative guidelines for disease prevention in a normal population, one would define a high proportion as 'at risk' " (Green & Vogt, 2016, p. 120).

As health risk assessment expands from a doctor's office to the consumer-oriented internet and from medical journals to *People Magazine*, it not only makes genomic medicine more commonplace, but also, we argue, underscores how it is being personalized in new ways. The individualized health subject has the "duty to procure information, and only those who act responsibly draw the correct, that is, risk-minimizing and forward-oriented, conclusions from this range of information" (Lemke, 2007, p. 161). Geneticist Michael Snyder claims that people are interested in learning about personal genomics, with basic concepts already "being taught in most high school science classes, as well as in some science museums and public forums such as libraries" (2016,

p. 143). The expansion of genomics into new domains underscores the emergence of a mutating, personalized biopolitics that is embedded in probability and based in possibility and potential. Potentiality, "an orienting concept in the life sciences" may also index a "gap" that exists "between what is and what might, could, or even should be" (Taussig et al., 2013, pp. 3, 5). The notion of duty means there is an opportunity for intervention as well as a moral claim to shape the future. Indeed, the moral claim that one *should* take action presupposes a particular kind of responsibilized, risk-managing subject, who understands their genetic responsibility and what it means to be a "genetic entrepreneur" (Harvey, 2009; also Rose, 2001; Konrad, 2003).

If one does not have insurance, access to medical care, or the time or finances to engage in early detection screening, assumptions are made that the individual "does not care about [them]self, and puts society in jeopardy" (Krupar, 2012, p. 55). Precision medicine's emphasis on prevention and universal pre-patienthood is a form of biopower that is "disciplining bodies … and inducing subjects to reveal—confess—their risks of … disease" and thus be condemned to a lifetime of constant monitoring (Krupar, 2012, p. 51). Such confessions can then be made into a public spectacle, exemplified in the Danish television show, *Gentesten ændrede mit liv* (The gene test changed my life), which features interviews with Danes who have undertaken genetic testing and counselling (Schmidt-Nielsen & Stage, 2023, p. 74). Such media spectacles become part of the cultural production of "this rising state of pre-patienthood in which health is constantly 'haunted' by impending illness" (Schmidt-Nielsen & Stage, 2023, p. 74).

As risk assessment and pre-patienthood become increasingly mundane, scholars also ask if there are other *responsible* ways of being. What about the right "not to know" about a risky genetic mutation (Gigerenzer & Garcia-Retamero, 2017)? How might researchers recognize the "lived value of social identity as temporalities of embodied prognosis" (Konrad, 2003, p. 42; also Jain, 2013)? In her call for "cancer glam", Shiloh Krupar offers a critical analysis of cancer detection through song and rhyme, in an attempt to "consider a different form of the topic … in solidarity with all those soon-to-be *lumpen*proletariat and with respect to all those that came before us" (2012, p. 47, emphasis in original). While acknowledging this is a "playful turn", Krupar emphasizes it is serious:

> to show you the world of early detection, risk management, and self-care imperatives
> the ways this obscures the exhaustion, injuries, and loss of autonomy
> the sad freedom to be susceptible and determine future well-being via cost-benefit abstractions
> the medical moralism of genetic testing and other actions that risky subjects must dutifully undergo to optimize life chances...
>
> (2012, pp. 47–48, sentence structure in original)

Screenings in particular have been shown to trigger experiences of risk, leading to emotions of uncertainty and vulnerability, and to questions about how we may combine "navigating and anticipating an uncertain (molecular) *future*" with "*present* patient care" (Swallow et al., 2024, p. 983, emphasis added). Along with the possibility that one's own genomic risk assessments emerge in multiple scores, we realize that "statistical measurements that [were] intended to tame randomness and provide certainty in managing risk" in fact may produce the opposite, what Chris Gillespie terms "measured vulnerability" (2012, pp. 37–39, 43). In a song called "Booby Traps", Krupar explores the "compilation of catch-22s, roundabouts, self-fulfilling prophesies, tautologies, and overall dissatisfactions" in the cancer early detection world, as we also recognize these practices may save lives. In this world, dutiful pre-patients engage in self-surveillance, "reproducing themselves as relentlessly diagnostic subjects" (2012, p. 59). The song "Booby Traps" begins:

> There once was a breast cancer diagnosis
> That never resulted in prognosis.
> Convinced of pathology,
> The patient underwent biopsy
> And wound up with fateus necrosis.
>
> The fateus necrosis raised suspicion
> The doctor advised "more detection!"

Ultrasound, MRI exam,
Never-ending mammogram,
A life sentence of medical intervention.

The patient cried "when am I cancer free?
I'm sick of this ongoing militancy!"
So she unhooked her breast
From the latest test
And left the department of oncology …
(Krupar, 2012, p. 57, sentence structure in original)

When senior scientists argue that due to the decreasing costs of genetic screening, everyone should get a full genetic panel test, or argue that cancer is a genetic disease and we need a public health intervention that makes genetic testing available and financially accessible to all, we recognize how expected and common the genome is becoming to definitions of how to live healthy lives. Calls for such broad application of such testing, along with popularized encouragement for all of us to get our cancer risk assessment scores, underscores that genetic risk is an increasingly accepted and ordinary way to govern life, despite its fraught and multiple calculations. Genomics are indeed mundane.

Acknowledgements We thank the anonymous reviewers (who are also chapter authors in this volume) and volume editors for their insightful comments. Monica DeHart and Jennifer Hubbert also offered helpful comments during the writing process. We also thank the STS reading group at UW Tacoma, and in particular Matthew Weinstein, for providing the intellecutal space for us to have the conversations that led to this chapter. Of course, we take responsibility for any errors.

References

"Accessing SEER Data." (n.d.). NIH - National Cancer Institute. Accessed December 8, 2024. https://seer.cancer.gov/data/access_seer_data.pdf

ACMG Board of Directors. (2021). Direct-to-consumer prenatal testing for multigenic or polygenic disorders: A position statement of the American

College of Medical Genetics and Genomics (ACMG). *Genetics in Medicine, 23*(11), 2027–2028.

Adams, V., Murphy, M., & Clarke, A. E. (2009). Anticipation: Technoscience, life, affect, temporality. *Subjectivity, 28*, 246–265.

Alaszewski, A. (2023). Risk and health. In A. Petersen (Ed.), *Handbook on the Sociology of Health and Medicine*. Edward Elgar Publishing.

Armstrong, D. (1995). The rise of surveillance medicine. *Sociology of Health and Illness, 17*(3), 393–410.

Armstrong, D. (2012). Screening: Mapping medicine's temporal spaces. In N. Armstrong & H. Eborall (Eds.), *The sociology of medical screening* (pp. 17–32). Blackwell Publishing.

Armstrong, N., & Eborall, H. (2012). The sociology of medical screening: Past, present and future. *Sociology of Health & Illness, 34*(2), 161–176.

Aronowitz, R. A. (2009). The converged experience of risk and disease. *The Milbank Quarterly, 87*(2), 417–442.

Aronowitz, R. A. (2015). *Risky medicine: Our quest to cure fear and uncertainty.* University of Chicago Press.

Beck, U. (1992). *Risk society: Towards a new modernity* (Vol. 17). Sage.

Bennett, L. R., & Manderson, L. (2024). Anthropological explorations of women's reproductive cancers. In C. C. Van Hollen & N. S. Appleton (Eds.), *A companion to the anthropology of reproductive medicine and technology* (pp. 381–396). John Wiley & Sons.

Beteul, E. (2021, December 15). Nucleus Genomics announces $3.5M seed round for a fresh take on genetic tests. *TechCrunch*. https://techcrunch.com/2021/12/15/nucleus-genomics-announces-3-5m-seed-round-for-a-fresh-take-on-genetic-tests/

Binuya, M. A. E., Engelhardt, E. G., Schats, W., Schmidt, M. K., & Steyerberg, E. W. (2022). Methodological guidance for the evaluation and updating of clinical prediction models: A systematic review. *BMC Medical Research Methodology, 22*(1), 1–14.

Brown, K. V. (2024, November 20). The cancer gene more men should test for. *The Atlantic*. Retrieved November 21, 2024, from https://www.theatlantic.com/health/archive/2024/11/brca-breast-cancer-men-prostate-pancreas/680698/

Brune, K. A., Lau, B., Palmisano, E., Canto, M., Goggins, M. G., Hruban, R. H., & Klein, A. P. (2010). Importance of age of onset in pancreatic cancer kindreds. *Journal of the National Cancer Institute, 102*(2), 119–126.

Capili, B., & Anastasi, J. K. (2021). Overview: Cohort study designs. *American Journal of Nursing, 121*(12), 45–48.

Chapman, C. R. (2023). Ethical, legal, and social implications of genetic risk prediction for multifactorial disease: A narrative review identifying concerns about interpretation and use of polygenic scores. *Journal of Community Genetics, 14*(5), 441–452. https://doi.org/10.1007/s12687-022-00625-9

Cooper, M. (2006). Pre-empting emergence: The biological turn in the war on terror. *Theory, Culture & Society, 23*(4), 113–135.

Ducournau, P., & Beaudevin, C. (2011). Of deterritorialization, healthism and biosocialities: The companies' marketing and users' experiences of online genetics. *Journal of Science Communication, 10*(3), 1–10.

Dumit, J. (2012). *Drugs for life: How pharmaceutical companies define our health.* Duke University Press.

Etienne, V. (2024, March 13). Olivia Munn's breast cancer risk assessment score 'saved my life'—What to know about the tool. *People Magazine.* Retrieved June 2, 2025, from https://people.com/breast-cancer-risk-assessment-score-what-is-it-olivia-munn-8608670

Evidence and recommendations: NHS population screening. (2015). Accessed April 22, 2025, from https://www.gov.uk/guidance/evidence-and-recommendations-nhs-population-screening

Gigerenzer, G., & Garcia-Retamero, R. (2017). Cassandra's regret: The psychology of not wanting to know. *Psychological Review, 124*(2), 179–196. https://doi.org/10.1037/rev0000055

Gillespie, C. (2012). The experience of risk as "measured vulnerability": Health screening and lay uses of numerical risk. In N. Armstrong & H. Eborall (Eds.), *The sociology of medical screening* (pp. 33–45). Blackwell Publishing.

Good, M.-J. D. (2001). The biotechnical embrace. *Culture, Medicine and Psychiatry, 25*, 395–410.

Green, S., & Vogt, H. (2016). Personalizing medicine in silico and in socio. *Humana.Mente Journal of Philosophical Studies, 9*(30), 105–145.

Gunter, C. (n.d.). Single nucleotide polymorphisms (SNPs). National Human Genome Research Institute. Retrieved June 5, 2024, from https://www.genome.gov/genetics-glossary/Single-Nucleotide-Polymorphisms-SNPs

Hacking, I. (2006). Genetics, biosocial groups and the future of identity. *Daedalus, 135*(Fall), 81–95.

Harvey, A. (2009). From genetic risk to post-genomic uncertainties: Nutrigenomics and the birth of the "genetic entrepreneur". *New Genetics and Society, 28*(2), 119–137.

Harvey, A. (2010). Genetic risks and healthy choices: Creating citizen-consumers of genetic services through empowerment and facilitation. *Sociology of Health & Illness, 32*(3), 365–381.

Heinsen, L. L., Wahlberg, A., & Petersen, H. V. (2022). Surveillance life and the shaping of "genetically at risk" chronicities in Denmark. *Anthropology and Medicine, 29*(1), 29–44.

Heyman, B., Alaszewski, A., & Brown, P. (2012). Health care through the "lens of risk" and the categorisation of health risks—An editorial. *Health, Risk & Society, 14*(2), 107–115.

Holland, W. (2009). Periodic health examination—A brief history and critical assessment. *Eurohealth, 15*(4), 16–19.

Hood, L., & Friend, S. H. (2011). Predictive, personalized, preventive, participatory (P4) cancer medicine. *Nature Reviews Clinical Oncology, 8*(3), 184–187.

Ilic, M., & Ilic, I. (2016). Epidemiology of pancreatic cancer. *World Journal of Gastroenterology, 22*(44), 9694–9705. https://doi.org/10.3748/wjg.v22.i44.9694

Jain, S. L. (2013). *Malignant: How cancer becomes us*. University of California Press.

Jansen, S. N. G., Kamphorst, B. A., Mulder, B. C., van Kamp, I., Boekhold, S., van den Hazel, P., & Verweij, M. F. (2024). Ethics of early detection of disease risk factors: A scoping review. *BMC Medical Ethics, 25*, 25.

Jasanoff, S. (2015a). Future imperfect: Science, technology, and the imaginations of modernity. In S. Jasanoff & S.-H. Kim (Eds.), *Dreamscapes of modernity: Sociotechnical imaginaries and the fabrication of power* (pp. 7–43). University of Chicago Press.

Jasanoff, S. (2015b). Imagined and invented worlds. In S. Jasanoff & S.-H. Kim (Eds.), *Dreamscapes of modernity: Sociotechnical imaginaries and the fabrication of power* (pp. 321–341). University of Chicago Press.

Jensen, L. G., Svendsen, M. N., & Snell, K. (2023). Strategies on personalized medicine and the power of the imagined public. *New Genetics and Society, 42*(1), 1–21.

Kaplan, J. M., & Fullerton, S. M. (2022). Polygenic risk, population structure and ongoing difficulties with race in human genetics. *Philosophical Transactions of the Royal Society B, 377*, 1–8.

Kerr, A., Chekar, C. K., Ross, E., Swallow, J., & Cunningham-Burley, S. (2021). *Personalized cancer medicine: Future crafting in the genomic era*. Manchester University Press.

Klein, A. P. (2021). Pancreatic cancer epidemiology: Understanding the role of lifestyle and inherited risk factors. *Nature Reviews Gastroenterology & Hepatology, 18*, 493–502.

Konrad, M. (2003). Predictive genetic testing and the making of the pre-symptomatic person: Prognostic moralities amongst Huntington's-affected families. *Anthropology & Medicine, 10*(1), 23–49.

Krogsbøll, L. T., Jørgensen, K. J., Larsen, C. G., & Gøtzsche, P. C. (2012). General health checks in adults for reducing morbidity and mortality from disease: Cochrane systematic review and meta-analysis. *BMJ, 345*, e7191.

Krupar, S. (2012). The biopsic adventures of mammary glam: Breast cancer detection and the practice of cancer glamor. *Social Semiotics, 22*(1), 47–82.

Kuchenbaecker, K. B., Hopper, J. L., Barnes, D. R., Phillips, K.-A., Mooij, T. M., Roos-Blom, M.-J., Jervis, S., van Leeuwen, F. E., Milne, R. L., Andrieu, N., Goldgar, D. E., Terry, M. B., Rookus, M. A., Easton, D. F., Antoniou, A. C., & the BRCA1 and BRCA2 Cohort Consortium. (2017). Risks of breast, ovarian, and contralateral breast cancer for BRCA1 and BRCA2 mutation carriers. *JAMA, 317*(23), 2402–2416.

Lemke, T. (2007). Susceptible individuals and risky rights: Dimensions of genetic responsibility. In R. V. Burri & J. Dumit (Eds.), *Biomedicine as culture: Instrumental practices, technoscientific knowledge, and new modes of life* (pp. 151–165). Routledge.

Lock, M. (2011). Genomics: Embodying molecular genomics. In F. E. Mascia-Lees (Ed.), *A companion to the anthropology of the body and embodiment* (pp. 278–294). Blackwell.

Mizrahi, J. D., Surana, R., Valle, J. W., & Shroff, R. T. (2020). Pancreatic cancer. *The Lancet, 395*(10242), 2008–2020.

Mohammadi-Shemirani, P., Sood, T., & Paré, G. (2023). From 'omics to multi-omics technologies: The discovery of novel causal mediators. *Current Atherosclerosis Reports, 25*, 55–65.

National Research Council. (2011). *Toward precision medicine: Building a knowledge network for biomedical research and a new taxonomy of disease*. The National Academies Press.

NCI. (n.d.-a). *BRCA gene mutations: Cancer risk and genetic testing*. https://www.cancer.gov/about-cancer/causes-prevention/genetics/brca-fact-sheet

NCI. (n.d.-b). *BRCA mutations: MRI, surgery reduce breast and ovarian cancer deaths*. https://www.cancer.gov/news-events/cancer-currents-blog/2024/brca-breast-ovarian-cancer-mri-surgery-fewer-deaths

Neagu, A.-N., Bruno, P., Johnson, K. R., Ballestas, G., & Darie, C. C. (2024). Biological basis of breast cancer-related disparities in precision oncology era. *International Journal of Molecular Sciences, 25*(7), 4113.

Nelson, S. C., & Fullerton, S. M. (2018). "Bridge to the literature"? Third-party genetic interpretation tools and the views of tool developers. *Journal of Genetic Counseling, 27*, 770–781.

Nolan, J. J., & Ormondroyd, E. (2023). Direct-to-consumer genetic tests providing health risk information: A systematic review of consequences for consumers and health services. *Clinical Genetics, 104*(1), 3–21.

Nordgren, A., & Juengst, E. T. (2009). Can genomics tell me who I am? Essentialistic rhetoric in direct-to-consumer DNA testing. *New Genetics and Society, 28*(2), 157–172.

Novas, C., & Rose, N. (2000). Genetic risk and the birth of the somatic individual. *Economy and Society, 29*(4), 485–513.

Nucleus Genomics. (2024). Nucleus. 2024. https://mynucleus.com/. Accessed June 16, 2024.

O'Malley, P. (2004). *Risk, uncertainty and government*. Taylor and Francis Group.

Perrin, E., Kalsbeek, W., & Scanlan, T. (Eds.). (2001). *Toward a health statistics system for the 21st century*. National Academy Press.

Prainsack, B. (2017). *Personalized medicine: Empowered patients in the 21st century?* New York University Press.

Prainsack, B. (2023). Sociologies of precision medicine. In A. Petersen (Ed.), *Handbook on the sociology of health and medicine* (pp. 439–454). Edward Elgar Publishing.

Pritchard, C. (2019). Commentary: Rename breast-cancer syndrome to help save lives. *Nature, 571*, 27–29.

Rabinow, P., & Rose, N. (2006). Biopower today. *BioSocieties, 1*, 195–217.

Raz, A., & Minari, J. (2023). AI-driven risk scores: Should social scoring and polygenic scores based on ethnicity be equally prohibited? *Frontiers in Genetics, 14*, 1–4.

Reardon, J. (2017). *The postgenomic condition: Ethics, justice and knowledge after the genome*. University of Chicago Press.

Rogers, A. (2025, June 5). *Researchers optimize genetic tests for diverse populations to tackle health disparities*. NIH National Human Genome Research Institute. https://www.genome.gov/news/news-release/researchers-optimize-genetic-tests-for-diverse-populations-to-tackle-health-disparities

Rose, N. (2001). The politics of life itself. *Theory, Culture & Society, 18*(6), 1–30. https://doi.org/10.1177/02632760122052020

Rose, N. (2007). *The politics of life itself: Biomedicine, power and subjectivity in the twenty-first century*. Princeton University Press.

Schmidt Nielsen, A.-K., & Stage, C. (2023). Genetic hauntings: Mediating pre-patienthood and haunted health on TV. *MedieKultur: Journal of Media and Communication Research, 39*(74), 127–145. https://doi.org/10.7146/mk.v39i74.137743

Snyder, M. (2016). *Genomics and personalized medicine: What everyone needs to know*. Oxford University Press.

Starr, P. (1982). *The social transformation of American medicine*. Basic Books.

Sunder Rajan, K. (2005). Subjects of speculation: Emergent life sciences and market logics in the United States and India. *American Anthropologist, 107*(1), 19–30. https://doi.org/10.1525/aa.2005.107.1.019

Surveillance, Epidemiology, and End Results Program. (n.d.). *National Cancer Institute*. Retrieved June 3, 2024, from https://seer.cancer.gov/

Svendsen, M. N. (2006). The social life of genetic knowledge: A case-study of choices and dilemmas in cancer genetic counselling in Denmark. *Medical Anthropology, 25*(2), 139–170. https://doi.org/10.1080/01459740600604422

Swallow, J., Broer, T., Kerr, A., & Cunningham-Burley, S. (2024). Laboratory practices, potentiality, and material patienthood in genomic cancer medicine. *Science, Technology, & Human Values, 49*(5), 967–988. https://doi.org/10.1177/01622439241243987

Taussig, K.-S., Hoeyer, K., & Helmreich, S. (2013). The anthropology of potentiality in biomedicine: An introduction to Supplement 7. *Current Anthropology, 54*(S7), S3–S14. https://doi.org/10.1086/671401

Turrini, M., & Bourgain, C. (2021). Appraising screening, making risk in/visible: The medical debate over non-rare thrombophilia (NRT) testing before prescribing the pill. *Sociology of Health & Illness, 43*(7), 1627–1640. https://doi.org/10.1111/1467-9566.13324

Verran, H. (2010). Number as an inventive frontier in knowing and working Australia's water resources. *Anthropological Theory, 10*(1–2), 171–178. https://doi.org/10.1177/1463499610370873

Verran, H. (n.d.). Narrating a number and staying with the trouble of value. In *Data Handbook 2*. Retrieved November 20, 2024, from https://datajournalism.com/read/handbook/two/assembling-data/narrating-a-number-and-staying-with-the-trouble-of-value

Wray, N. R., Lin, T., Austin, J., McGrath, J. J., Hickie, I. B., Murray, G. K., & Visscher, P. M. (2021). From basic science to clinical application of polygenic risk scores: A primer. *JAMA Psychiatry, 78*(1), 101–109. https://doi.org/10.1001/jamapsychiatry.2020.3049

Yanes, T., McInerney-Leo, A. M., Law, M. H., & Cummings, S. (2020). The emerging field of polygenic risk scores and perspective for use in clinical care. *Human Molecular Genetics, 29*(R2), R165–R178. https://doi.org/10.1093/hmg/ddaa148

8

The Analyst, the Interpreter, and the Intruder: Genomics Labor Division and the Textures of the Clinical Genetics' Specialty Controversy in Spain

Violeta Argudo-Portal

Introduction

In 2022, during an interview with Gregorio, a former breast surgeon who is currently a medical geneticist and cancer genetic counselor, and the leader of a genetics lab at a public Spanish hospital, he explained how he has modulated his approach to genetic test interpretation in the field of cancer susceptibility genes over the last two decades. While at first he bought into the risk estimations for susceptibility genes, and it was common for him to communicate cancer risk estimate percentages, in the last decade, the way he interprets and communicates test results has shifted, becoming more modest. "We are not at the start of the 2000s. Now we know that we need to keep adjusting risk estimates and following up on genomics updates day by day. We need to explain this to the patient" (Interview, 2022). His shift underscores a broader issue: Cancer predisposition test results are an ongoing probabilistic craft rather than fixed

V. Argudo-Portal (✉)
Department of Social Anthropology, University of Barcelona, Barcelona, Spain
e-mail: vargudoportal@ub.edu

V. Argudo-Portal et al. (eds.), *Mundane Genomics*,
https://doi.org/10.1007/978-981-95-4550-6_8

facts. Risk estimates based on genetic tests are unstable in most cases; therefore, he has decided to avoid the formula: a positive test for the X variant equals the X risk estimate. In the past, during the genetic boom of the early 2000s, he linked a predisposition test result to a percentage when talking with a potential patient, but now it seems daring to him. After all, that person will walk out of the room with a risk percentage following them, no matter how provisional that number is in most cases.

For him, geneticists, when talking with patients, must have the knowledge and skills to be aware that test results interpretation goes beyond a negative or positive result to the susceptibility genes from the tested panel. He noted, "[T]hat is why you cannot rely on sending your sample with a carrier and then waiting for some lab to send you a report saying if you have this or that predisposition; our legislation is very clear in that sense. You need a professional with the knowledge to interpret and communicate what a test means." Gregorio draws attention to a key issue: the relevance of considering who is doing the interpretation work of the test, the need for genomics literacy, and the question of whether some genetics tests are interpreted at all. Under Spanish Legislation (14/2007), genetic tests using medical means are only allowed if "done by a qualified professional [...] offering proper genetic counseling"; that is clear. However, two seemingly mundane questions require examination: Who counts as a qualified professional? How is genomics ordered in healthcare to ensure a modest interpretation? These are particularly salient questions if we consider that Spain is the only country in the European Union without a recognized specialty in clinical genetics (although, as we will see, it is neither clear nor agreed what clinical genetics means), due to a decades-long discussion on the skills and knowledge required to work on genomics in the healthcare system, professionals' jurisdictions, roles in the clinic, and different ways of understanding contemporary genetics practice—issues that are all at stake.

This chapter delves into the controversy around professional jurisdictions in clinical genetics in Spain. This is used as an entry point to unpacking the expertise arrangements required to interpret genetic test results in everyday practice and the types of genomics understanding embedded in such explanations—with a focus on the field of genetic predisposition and polygenic risk scores. The controversy around the genetics specialties in Spain makes it a rich context for exploring genomics healthcare politics, addressing issues such as professionals' literacy, healthcare labor division, and ordering. The chapter offers a window onto the Spanish context, aiming to spur general reflection regarding the integration of genetics into healthcare, which affects professionals, national healthcare systems, and patient care. It draws attention to how genomics shapes and is being shaped by contemporary healthcare systems, focusing on workforce orderings and understandings of what it takes to make genomics care doable.

The issue of professional jurisdictions and the division of expert labor is a classic inquiry in the social sciences (Abbott, 1988). Such a body of literature has been extended to the study of expertise (Collins & Evans, 2007; Grundmann, 2017), which includes lay people as experts (Wynne, 1989), where expertise is not a formal attribution but a relational craft, as anthropology of science has also illustrated (Arregui, 2014). In a nutshell, we could say that while the sociology of professions reduced expertise to experts and was linked to organizational and institutional credentialing, the sociology of expertise refers to the arrangements that conform to the "capacity to perform a task better and faster than others," and attention is directed to what experts do rather than attending only to their professional jurisdictions (Eyal, 2013, p. 877). Of course, both sociologies are in dialogue. However, for Eyal, it is critical to consider different modes of analysis when approaching experts and expertise: Understanding experts requires unpacking jurisdictions, that is, "who has control and to what degree over a set of tasks," and expertise analysis requires posing the question of "what arrangements must be in place for a task to be accomplished" (Eyal, 2013, p. 872), which places the focus on arrangements for genomics integration into healthcare.

This work is in indirect dialogue with the literature on expertise and directly inspired by Fortun's (2023) insights on the quotidian meanings, practices, and techno-politics of genomics, notably, his notion of *genomics scrupulousness,* which offers a better understanding of professionals' tensions regarding how to order genetics in healthcare. For Fortun (2023), scrupulousness is a mode of care, or *travail éthique* in Foucauldian terms (Foucault, 2020), situated between excess and never enough: "attending to the smallest trouble or doubt threatens, always and immediately, to be indistinguishable from an obsessive paralysis of over-refinement–what you needn't and shouldn't worry or care about. To be scrupulous is to risk being…scrupulous" (Fortun, 2023, p. 145). Scrupulousness "is something that has to be learned, practiced, cultivated in a collective space, a pedagogical place that cultivates tacit and embodied knowledge along with book knowledge" (Fortun, 2023, p. 153). As we will see throughout the chapter, scrupulousness and what I call *genomics modesty* as ambivalent modes of care, both attentive to oneself and others' practices, are key to examining the textures of the controversy around the genetics clinical specialty in Spain. For such exploration, hinging between too much detail and not enough when approaching genetic tests interpretation, there is another key element pivoting between the mundane and the extraordinary throughout this chapter with regard to the arrangements that make genomics labor (un)doable and require that attention be paid to mundane healthcare orderings.

To explore the dynamics of genomics' everyday practice and tensions in terms of healthcare reordering, I am inspired by the work of anthropologist Veena Das in *The Textures of the Ordinary* (2020). She offers an exquisite reflection on a lifetime of ethnographic work, read through Wittgenstein, which is particularly useful for establishing some key aspects of how I approach mundane genomics throughout this chapter. Das (2020, p. 6) observes that "[t]he sense the everyday will be returned but in a deformed way draws us into a wider discussion of the relation between what is ordinary and what is extraordinary in thinking of everyday life." Medical specialty feuds and disputes, particularly in the context of genetics, might be seen as part of the ordinary organizational and professional jurisdictions that structure healthcare. At the same time, it is crucial to acknowledge the extraordinary importance of examining the seemingly mundane aspects of how things are done and arranged in practice; whether due to or despite formal systems of order, it is crucial to discuss their pitfalls, promises, and challenges. We then find that the ordinary constantly shifts and shapes our understanding of the extraordinary, and the mundane might then be understood as made of ongoing shifts between the ordinary and the extraordinary, much as Fortun's (2023) "genomics scrupulousness" is made of excess and never enough.

More than three decades after the genomics gold rush, multiple discussions on the different genomics challenges in healthcare have arisen, such as inequalities in genetics services access (Lopes-Cendes & De Oliveira, 2025), genetics naming practices (Navne, 2023), lab-to-clinic translations (Harvey, 2011; Horgan et al., 2024), and the issue of genomics integration into healthcare; independently of whether there is a clear specialty ordering or a genomics strategy such as in the UK (NHS, 2022), there is a general interrogation of such integration (Battista et al., 2012; Burns et al., 2019; Marchant et al., 2020; Stark et al., 2019). In this work, I explore ordinary healthcare organization around specialties, genomic uncertainties, unbalanced relations between technological possibilities and professionals' knowledge, and the multiple genomics understandings comprising the textures of mundane genomics in Spain.

I first present the materials and methods upon which this chapter draws. I then outline the Spanish context by offering a selective genealogy of genetics in Spain in three acts to situate better why it is the only EU country without a clinical genetics specialty, and cast light on associated controversies. In the following sections, I delve into the textures of the controversy connected to clinical genetics in Spain through the lens provided by three figures: the analyst, the interpreter, and the intruder. The analytic approach reveals how a more scrupulous reading of the discussions at play in the controversy over genetics specialties—which tends to be read as a mundane matter of just another medical specialty dispute—highlights the division of genomics labor and different understandings of genetics tests and genomics in healthcare. These three figures raise questions about the labor division in healthcare and whether these tensions and different epistemic rationalities are what make genetics in healthcare doable, keep it "scrupulous" (Fortun, 2023), and cultivate a *genomics modesty*, rather than considering such arrangements dismissible. This work aims to shed light and activate a discussion on the challenges of genomics integration into national healthcare systems, focusing on the ordering of its practice, which calls attention to the politics of care and knowledge.

Methods and Materials

The chapter's research stems from a bigger project exploring predictive genetic tests in Spain (2021–2025).[1] It draws on intersecting qualitative methods: semi-structured interviews, observation, document analysis, and a descriptive survey. Ten interviews with healthcare professionals in the public and private sectors working on genetics in reproduction (some of which were conducted with the project members Vincenzo Pavone and principal investigator Mauro Turrini), and 10 with healthcare professionals in the national healthcare system working on cancer predisposition, were conducted between 2021 and 2023. The interviews were fully transcribed and thematically coded. In the course of research, I identified a transversal issue overlooked in the Spanish context: the lack of both a clinical genetics specialty and consensus regarding who and what it takes to interpret a predisposition genetic test nowadays. To delve more deeply into these issues, I conducted a document analysis of secondary sources to support analysis of the history of human genetics in Spain, which includes a review of the digitalized minutes of the meetings of genetics associations, professional reports on the history of genetics in Spain, and a media follow-up on the clinical controversies around the genetics specialty (2010–2024).

As I encountered difficulties addressing this controversy through interviews, a descriptive survey was conducted in 2024 to collect professionals' approaches to the skills required to read genetics tests and the clinical genetics specialty controversy in Spain. The survey received 24 responses from the 100 professionals to whom it was sent. However, it is essential to note that between 2023 and 2024, some collectives I contacted were afraid that mass dissemination of the survey could feed the controversy on social networks. This was at a particular moment when the situation was tense as a response was awaited from the European Parliament on a petition to have clinical genetics recognized as a specialty. Moreover, it was a period of suspense for the genomics community with regard to the Health Ministry Decision on which genetic tests would be included in the national healthcare system. Attending to their concerns about my

[1] The Spanish Research Council Ethics Committee approved the research for this project.

making too much noise with the survey and reactivating tensions at a sensitive time, I decided first to postpone conducting it and finally not to disseminate it publicly. Following requests from the community studied, the survey was distributed only by one-to-one email to professionals working on genetics in the public and private healthcare sector and commercial labs covering the country's 17 regions. Despite the survey's consequent limitations, it qualitatively served to add 24 descriptive testimonies on the controversy, providing the density I sought to complement the interview data. Finally, note that all names in this chapter are pseudonyms.

A Non-existent Speciality: A Selective Story in Three Acts

The controversy over the lack of a clear ordering of genetics in the Spanish national healthcare system, focusing on the lack of a clinical genetics specialty, has had several public and media mobilizations in recent years. A group of about 40 people, some dressed in white coats, others in street clothes with long winter coats, gathered in front of the Spanish Ministry of Health in Madrid on January 23, 2024, carrying banners with the slogan #EspecialidadGenéticaYa (#GeneticsSpecialityNow). Different rallies of this type occurred throughout the first half of 2024, both in front of the Ministry of Health and in various other cities around the country, coinciding, for example, with the Day of Rare Diseases or the International Day of DNA. The platform for creating the multidisciplinary specialty of clinical genetics has articulated the calls for these rallies, bringing together genetics professionals and the Spanish College of Biologists, as well as patient associations (EGEREC, Fragile X, 22q, SIMA Marfan, Klinefelter, ALS, Tatton Brown Rahman, Epilepsy, Ehlers Danlos, Familial Spastic Paraparesis, Huntington's, Parkinson's or Metastatic Cancer). As the platform slogan indicates, it arises from the concern and demand of some professionals to recognize the specialty of clinical genetics in Spain. Their statement explains that their activities aim to "raise social awareness of the urgent need for the creation of the

Multidisciplinary Specialty of Clinical Genetics, due to the damage this causes to the National Health System, Spain being the only European country that lacks such a specialty."

The platform has various profiles on social networks that have promoted rallies, circulated a manifesto, and participated in press releases pointing out the "urgency of the genetic specialty" in Spain, and how, once again, Spain is trailing the rest of Europe. Indeed, the narrative of Spain dropping behind the EU is vital to mobilizing the demand. How can it be that Spain is the only country in the EU without a recognized specialty in clinical genetics in its healthcare system? This narrative extends to the Spanish Association of Human Genetics (AEGH), which, following the platform last year of mobilizations, published a press release on its website stating, "The AEGH joins any action and mobilization driven for the recognition of the Health Specialty in Spain, the only European country with this lack."

All this fuss over the genetics specialty is not new. Indeed, Pàmpols et al. (2016, p. 385) note that "the claim for accrediting health professionals in genetics specialties in Spain is as old as AEGH [...]. Nevertheless, strong conflicts arose between physicians and nonphysician professionals, and the genetics specialty was paralyzed." We need to go back to the dawn of the end of the dictatorship in 1974, when the AEGH was created. Acknowledging that the genealogy of the frictions in the specialty has been sustained for more than four decades—an unusual situation in the Spanish context—prompts us to consider whether four decades of such exceptionality becomes part of the textures of the mundane. In what follows, I summarize the genealogy of the controversy and tensions around the specialty chronologically in three main acts.

Act 1: Fragmentations

The formalization of professional and scientific genetics associations in Spain must be situated as part of an ongoing collaboration in the 1960s between Portuguese and Spanish researchers in genetics who met for different symposiums, with a strong presence of studies presented under the umbrella of cytogenetics (Candela Castillo & García González, 2003). In

1972, the Spanish Society of Genetics (Sociedad Española de Genética) was created, followed in 1974 by the Spanish Association of Human Genetics (Asociación Española de Genética Humana), and in 1978, as part of the Spanish Association of Pediatrics (Asociación Española de Pediatría), a section named the Spanish Society of Clinical Genetics and Dysmorphology (Sociedad Española de Genética Clínica y Dismorfología) was founded. In 1988, the Spanish Society of Environmental Mutagenesis and Genomics was inscribed. All three associations remain active today and illustrate the tensions in the field.

The Spanish Society of Genetics is defined as a scientific association. The archive of the minutes of its meetings demonstrates that the association was driven mainly by researchers based at universities and research centers and devoted considerable effort to creating a journal (*Genética Ibérica*, 1985–1989),[2] offering training, and promoting changes in naming higher education subjects to incorporate the presence of genetics. The minutes of the Spanish Association of Genetics, on the other hand, recursively present its aim to agglutinate other associations to produce a field of genetics in Spain and unify association fees, particularly during the genetics gold rush of the 1990s (Candela, 2003). However, these efforts failed. Nowadays, the Spanish Society of Genetics oversees the National Genetics Prizes and has research streams in cytogenetics, population and evolution genetics, and animal and plant genetics. The Society is characterized by the strong presence of university professors and researchers.

The involvement of the Spanish Association of Human Genetics—founded mostly by physicians in the 1970s—with the European Society for Human Genetics (established in 1967) has been crucial, bringing claims legitimized at the EU level to the Spanish Ministry of Health. The last data from the Spanish Association of Human Genetics note that of the 370 members, 200 are biologists, 97 physicians, 34 pharmacists, and 39 from other professions (Pàmpols et al., 2016). The association has advocated strongly for the specialty of clinical genetics since its inception

[2] For some, the journal's short life (1985–1989) was indicative of the professional jurisdictions' tensions and association atomization in the country, in contrast to other journals such as *Heredity* (American Genetic Association) or *Genomics* (Canada) that counted on the support of their countries societies and remain active until today. For a more detailed description of this issue, see Candela Castillo and García González (2003).

and continues to do so today, a position explained by strong representation of biologists in the association, as we will see;[3] indeed, the specialty could help biologists find a permanent position in the healthcare system organization. Finally, the Spanish Society of Clinical Genetics and Dysmorphology was primarily formed by pediatricians when newborn blood-spot screening was formally scaled up nationally in the 1980s in Spain (however, newborn screening was introduced in some regions in the late 1960s). This association remains mainly constituted by pediatricians. The coexistence of these three associations reveals long-lasting frictions and atomization in the field of human genetics in Spain, illustrating the different professional jurisdictions that seem to have grown apart since the formalization of the field in the country four decades ago.

Act 2: The Rise and Fall of Specialty Recognition

We can identify two key moments in the rise and fall of specialty recognition. The first was in 1989 when the National Commission for Medical Specialties agreed to include the specialty as requested by the Spanish Association of Human Genetics; however, due to discrepancies between physicians and other professionals, it was not formalized. Then, right on top of the genetic gold rush, in 2001, the European Society for Human Genetics (ESHG) promoted the formal and international recognition of genetics as a medical specialty at the EU level. This call for recognition was worked around in the previous years, when in 1996, the ESHG, the American Society of Human Genetics, and the Human Genetics Society of Australasia founded the International Federation of Human Genetic Societies. This gathering resulted in an international meeting in 2001, feeding the first steps for specialty recognition in the EU.[4] However, it was not until 2011, when the Commission passed EU No 213/2011 amending Directive 2005/36/EC that "clinical/medical" genetics was officially recognizing as a specialty, enforced the following day when the

[3] In Spain, these positions are called BIR, Resident Biologist Intern, in contrast to medical positions MIR, Resident Medical Intern.

[4] For a complete chronology of the European Society for Human Genetics meetings from 1967 to 2024, see: https://www.eshg.org/conferences/past-eshg-meetings

regulation was passed on March 3, 2011. Interestingly, the regulation recognized the specialty of "medical genetics," but the ESHG on its website refers to "medical/clinical genetics."

When reading about genetics specialties at the EU level, a slash appears between medical and clinical genetics, showing an ambiguous and contested emic differentiation between doctors (medical genetics) and professionals working in the clinic with training other than medicine—a slash that produces a grouping in which professionals with a medical degree are differentiated from other professionals (nurses, laboratory technicians, or, in the Spanish case, graduates in genetics, biology, chemistry, biotechnology, and so on). The use of the slash is mostly diluted, but still present. In 2025, the European Certificate of Medical Genetics and Genomics still refers to medical/clinical genetics and genomics training in its body text, although the training requires a prior medical qualification. Therefore, the slash still appears sometimes, and the clinical label is used for medical and laboratory genetics.[5] Throughout this chapter, I have decided to retain this form rather than inducing a clarity not present in the field by using, for instance, only the terms laboratory and medical genetics. I think this messy naming and slash use matters, as it illustrates the frictions of the field and the disputes about what clinical means in genetics and genomics nowadays. After this clarification, let us return to 2012.

The European Board of Medical Genetics was formed in 2012, and under it, the specialty of Clinical Laboratory Geneticist was created for those providing genetics laboratory services. At the EU level, laboratory genetics usually refers to lab professionals (biologists, chemists, pharmacists, and other qualifications linked to sample lab processing and analysis). In contrast, medical genetics refers to physicians, and genetics counseling remains a blurry field. Then, with the recognition of the specialty, rapid demarcations had to be drawn between doctors and other professionals. The so-called multidisciplinary approach is fragmented into three professional branches. The European Board of Medical Genetics currently differentiates between clinical laboratory geneticists, genetic counsellors, and medical geneticists.

[5] See: https://www.uems-ecmgg.org/Home.html

Amidst this rise in genetics recognition at the EU level, we can identify the second key moment in the rise and fall of the specialty. A Decree was passed in 2014 in Spain, creating the "multidisciplinary specialty of Clinical Genetics," which physicians, pharmacists, biologists, and chemists could access. However, legal recognition of the specialty lasted fewer than two years. In 2016, the Decree that recognized the specialty was nullified due to the lack of a complete analysis of the economic and budgetary impact of the proposal, thus returning to the same situation as two decades earlier.

Act 3: Almost There, But No

Although the claim for specialty recognition, with its disputes and press releases, has been latent in the field since the decree was nullified in 2016, it remained dormant in terms of the public eye until 2024. During 2024, the rallies and social networks reactivated the public demand to have the specialty recognized, coinciding with the publication of the national healthcare system's catalog of genetic and genomic tests and services, establishing the portfolio of tests and services agreed to be guaranteed to all. In June 2024, the Ministry of Health approved a catalog of genetic tests and a counseling committee. The first prioritized areas of the catalog were: adult oncohematology, pediatric oncohematology, pharmacogenomics, heart disease and circulatory system disorders, ophthalmological diseases, hereditary and mitochondrial metabolic diseases, neurological and neuromuscular diseases, and neurodevelopmental disorders, including neurocognitive deficits, followed by the inclusion of tests and services on skin, liver, respiratory diseases, solid tumors, and hereditary adult cancers.[6] Some expected the specialty's recognition to come to fruition along with the catalog's official publication, but it did not.

The Health Ministry has prioritized setting the geographic scene of services, including mapping the situation regarding genetic services and professionals in Spain, and has left specialty recognition for later. In a way, expertise arrangements and service access have been prioritized over

[6] The catalog can be accessed here: https://cgen.sanidad.gob.es/#/

decisions on professional jurisdictions. However, for all the survey respondents, the main concern regarding access to genetics services revolves around the lack of professionals with the required training; as such, prioritizing ordering the offer of services over tackling professional training makes no sense for the patients nor the expenditure of the national healthcare system. In line with health policy literature that points out that the effects of clinical genetics services are still uncertain globally (Payne et al., 2007), and there is not a criterion for assessing their outcomes, one survey respondent argues:

> The Ministry's recent actions, aimed at establishing a portfolio of services to disseminate knowledge and offering tests without territorial restrictions, may lead (is already leading), in the absence of trained professionals (clinical geneticists and laboratory specialists in molecular genetics), to uncontrolled spending without optimizing the clinical utility of genetics for healthcare and public health. (Survey response, 2024)

In this context, in September 2024, the platform for the recognition of the specialty in Spain resorted to the arbitration of the European Parliament, demanding "the recognition and official regulation of the clinical genetic multidisciplinary specialty in Spain." In November, the petition received the Parliament's response: "It is up to the Member State to decide on whether to regulate a medical specialty or not."[7] Thus, Spain remains with blurry jurisdictions around who counts as a clinical genetics expert and whether that should refer to a medical specialty. What is at stake and remains unanswered is whether there is a need to institutionalize genomics in healthcare through fragmented specialties or if other forms of ordering and work are preferred—and why, by whom, and which understanding of genomics in healthcare is at play. In the next section, I point out that under the professional jurisdictions' feuds, and expert labor division, there are also critical arguments on what it takes to read a genetic test in the clinical sphere, with some professionals drawing a differentiation between genomic analysis and genomics

[7] See the response to Petition NO 0702/2024: https://www.europarl.europa.eu/doceo/document/PETI-CM-766823_EN.pdf

interpretation—aspects that are key to better understanding the tensions in such genetics expertise arrangements.

The Analyst, the Interpreter, and the Intruder

The current genetics practice in Spain could be summarized by emulating the title of a 1960s spaghetti western film: "The Analyst, the Interpreter, and the Intruder." In this analytic section, I show how different professionals working on genetics in Spain explain why it is the only country without a clinical genetics' specialty. Then, I unpack how discussions of the clinical genetics specialty, and their mundane character as yet another professional jurisdiction dispute in the medical sciences, must transcend the latter reading; I also pay close attention to the narratives in order to unfold the kind of genetics envisioned, and what are they talking about when talking about genetics in healthcare. In a way, I argue that the specialty issue has turned out to be a Gordian knot that also addresses the difficulties of ordering genetics and the coexistence of fragmented knowledge in healthcare settings.

Labor (Un)Division

The decades-long wait for the recognition of the specialty has amplified disputes over professional boundaries, heightened struggles, and deepened tensions over the role clinical genetics should play in the Spanish healthcare system. Central to these debates are questions about how genomic predispositions should be interpreted in healthcare, who is best suited to undertake such a task, and how the division of labor connected to clinical genomics should be ordered. A survey notes, "It has taken so long for the specialty to develop that it has generated intrusiveness, atomization, silos and, therefore, lobbies against it" (Survey response, 2024). Others have repeated the same thing in the interviews and survey responses when asked about the specialty. For them, the decades-long wait has seen increased fragmentation of the geneticist's profession (lab analyst, medical genetics, genetics counseling, genetic bioinformatician, biomedicine researcher, and so on) along with legislative sloppiness, which has turned recognition of the specialty from a mundane bureaucratic procedure into an extraordinary, titanic task. As we will see, it is a titanic task because it requires figuring out who can work in genomics in healthcare, with what training, why, and for what purpose, which in turn requires tackling the quotidian labor that makes genomics (un)doable and addressing the overall role of genomics in healthcare.

During this wait, much has happened in the field of genomics research that has led to an expansion of genetics uncertainties and discussions on genomics actionability in the clinic (Bonache et al., 2018; Cambrosio et al., 2018; Kerr et al., 2019). This shows that the problem of integrating genetics into the national healthcare systems has not been solved, and that the complexities of such integration would not be solved by simply establishing a specialty, as the boundaries between molecular biology and medicine keep being reworked (Keating et al., 2016; Rheinberger, 2009), requiring broader reflection. Addressing clinical genomics opens a discussion about medical training, the reshaping of the 1990s approach to genetics—as it is no longer the same as at the end of the twentieth century and the beginning of this millennium—securing funding, making political decisions on the technological possibilities, and technoscientific market demands, in addition to tackling questions about postponed healthcare transformations that require rethinking healthcare workforce roles and ordering. That is why, for Dr. García, what has led Spain to this situation is multifactorial, and includes professional corporatist interests, the complexity of incorporating genetics into university training, and the need to inject investment to provide services and create regular and stable positions. In a way, the ordering of the mundane interferes with the expectations of genomics; professionals need to be trained, healthcare reordered, and a different way to approach diseases and genetics needs to be reconsidered and worked through.

In addition to the professional silos and atomization or conglomerate of causes, another key issue arose in the materials: the proposal of the Spanish Association of Human Genetics for a single specialty of clinical genetics that would include physicians, biologists, and chemists. This proposal is seen as inadequate and far from what has been approved at the

EU level, where there is a distinction between clinical laboratory genetics and medical/clinical genetics and genomics. When asking how it is that Spain has not recognized genetics as a specialty in the national healthcare system, thereby regulating its training and general ordering, a physician in his seventies notes:

> I am a physician. I started working in the genetic-clinical field in 1985. I have been working as a geneticist for almost 40 uninterrupted years in public and private health care. I have detected two explanations that, acting synergistically, contribute to this unusual fact [lack of genetics specialty in Spain]: the spurious interest of certain professional "lobbies" that the specialty does not go ahead and an internal struggle not to assume the existence (as in Europe) of two specializations: Clinical Genetics (Physicians) and Laboratory Genetics (other health personnel, such as biologists and other specialties). The impossibility of creating a specialty combining clinical and laboratory genetics hinders its recognition. Clinical genetics is centered on the patient and their disease and corresponds to the physician. Laboratory genetics is centered on the test and the sample, and corresponds to the other specialties: clinical analysis, biologists, molecular biologists, chemists, pharmacists, etc. However, the two specialties are necessary and must work together. (Survey response, 2024)

In this context, the Spanish Association of Human Genetics proposal advocates a single specialty, so different life sciences professionals could fall into the category of "clinical genetics" in addition to physicians; indeed, the platform for the specialty calls for a "multidisciplinary clinical specialty." For some, this approach is a strategic move for all those professionals with degrees other than medicine—who usually have a challenging time finding stable positions in the healthcare sector—to access the specialty. Offering an antagonistic reading of the situation, such a proposal tends to be presented as "stealing medical specialties" in physicians' newspapers.[8] Beyond this lobbyist reading, we can observe that while the unique clinical genetics specialty has greater media coverage and a platform for mobilization with a manifesto, the other proposal advocating

[8] See the example mentioned above: https://cgcob.es/wp-content/uploads/2024/11/RS-137-24-CG-2024-11-18-Articulo-BIR-CGCOB-Presidenta-Comision-de-Sanidad-Isabel-Tejada.pdf

two specialties—laboratory and clinical genetics—seems to exert resistance by other means. In this dispute, however, it is not only the "making a living" of the more precarious workers of the healthcare system (biologists, chemists, lab technicians) that is at stake, but also a mode of ordering genetics in healthcare. When molecular relations are extended into the mundane requirements of ordering healthcare services and tasks, overlooked aspects of expert labor division or what it takes to interpret a genetic test result for a patient are brought into the spotlight.

The Analyst and the Interpreter

Alfredo works at a public hospital, where his primary responsibilities include interpreting genetic test results and performing administrative tasks. He opted to respond to the survey instead of participating in an interview, arguing that it was an important topic, and he preferred to have time to write down his responses. He outlined three key points he believes are crucial for understanding the lack of ordering surrounding clinical genetics in Spain.

> 1) There is self-interested confusion on the part of some clinical analysis specialties in considering genetics as just another form of clinical analysis when the big difference is that these are non-quantitative studies with many cases in which the results are highly complex and cannot be *interpreted* without a deep knowledge of genetics. These clinical analysis specialties pretend to believe that because the term "clinical genetics" contains the word clinical it must be clinical analysis.
>
> 2) In our country, there is tremendous intrusiveness and boldness on the part of professionals not trained in genetics who are working in variant *interpretation*, report preparation, and genetic counseling without training in the field.
>
> 3) Genetics requires years of training and experience; it cannot be acquired with a few months of training or just one intensive course. Patients requiring genetic studies deserve specifically and extensively trained professionals. (Survey response 2024, emphasis added)

The points raised are dense and illustrate how disputes over the lack of ordering of the geneticists' profession are not only a matter of professional jurisdictions but also about coming to terms with assumptions of what "clinical" means, what kind of studies are genetic studies, and the mismatch between experts and expertise; thus, the very meaning of "clinical" is at stake. Alfredo clearly distinguishes between lab analysts and medical geneticists and follows the same reasoning that structures the two specialties at the EU level. However, Alfredo's claims not only demarcate clinical analysists as professionals in the wet or dry lab and clinical geneticists in the clinic, but also raise the point, and as others have also done, of a distinction between genetic analysis and genetic interpretation The lab specialty, be it wet or dry, would be conducting genetic analyses, while the task of medical genetics would be interpreting.

In this regard, anthropologists' discussions about the differences and links between analysis and interpretation (Geertz, 1973; Holbraad et al., 2018) help unpack the labor division as a particular relation forged with genomics, while giving meaning to what genomics does, is, and requires in healthcare settings. While analysis refers to breaking down data to identify correlations, patterns, or trends, interpretation synthesizes and assigns meaning and relevance to results. For Alfredo, a genetic test—based, for instance, on a demarcated pool of data drawn from a gene panel or SNP arrays—cannot be reduced to a quantitative report; genetics in healthcare requires interpretation. Clinical analysis as a diagnostic method, based on different sample processing and measurements combined with wet and dry laboratories, is insufficient. That is why Alfredo notes that genetic studies are "non-quantitative." In this vein, he argues that genetic expertise in the clinical sphere cannot equal report production from automated sequencing. The professionals' capacity to interpret genetics is paramount for genetics to land in the clinic, as it incorporates a larger framework to give narrative and meaning to genetic test results. In a way, Alfredo's statements illustrate a concern in line with Keating and Cambrosio's (2003) when they observe in biomedicine etymology an "inversion of the relationship of biology to medicine." Thus, for Alfredo, including laboratory analysts in the specialty of clinical genetics is another step toward biology (and bioinformatics) cornering medical practice.

Gregorio, a geneticist expert in hereditary breast cancer, whose interview opened this chapter, also discussed the significance of going beyond analysis results—the positive or negative to x gene or variant—and taking on interpretation tasks. When I visited his hospital lab and talked about predisposition testing for breast cancer, he highlighted how he had changed his relationship to genomics test results in the last two decades, and stressed the relevance of geneticists in the clinic interpreting the results and not only reproducing analytics results. For him, having a broader picture and understanding of contemporary genomics is key for interpretation. Test results cannot be treated as mere lab analysis or translated to the patient as definitive; they require interpretation. Gregorio explained:

> We know now that the field changes so fast and that, for instance, we were giving really high predisposition risk scores due to the type of patient pool we were working on. Interpretation is key, and explaining to the patient that genomics is in constant reformulation—even more, when we talk about risk estimates for cancer predisposition [...] that is why I do not like talking about predisposed people. We talk about higher or lower risk considering the data and knowledge we have now, or I talk about predisposition genes, not [predisposed] patients. That is why it is essential to interpret the test results and not just say, positive or negative to that [gene or variant]. (Interview, 2022)

Inserted in this differential logic about what analysis and interpretation do is that medical geneticists understand "the clinical" as a space in which clinicians, as interpreters, give meaning to genetics, and do not subsume the clinical under the molecular in order to prevent the medical explanation being reduced to a biological one (Keating & Cambrosio, 2003). Alfredo's second point adds a third actor to the analyst and the interpreter: the intruder, to whom we turn in the next section.

The Intruder

Throughout my research on genetics in Spain, the shadow of the intruder was invoked multiple times. Intrusion appeared in almost all the conversations I had on the profession of genetics. Depending on where the intrusion was detected, the perpetrator shifted from one professional to another, making all professionals involved in genetics nowadays potential intruders, which has to do with a quest for genomics scrupulousness (at least for some).

When talking with María, who gained a BA in Genetics from the only official degree in the country, anyone without the degree was an intruder in the field. When discussing with non-physician professionals working on genetic test processing, such as biologists or biochemists, they wonder how they could be regarded as intruders in the field when they are the ones "carrying the burden of genomics in many hospitals with their labor in the lab" (Survey response, 2024). As such, biologists, biochemists, or biomedicine researchers note that they cannot be intruders as genetics requires their expertise and that "not all physicians have a flexible enough understanding of the disease, except for pediatricians and oncologists that have a broader approach to disease" (Survey response, 2024). Thus, their perspective from outside the medical realm is valuable; for them, then, genomics requires scrupulousness and simultaneously a "flexible understanding of the disease." When talking with physicians, if a laboratory analyst interprets a test and writes a report that is seen as an intrusion, as is getting into the clinical sphere when giving meaning to genetic test results.

At the same time, genetics should not be an isolated specialty for some physicians. Álvaro, a physician who has been involved in healthcare reproductive genetics in Spain since the 1980s, told me, "All medical practice today is in connection to genetics. Today, cardiology cannot be understood without genetics, nor can oncology, and neither can assisted reproduction. That is the challenge" (Interview, 2022). For him, the issue is the scarcity of professionals with enough knowledge to practice genetics in the clinic, regardless of their university degree.

Álvaro notes that professionals with the knowledge and abilities to be interpreters are scarce. Indeed, when asked in either interviews or the survey why there are access inequalities to genetics services in the Spanish healthcare system, the primary response was: lack of professionals. Then, there is a paradox: While genomics seems to be permeating medical practice in a transversal manner, the "qualified professional" figure remains blurry, scarce, and hard to grasp. This point shows that technological possibilities and decades of experimental genomics research do not automatically become a pool of professionals with enough training and knowledge to integrate genomics into healthcare. Yet the mundane work of ordering healthcare training is widely overlooked and assumed to happen out of nowhere.

The mundane textures of the specialty controversy involve tensions around tasks, professional jurisdictions, and how genomics is understood as a field of simultaneous scrupulousness and transversality. For Santos, who has been working in a hospital on genetic predisposition for two decades, these tensions and lack of formal ordering of genomics in healthcare settings in Spain come from a rigid misunderstanding of contemporary genetics and the national healthcare specialties system.

> Genetics is a transversal discipline that impacts all clinical specialties. It is possible that they had a partial, segmented, pathology-dependent vision and did not consider genetics' transversal value. Subsequently, they did not know how to adapt it since the Spanish system of specialties is very robust and hermetic, so I understand that it has been difficult for them to open to a more panoramic vision. (Survey response, 2024)

For Santos, limiting genetics to a pathology-dependent approach controlled by a rigid specialties order is problematic as it does not follow the transversality of genetics. Understanding the tensions around mundane genomics labor and the specialty requires looking into the issue of healthcare organization, which is highly fragmented into specialties, and, of course, under such logic, the battle revolves around how to make clinical genetics another isolated new area and who can join in. Such fragmentation shapes the ordinary functioning of hospitals, diagnostics, and patient care in Spain, where medical specialties remain strong, and multidisciplinary teams are an exception, primarily present in oncology units. What is clear, as has been shown at a global level (Battista et al., 2012), is that much is still to be done to forge multidisciplinary specialists, provide resources for communities of practice (Long et al., 2019; Nicolini et al., 2016), and rearticulate tasks and roles in a healthcare system with little transformation since the 1980s. Ultimately, this issue pivots around who can be qualified as a clinical geneticist without questioning whether such an ordering as a specialty works for genomics in the current healthcare context.

Taking a radical and normative stance, one might argue that due to the lack of formal specialty and normative ordering in healthcare, the "qualified professional" to which the legislation refers does not exist, making anyone an intruder. Of course, this is a provocative stance as expertise transcends professional crediting or jurisdictions (Grundmann, 2017). In the Spanish case, genomics work is done by a quotidian arrangement of different professionals despite lacking a formal specialty. However, there is a missing question about the mundane textures of the specialty controversy in Spain, which has less to do with professional jurisdictions than with how genomics is understood and whether that clashes with healthcare order. How will genomics be ordered in the Spanish healthcare system without coming to terms with what clinical genomics is, after all? In this regard, Julia, who works at both a private and public hospital, points out that it is crucial to consider the ongoing transformations in genomics:

> Technical advancement has caused a migration from a targeted study, i.e., direct linkage between phenotype and genetic study, to a much broader scope, i.e., from suspect phenotype to many candidate genes or variants. This has led to the production of a more exploratory approach versus a more confirmatory approach. (Survey response, 2024)

The interpreter role must also be situated in how the capacities of sequencing technologies are shaping diagnostics, given that genomic detail seems to be moving toward an exploratory approach rather than the promised precision. Clinical genomics, as a plastic field constantly realigning knowledge and professionals, with few actionable results and temporary risk estimations, is what, for some, makes moving from a clinical analysis reading to test interpretation crucial. Such an approach feeds scrupulousness not only as attention to detail but also by addressing genetics test results and sequencing reports differently, not as confirmation but as an exploratory realm.

Instead of offering resolution, genetic test results open explorations of different sorts. As Marres and Stark (2020) note, more important than results or a failure to offer resolutions is what test results put to the test. On this occasion, the types of knowledge, tasks, and professionals required are still being tested when discussing arranging genomics in the clinic. This open discussion on what is needed offers an opportunity to craft *genomics modesty*, a much slower and more reflective approach to genomics science and, as such, to its integration into healthcare. If something has been learned from the genetic hype and promises, modesty is a needed companion in technoscience. Overall, the analyst, the interpreter, and the intruder are part of how genomics is being arranged in quotidian practice, and they reflect ongoing transformations in the field that, for now, require the tension of the combination. The three figures illustrate the lack of closure in conversations regarding the role of genomics in healthcare and its ordering.

Final Remarks

By unpacking the textures of the specialty controversy, I have examined the distinctions between the analyst and the interpreter that relate to the division of labor, particularly when assessing genetic predisposition, and how the diverse arrangement of professionals who only partially fit the category "geneticist" (lab analysts, bioinformaticians, surgeons, genetic counselors, nurses, etc.) makes genomics doable in practice despite multiple scientific and professional tensions. In a way, the intruder narrative is paradoxical in Spain, as there is no straightforward ordering of genomics professions; however, genetics services in hospitals continue daily amidst the healthcare services' heavy workloads and patients' lives and worries. In such a context, the disputes between the interpreter, the analyst, and the intruder become mundane but should not for that reason be ignored. I have pointed out how mundane issues of ordering healthcare turn out to be extraordinary endeavors, and the reverse.

The friction and reworking of the expert labor division are part of the mundane arrangements in genomics hospital routines. Beaudevin, Peerbaye, and Bourgain's (2019) study of the various frictions around tumor genetics in a French university hospital, unpacks the network of professionals and the diversity of meanings and uses ascribed to tumor profiling. It highlights that on an ordinary basis, genomics labor at a hospital might differ from genomics work in experimental settings and that fragmented expertise outside the jurisdiction of "clinical geneticists" (experts in "true genetics") might be the norm (Beaudevin et al., 2019, p. 656). They go further, noting that such fragmentation and "frictions are, therefore, arrangements that distribute and order different professional and epistemic rationalities to make tumor genetics doable" (p. 655). Rather than trying to dissolve the tensions present in the field between analysts, interpreters, and intruders, then, we might as well accept that these arrangements are not always fed by affinities (scientific or professional) but composed of frictions which are indeed what sustain genomics in clinical practice daily, including the Spanish context so far.

The analyst, the interpreter, and the intruder speak about what Puig de la Bellacasa, in her work and revision of feminist ruminations on knowledge politics, care, and relations, refers to as "dissenting-within," that is, attending to "how cuts foster relationships, rather than by how they disconnect worlds" (2017, p. 78). Differences, then, are key to understanding interdependencies, as are frictions (Tsing, 2005). I wonder then, if frictions of this kind in healthcare cultivate some forms of *genomics modesty*, where different professionals are attentive to what others do, involved in an ongoing training based on scrupulousness, and aiming to ensure the best possible healthcare in a highly uncertain field and fragile public healthcare landscape. Instead of assuming that easing tensions between professional jurisdictions and establishing clear demarcations to resolve the controversy are what are required at this point to facilitate genetics integration into healthcare, one might consider whether the quotidian practice of intense scrupulousness and tensions between different professionals' standpoints is precisely what fosters a much-needed sense of modesty and critical reflection on genomics in healthcare settings.

Acknowledgments I would like to thank all the healthcare professionals and genetics associations that have taken the time out of their busy schedules to participate in this research, enabling me to collate their viewpoints and concerns. This work is the product of the IfGene research project, "Projecting the future of genetic prediction" (PID2020-115899GB-I00, PIs: Mauro Turrini and Rubén Blanco), funded by the Spanish Ministry of Science and Innovation

(MICIU) through the Agencia Estatal de Investigación (AEI) (MICIU/AEI/10.13039/501100011033), with the support of the European Regional Development Fund (ERDF). The research was also developed thanks to the Juan de la Cierva Fellowship (Grant FJC2021-046469-I funded by MCIN/AEI/10.13039/501100011033 and the European Union "NextGenerationEU"/PRTR, PI Violeta Argudo-Portal).

References

Abbott, A. (1988). *The system of professions: An essay on the division of expert labor*. The University of Chicago Press.

Arregui, A. G. (2014). 'Teoría', 'Inteligencia' y 'Ciencia' como Índices Bajo Amazónicos del Conocimiento Incorporado. *Amazônica - Revista de Antropologia, 6*(1), 90–108.

Battista, R. N., Blancquaert, I., Laberge, A.-M., Van Schendel, N., & Leduc, N. (2012). Genetics in health care: An overview of current and emerging models. *Public Health Genomics, 15*(1), 34–45. https://doi.org/10.1159/000328846

Beaudevin, C., Peerbaye, A., & Bourgain, C. (2019). 'It has to become true genetics': Tumour genetics and the division of diagnostic labour in the clinic. *Sociology of Health & Illness, 41*(4), 643–657. https://doi.org/10.1111/1467-9566.12844

Bonache, S., Esteban, I., Moles-Fernández, A., Tenés, A., Duran-Lozano, L., Montalban, G., Bach, V., Carrasco, E., Gadea, N., López-Fernández, A., Torres-Esquius, S., Mancuso, F., Caratú, G., Vivancos, A., Tuset, N., Balmaña, J., Gutiérrez-Enríquez, S., & Diez, O. (2018). Multigene panel testing beyond BRCA1/2 in breast/ovarian cancer Spanish families and clinical actionability of findings. *Journal of Cancer Research and Clinical Oncology, 144*(12), 2495–2513. https://doi.org/10.1007/s00432-018-2763-9

Burns, B. L., Bilkey, G. A., Coles, E. P., Bowman, F. L., Beilby, J. P., Pachter, N. S., Baynam, G., Dawkins, H. J. S., Weeramanthri, T. S., & Nowak, K. J. (2019). Healthcare system priorities for successful integration of genomics: An Australian focus. *Frontiers in Public Health, 7*, 41. https://doi.org/10.3389/fpubh.2019.00041

Cambrosio, A., Keating, P., Vignola-Gagné, E., Besle, S., & Bourret, P. (2018). Extending experimentation: Oncology's fading boundary between research and care. *New Genetics and Society, 37*(3), 207–226. https://doi.org/10.1080/14636778.2018.1487281

Candela, M. (Ed.). (2003). *Los orígenes de la genética en España.* Sociedad Estatal de Conmemoraciones Culturales.

Candela Castillo, M., & García González, A. (2003). La Sociedad Española de Genética. In Candela, M. (Ed.). *Los orígenes de la genética en España.* Sociedad Estatal de Conmemoraciones Culturales.

Collins, H., & Evans, R. (2007). *Rethinking expertise.* The University of Chicago Press.

Das, V. (2020). *Textures of the ordinary: Doing anthropology after Wittgenstein.* Fordham University Press.

Eyal, G. (2013). For a sociology of expertise: The social origins of the autism epidemic. *American Journal of Sociology, 118*(4), 863–907. https://doi.org/10.1086/668448

Fortun, M. (2023). *Genomics with care: Minding the double binds of science.* Duke University Press. https://doi.org/10.1215/9781478024521

Foucault, M. (2020). *The history of sexuality. Volume 2: The use of pleasure* (R. J. Hurley, Trans.). Penguin Classics.

Geertz, C. (1973). *The interpretation of cultures: Selected essays.* Basic Books.

Grundmann, R. (2017). The problem of expertise in knowledge societies. *Minerva, 55*(1), 25–48. https://doi.org/10.1007/s11024-016-9308-7

Harvey, A. (2011). From lab to lifestyle: Translating genomics into healthcare practices. *New Genetics and Society, 30*(4), 309–327. https://doi.org/10.1080/14636778.2011.592005

Holbraad, M., Green, S., Corsín Jiménez, A., Das, V., Bird-David, N., Kohn, E., Hage, G., Bear, L., Knox, H., & Kapferer, B. (2018). What is analysis? *Social Analysis, 62*(1), 1–30. https://doi.org/10.3167/sa.2018.620103

Horgan, D., Pesapane, F., Van Der Buckle, M., De Maria, R., Dube, F., Singh, J., Ługowska, I., Bayle, A., Hofman, P., Malapelle, U., Hills, T., Capoluongo, E. D., & Subbiah, V. (2024). From theory to practice: Implementing next-generation sequencing and public health genomics in healthcare systems. *Critical Reviews in Oncology/Hematology, 201*, 104433. https://doi.org/10.1016/j.critrevonc.2024.104433

Keating, P., & Cambrosio, A. (2003). *Biomedical platforms: realigning the normal and the pathological in late-twentieth-century medicine.* The MIT Press.

Keating, P., Cambrosio, A., & Nelson, N. C. (2016). "Triple negative breast cancer": Translational research and the (re)assembling of diseases in post-genomic medicine. *Studies in History and Philosophy of Science Part C: Studies in History and Philosophy of Biological and Biomedical Sciences, 59*, 20–34. https://doi.org/10.1016/j.shpsc.2016.05.003

Kerr, A., Swallow, J., Chekar, C. K., & Cunningham-Burley, S. (2019). Genomic research and the cancer clinic: Uncertainty and expectations in professional accounts. *New Genetics and Society, 38*(2), 222–239. https://doi.org/10.1080/14636778.2019.1586525

Long, J. C., Pomare, C., Best, S., Boughtwood, T., North, K., Ellis, L. A., Churruca, K., & Braithwaite, J. (2019). Building a learning community of Australian clinical genomics: A social network study of the Australian Genomic Health Alliance. *BMC Medicine, 17*(1), 44. https://doi.org/10.1186/s12916-019-1274-0

Lopes-Cendes, I., & De Oliveira, T. C. (2025). Inequalities and inclusion in genomics applied to healthcare: A Latin American perspective. *Annual Review of Genomics and Human Genetics*. https://doi.org/10.1146/annurev-genom-111224-100329

Marchant, G., Barnes, M., Evans, J. P., LeRoy, B., & Wolf, S. M. (2020). From genetics to genomics: Facing the liability implications in clinical care. *The Journal of Law, Medicine & Ethics, 48*(1), 11–43. https://doi.org/10.1177/1073110520916994

Marres, N., & Stark, D. (2020). Put to the test: For a new sociology of testing. *British Journal of Sociology, 71*(3), 423–443. https://doi.org/10.1111/1468-4446.12746

National Health Service England. (2022). *Accelerating genomic medicine in the NHS*. A strategy for embedding genomics in the NHS over the next 5 years. Publication reference: PR1627. https://www.england.nhs.uk/long-read/accelerating-genomic-medicine-in-the-nhs/

Navne, L. E. (2023). Making sense of new disease categories: Naming, spatialising, and serialising in genomic medicine. *Medicine Anthropology Theory, 10*(3), 1–28. https://doi.org/10.17157/mat.10.3.6686

Nicolini, D., Scarbrough, H., & Gracheva, J. (2016). In E. Ferlie, K. Montgomery, & A. R. Pedersen (Eds.), *Communities of practice and situated learning in health care* (Vol. 1). Oxford University Press. https://doi.org/10.1093/oxfordhb/9780198705109.013.20

Pàmpols, T., Ramos, F. J., Lapunzina, P., Gozalo-Salellas, I., Pérez-Jurado, L. A., & Pujol, A. (2016). A view on clinical genetics and genomics in Spain: Of challenges and opportunities. *Molecular Genetics & Genomic Medicine, 4*(4), 376–391. https://doi.org/10.1002/mgg3.232

Payne, K., Nicholls, S. G., McAllister, M., MacLeod, R., Ellis, I., Donnai, D., & Davies, L. M. (2007). Outcome measures for clinical genetics services: A comparison of genetics healthcare professionals and patients' views. *Health Policy, 84*(1), 112–122. https://doi.org/10.1016/j.healthpol.2007.03.005

Puig de la Bellacasa, M. (2017). *Matters of care: Speculative ethics in more than human worlds*. University of Minnesota Press.

Rheinberger, H.-J. (2009). Recent science and its exploration: The case of molecular biology. *Studies in History and Philosophy of Science Part C: Studies in History and Philosophy of Biological and Biomedical Sciences, 40*(1), 6–12. https://doi.org/10.1016/j.shpsc.2008.12.002

Stark, Z., Dolman, L., Manolio, T. A., Ozenberger, B., Hill, S. L., Caulfied, M. J., Levy, Y., Glazer, D., Wilson, J., Lawler, M., Boughtwood, T., Braithwaite, J., Goodhand, P., Birney, E., & North, K. N. (2019). Integrating genomics into healthcare: A global responsibility. *The American Journal of Human Genetics, 104*(1), 13–20. https://doi.org/10.1016/j.ajhg.2018.11.014

Tsing, A. (2005). *Friction: An ethnography of global connection*. Princeton University Press.

Wynne, B. (1989). Sheepfarming after Chernobyl: A case study in communicating scientific information. *Environment: Science and Policy for Sustainable Development, 31*(2), 10–39. https://doi.org/10.1080/00139157.1989.9928930

Part III

Consumers and Identities

9

"I Didn't Want to Buy More Tickets for that Lottery": Metaphors for Genetic Testing in a Qualitative Study about Direct-to-Consumer Genetic Tests

Pablo Santoro and Rubén Blanco

Introduction

Direct-to-Consumer Genetic Testing (DTCGT) has become part of the testing ecosystem through which contemporary individuals construct their medical subjectivities. Early promises of personal genomics and consumer empowerment were initially made by the first wave of DTCGT companies, such as Sciona, during the bioeconomic bubble following the completion of the Human Genome Project in the early 2000s. While these have not been fulfilled, tens of millions of people have undergone DTCGT in a market that has significantly expanded during the last decade (Hazel et al., 2021). The market is led predominantly by US companies like 23andMe, but by 2016, over 246 DTCGT firms were operating worldwide (Hogarth & Saukko, 2017). Although ancestry testing remains the most successful branch of this industry, health-related tests

P. Santoro (✉) • R. Blanco
Departamento de Sociología: Metodología y Teoría, Universidad Complutense de Madrid, Madrid, Spain
e-mail: psantoro@cps.ucm.es; ruben.blanco@cps.ucm.es

V. Argudo-Portal et al. (eds.), *Mundane Genomics*,
https://doi.org/10.1007/978-981-95-4550-6_9

have also gained significant traction. These are often marketed as tools to empower patients, democratize research, and personalize healthcare (Prainsack, 2011), supported by claims to provide insights into disease predispositions, pharmacogenomics, hereditary mutations, physical traits, food intolerances, and even psychological characteristics. Critics, including healthcare regulators, clinicians, and policy advisors, have repeatedly warned of the risks associated with disseminating genomic information without expert oversight, citing concerns about the clinical reliability and utility of these tests, and about potential misinterpretations by users (Turrini, 2018). Nonetheless, despite opposition from medical associations, the health-related DTCGT sector remains a vibrant, although niche industry exhibiting continued global expansion. In Spain, the context of our study, start-ups including 24Genetics, ADNtro, and TellMeGen have emerged in recent years, focusing primarily on health and nutrition tests, while most Spanish health insurance companies and diagnostic labs now include genetic testing in their portfolios alongside other private genetic services, such as IVF preimplantation diagnostics and DNA paternity tests (Orihuela & Turrini, 2022).

Extensive scholarly work has examined the social dimensions of health-related DTCGT. Beyond legal and bioethical discussions, social scientists and public health researchers have investigated the motivations and experiences of DTCGT users, and the general public's attitudes toward these tests. Early studies of DTCGT users presented a complex picture (McGuire et al., 2009; Kaufman et al., 2012; Vayena et al., 2012; see Covolo et al., 2015, for a systematic review), showing how medical, personal, and social dimensions intermingle in decisions to undergo genetic testing (Turrini & Prainsack, 2016). Since then, much of the research has focused on specific descriptive and behavioral questions, such as levels of trust in DTCGT companies, privacy concerns, consultations with physicians before or after taking the test, and possible behavioral changes post-testing (Mavroidopoulou et al., 2015; Roberts et al., 2017; Oliveri et al., 2022; Ongaro et al., 2022). While efforts to quantify and measure these aspects have yielded valuable insights, findings are often contradictory, partly due to variations in national contexts, sample populations, and methodologies. For instance, some studies highlight widespread skepticism and significant privacy concerns (Schaper et al., 2019), whereas others emphasize optimism and a tendency to downplay risks (Vayena et al., 2014).

One limitation of these studies is that they often fail to interrogate the tests themselves—how people conceptualize and understand what DTC tests *are* and what they *do*. Other approaches, particularly qualitative ones, have explored not just attitudes or motivations but also the broader discourses, meanings, and practices surrounding these tests and their use. For example, research on ancestry test users has shown how these tests reshape understandings of kinship, genealogy, and racial identity, but only through a nuanced understanding of the knowledge provided by ancestry companies (Strand & Källén, 2021; Roth & Ivemark, 2018). Similarly, studies on health-related DTC tests reveal pragmatic and context-specific uses, such as Chinese parents' engagement with "talent tests" (Au, 2022) or the tensions caused by integrating commercial genotyping into the UK's NHS (Finlay, 2017).

In this chapter, we align with these latter approaches and with the new sociology of testing proposed by Marres and Stark (2020). Instead of asking about the validity of these tests or focusing solely on the personal motivations of users or behavioral changes post-testing, we examine how laypeople understand health-related DTC tests: in other words, what kind of information they believe these tests provide, how they perceive its utility, and how they interpret key concepts such as "genetic predisposition." We are especially interested in this latter concept; indeed, marketing campaigns often emphasize the ability of such tests to measure a "predisposition" or "susceptibility" to various health conditions. Meanwhile, scholars have noted that second-wave DTC tests have shifted from monocausal and deterministic views of genetic causality toward more nuanced, polygenic interpretations (Hogarth & Saukko, 2017), influenced by advances in genomics, epigenetics, and other "new biologies" (Blackman, 2016), as well as by the regulations introduced by the US FDA and the European Union since 2012 (Curnutte, 2017). While genomic sciences increasingly seem to emphasize biosocial complexity and challenge reductionist conceptions of genetics (Reardon, 2017), the extent to which patients and the general public adopt these frameworks when interpreting information about genetic predisposition remains unclear—particularly in the context of consumer products such as DTC tests, where no medical supervision is involved.

Drawing on data from nine focus groups conducted in 2023 and 2024 with DTCGT users and members of the general public in Spain, this chapter presents some of the participants' perspectives on genetic testing. When discussing DTC genetic tests, participants also shared ideas and views about what it means to be "genetically predisposed" to a particular disease or health condition. As the analysis demonstrates, different interpretations of the concept of genetic predisposition often shaped and explained participants' varying attitudes toward genetic testing. To examine these perspectives and interpretations, we adopt metaphorical analysis as a conceptual tool—a widely used approach in the social studies of genetics, as we explain in the following section.

Genetics and Metaphors

Since at least the mid-twentieth century, philosophers and linguists have recognized that metaphor—the understanding and experience of one thing in terms of another—is not merely a stylistic device but an essential component of language, thought, and knowledge (Black, 1955; Lakoff & Johnson, 1980). Metaphors are so fundamental to our cognitive processes and our understanding of the world that some philosophers of language have argued that all language is metaphorical to some degree (Hesse, 1988). Despite the common misconception of science as a purely denotative domain, scientific language is no exception; metaphors are pervasive in scientific discourse and are, arguably, especially abundant in the life sciences—a field built, from its inception, on crucial metaphors such as the *organism-as-machine* and *natural selection* (Kampourakis, 2020; Reynolds, 2022). In genetics specifically, metaphors have been employed (and misused) extensively, both in the construction of theoretical frameworks and the orientation of research, and in the communication of genetic science to the public.

The literature on the role of metaphors in genetics is extensive. Classic studies by Dorothy Nelkin and Evelyn Fox Keller examined the pivotal role that metaphors have played in twentieth-century genetics since its very foundation. Fox Keller (1995, 2020) argued that the very notion of the gene was constructed through two distinct metaphors: the gene as *the atom of life* and the gene as *agent*. Similarly, Nelkin (2001) discussed essentialist metaphors from the eugenics movements of the 1920s and 1930s that relied on comparisons with animal breeding, and also addressed the image of genes as an *immortal sacred entity* analogous to the Christian soul. This, she argued, emerged following Crick and Watson's discovery of DNA in 1953 and provides the background for the innumerable media images that endow genes with social attributes.

In the latter half of the twentieth century, the field became dominated by what Stelmach and Nerlich (2015, p. 198) describe as *grand metaphors*, fundamental conceptual images similar to Lyotard's *master narratives*. These metaphors, including the *book of life*, the *blueprint*, the *map*, and the *code*, profoundly shaped our understanding of genes and genetics. The interplay, exchanges, and tensions among these metaphorical images not only served to popularize genetics in the media and popular discourse but also actively guided scientific work in genetic research. It is, therefore, unsurprising that the paradigm shift in genetics following the completion of the Human Genome Project (HGP)—and the recognition that far less was known about the genome than initially anticipated—unsettled these dominant metaphors and gave rise to new conceptual frameworks and images (Stelmach & Nerlich, 2015).

In recent years, scholars have examined the emergence of new metaphors in post-genomic sciences: for example, metaphors of *music*, *switches*, and *markers* have been used to describe processes in epigenetics (Stelmach & Nerlich, 2015). Sensationalist metaphors, such as *chain reaction* or *genetic bomb*, have been employed by the media to explain gene drive technology (Nerlich & Stelmach, 2022). In developmental genetics, metaphorical concepts like *folding*, *networks*, and *plasticity* have gained prominence (Pigliucci, 2010). Similarly, critiques of the machine metaphor in synthetic biology have led to the adoption of alternative metaphors in the field, such as *living machines* or the *genome as recipe* (Vaage, 2020).

However, most of these studies focus either on the heuristic use of metaphors by scientists or on their rhetorical and didactic functions in media and public science communication. Far less research has explored the use of metaphors by laypeople when they engage in "gene talk" (Fox

Keller, 2000), with a few exceptions. In two US-based studies, Condit (2009) and Gronnvoll and Landau (2010) examined lay metaphors for genes, uncovering how non-experts conceptualize genetic information through metaphorical images such as "the gene as virus" or "the gene as bomb." Both studies showed that lay metaphors did not always align with those provided by scientists or the media, and that the emotional resonance of metaphors was as important as their cognitive power. There is also limited research on metaphors specifically related to personal genomics and genetic tests. Saukko (2017) analyzes the metaphorical shift in DTC companies' communications—from Sciona's portrayal of genes as information to 23andMe's framing of genes as big data—and Martin and colleagues (2024) investigate how women tested for BRCA mutations make sense of their experiences through different metaphors. Beyond these studies, however, potential metaphorical understandings of DTCGT remain largely underexplored.

Methods

As part of a larger joint project exploring different practices, debates, and applications of genetic susceptibility testing within and beyond the clinical setting in Spain, we conducted nine focus groups to investigate public perceptions of DTCGT. While health-related tests were our main focus, we considered that including other views and experiences—pertaining to other types of DTCGT and genetic testing in clinical settings, and those of people without direct experiences of testing—could enrich the analysis. The study involved a total of 33 participants and was carried out in two phases.

The first phase, conducted between May and June 2023, comprised two focus groups of members of the general public, and four triangular focus groups of individuals with specific profiles deemed relevant to the research: "genetic patients" (people who had undergone a clinician-ordered genetic test), adoptive parents, and DTCGT users. Triangular focus groups are a qualitative technique involving just three participants, creating a hybrid format that combines elements of in-depth interviews (with a biographical focus) and traditional focus groups (focused on social representations and discourses) (Ruiz, 2012). The second phase of research, conducted between June 2023 and July 2024, targeted only users of DTCGT and included three additional triangular focus groups, each focused on a different type of test: health, ancestry, and "lifestyle" (diet and sports). Table 9.1 presents the characteristics of each focus group.

All participants resided in Madrid and were recruited through a professional service, ensuring diversity in terms of age, gender, and educational background. Meetings took place in a neutral, rented space, since we did not want discussion to be influenced by a medical or academic setting. A male moderator with extensive experience in focus group methodology and qualitative health research conducted all sessions, which lasted an average of 120 minutes. The two authors were present and participated in the moderation. Two distinct discussion guides were used. For groups with DTCGT users, the guide focused on reconstructing the decision-making process behind taking the test, and exploring participants' views, emotions, and reflections on the experience. For the

Table 9.1 Characteristics of focus groups

	Profile	*Age range*	*Sex*	*Type of DTCGT or genetic test*
FG1	General public—*older age, lower socioeconomic status*	65–75	4 women, 3 men	Non-tested
FG2	General public—*younger age, higher socioeconomic status*	25–35	3 women, 3 men	Non-tested
TFG1	Genetic patients	40–55	2 women, 1 man	Genetic test in clinical setting
TFG2	Adoptive parents	30–50	2 women, 1 man	DNA test
TFG3	DTCGT users	25–40	2 women, 1 man	DTCGT health
TFG4	DTCGT users	30–40	1 woman, 2 men	DTCGT ancestry, sports
TFG5	DTCGT users	40–70	2 women, 1 man	DTCGT health
TFG6	DTCGT users	30–40	3 men	DTCGT health, nutrigenomics
TFG7	DTCGT users	40–50	2 women, 1 man	DTCGT ancestry

other groups, the guide addressed broader topics, including participants' knowledge and opinions about DTCGT, ideas about genes and genetic predisposition, and general perspectives on genetic science and genetic testing. Both guides were designed to allow flexibility, enabling participants to steer the discussion toward the topics they deemed important. All focus groups were audio-recorded, and the recordings were transcribed verbatim in Spanish. (For this chapter, quotations have been translated into English.) Participants received a reimbursement of €50. Informed consent was obtained from all participants, and ethics approval was granted by the CSIC Ethics Committee (reference: 212/2021).

As is the case with other studies examining the metaphorical understanding of genetics by laypeople (Condit, 2009; Martin et al., 2024), it was during the preliminary analysis of data that we realized that participants frequently used metaphors to articulate their ideas about genes, genetic predisposition, and the experience of being tested. Metaphorical language not only facilitated the expression of their views on DTC tests but also illuminated their attitudes toward genetic testing in general and DTC tests in particular. Once we decided to focus on the study of metaphors as a key aspect of our analysis, we used ATLAS.TI to code the transcripts—identifying instances of metaphorical talk—and to organize emerging themes and subthemes.

In the analysis that follows, we examine laypeople's use of metaphors to discuss genes, genetic predisposition, and genetic tests, with particular attention to the emotional implications of the different metaphors and how they relate to attitudes and behaviors toward DTCGT. If we accept that metaphors have performative effects—as Fox Keller (1995) argues—and that "metaphors affect the ways we perceive, think, and act, shaping our understanding of events and structuring our attitudes" (Nelkin, 2001, p. 556), then analyzing metaphorical talk can become a way to make explicit the various attitudes people hold about genetic testing. Importantly, our approach does not seek to evaluate metaphors as "better" or "worse," nor to classify them as "alive" or "dead," unlike other studies of lay metaphors (Gronnvoll & Landau, 2010). Instead, we use them as tools for understanding how individuals make sense of genetic tests and their implications.

We now present the three most prominent metaphorical clusters identified in our fieldwork: (1) *metaphors of chance*, wherein genetic predisposition is mainly understood as a "lottery" and taking a DTC test is viewed as akin to "gambling"; (2) *metaphors of connection*, wherein genetic predisposition is interpreted through the image of the "chain" and the lens of heredity and kinship, and tests are viewed as making or unmaking ties in different ways; and (3) *metaphors of activation*, which refer to genes as "switches" that can be turned on or off, and to tests as sources of "light" and self-knowledge that can "activate" changes in lifestyle.

Metaphors of Chance: Lottery and Gambling

The first group of metaphors used by participants revolved around images of chance, gambling, and betting, both in relation to the understanding of genes and genetic predisposition and to the comprehension of genetic tests. This type of metaphor—"lottery," "roulette," "raffle," "game"—indicates the association of genetic testing with randomness and fortune.

Starting from the notion that genes are like pre-existing dormant diseases—an idea that also appears in other studies of lay metaphors for genetics (Gronnvoll & Landau, 2010)—genetic predisposition appears here as a matter of luck, where individuals are presymptomatic patients unaware of the risks they may carry. Genetics, in this sense, is often framed as an individual lottery; even though genes are inherited from one's parents, they are perceived as the specific "hand you've been dealt" in the "game of life." Participants frequently used expressions such as "*lo que te ha tocado*" ("what fate has given you") or "*las papeletas que te han tocado*" ("the tickets you've been dealt") to describe this randomness.

> Yeah, I think it's more like, genetics is more about the specific gene itself, like whether a particular gene has mutations or not, and the variations each individual has. But the genome, the human genome, is the whole set of it, and then it has different variations, and, well, you just get the one you ended up with in the lottery. (TFG4, Ancestry and Sports DTCGT users)
>
> Usually you think: "Well, that's not going to happen to me" [getting a serious disease like cancer]. Because it doesn't affect you so closely. So it's

> not until you see it getting close that you say, "Wow, I really have to be a bit careful with this." It's like you never think, "Oh, well, I could also get the lottery." (TFG3, Health DTCGT users)

Genetics is thus depicted as something governed primarily by chance and luck: Either one is fortunate and remains healthy, or unlucky and destined to develop certain diseases. This portrayal of genetic predisposition as a lottery, where good and bad health outcomes are distributed arbitrarily, is consistent with the fatalistic outlook reflected in the use of these kind of metaphors, which evokes a somewhat mythological image of luck—the Wheel of Fortune—whose designs are seen as unpredictable and beyond human control. In this way, a sense of deterministic fatalism emerges, with a predisposition to certain conditions or diseases understood as something that cannot be changed, independent of behavior, and marked by uncertainty. In this metaphorical field, the probability signaled by tests is akin to betting.

> - Suppose the test tells you that you have a 60% chance of getting cancer next year, what do you say?
> - I'd go to the doctor tomorrow.
> - Okay, you go to the doctor tomorrow, and the doctor says: "Alright, but we don't know. It's up to you. We can operate now, or not." Or they tell you that in two years, you may have Alzheimer's, or that you're going to lose your hair. So, you go to the doctor. But you've got a 60% chance—you don't know if it's going to happen or not. There's still a 40% chance it won't.
> - Yeah, but, you know, if someone wants to take the gamble, let them. It's like playing roulette—if you want to take a risk, go for it. Personally, I'd rather live my life. (FG2, General public)

In this context, the tests themselves are often likened to a lottery, a roulette, or a game of chance, as they inherently involve an element of risk rather than certainty. Taking a test for a multitude of diseases is perceived as akin to holding more lottery numbers, increasing the likelihood of receiving a positive result for at least one condition. For this reason, several participants who had taken a DTC health test chose to focus only on the specific disease about which they were concerned, rather than investigating all the potential health conditions that the test is supposed to detect.

> Mod:*And why did you just get tested for only one disease? I mean, in the test pack, you can get a whole bunch.*
> - Yeah, but I thought that was a bit too much, in the sense that you could end up worrying about so many things. I don't know, taking a test for who knows how many diseases didn't seem very practical to me, because with that, you're buying more tickets in the lottery and one of them might come back positive. (TFG3, Health DTCGT users)
> I almost decided not to go through with the prenatal test, not because of the cost—though at some point, I also thought it was totally overpriced—but because if the public healthcare system was already testing for the three most common conditions and it came back negative, then you could just live happily, you know? But if this other test was going to look for so many more, I remember thinking, "One of them will come up, something will." I do remember that feeling, for example, and thinking, "Hmm, I'm not sure if I'm convinced I should do it, because by testing for so many, there's a higher chance that one will come up." (TFG5, Health and Nutrigenomics DTCGT users)

According to this logic, and perhaps paradoxically, the greater the certainty attributed to genetic testing, the less useful these tests may seem. Knowing that one has a higher probability of developing a serious disease appears pointless when there is nothing one can do—when it ultimately comes down to chance and fate. Participants who were more reluctant to get tested frequently emphasized that a positive result might lead only to psychological consequences, such as excessive worry and anxiety about the future. The rhetorical question, "Would you want to know the date of your death?" was used in different focus groups to argue that increased knowledge about personal genetic predispositions does not necessarily lead to better health; rather, it may foster hypochondria and hinder one's ability to enjoy life.

> Well, the test can come back positive or negative, and in the end, it's like a fifty-fifty chance. And there are plenty of people who, even if they know about these tests, and are interested, and care about their health, they might say, "Look, I'm not ready to face this kind of result, so to avoid facing it, I'd rather not take the test, I'd rather live with that uncertainty for the rest of my life, and that's it." Because, of course, if you're suddenly told that you're highly likely to develop a serious illness, not everyone is prepared to know or to handle that. (TFG3, Health DTCGT users)

Metaphors of Connection: Chains and Mirrors

The second set of metaphors that emerged from participants' discourses presents a very different set of connotations. Instead of viewing genes merely as dormant diseases and genetic predispositions as matters of individual fate or chance, metaphors such as the "chain" or the "tie" emphasize the importance of genetic heredity and family continuity. These metaphors underscore the biological makeup one shares with one's parents and, more broadly, with all one's genetic ancestors. Genetics is what "gets passed down," what binds you to those who came before you. The chain metaphor, therefore, highlights the "hereditary," where "one comes from," and in group discussions, it was more closely linked to the image of DNA (which, not coincidentally, has a chain-like structure itself and is often associated with paternity tests) than to the notion of the gene.

> I think DNA is the remnants of where we come from; it is like a chain where everything we are is written. [...] I think that genes are what get inherited, you know what I mean? For example, I have this from my mother, this or that gene. For anything. For bad temper, diseases, whatever you want. I think that's what gets passed down. And, of course, what gets passed down is the same as what she has. Because it's within the chain. (FG1, General public)

While the lottery metaphor emphasizes the luck peculiar to the individual, the chain metaphor points to family, kinship, and continuity. As evidenced by the image of "remnants" in the quote above, it projects a genetic predisposition toward the past. It is, of course, the idea of heredity and family lineage that unites ancestry tests and health-related DTC tests; both serve as means of making visible the link that "ties you to history" and to your ancestors.

> - What the test tells you, in the end, is where you come from. Back then, you didn't exist, you weren't even a thought, but things were already happening that might affect you now, or have affected you since you were born, I don't know. It's fascinating because it's a way to understand where you come from. And maybe part of your personality has been shaped by the people who lived before you, who passed down traits or genetic predis-

positions from one generation to the next. Like, you have your grandfather's eyes and all, but maybe you've inherited more than just his eyes.
- It's like knowing you're not alone in the world.
- It ties you to history. (TFG7, Ancestry DTCGT users)

In our fieldwork, however, and with respect to the understanding of a genetic predisposition toward certain disease or health conditions, the chain metaphor tended to be symbolically restricted to family. In this sense, several participants who had taken a health-related DTC test did so because they were concerned about diseases running in their family, either genetic conditions that had been recently detected in a relative or serious diseases (cancer, Alzheimer's) thought to be hereditary. While screening programs identify relatives with the potential to develop serious genetic diseases, three of our interviewees did not qualify for those programs, either because they were more distant relatives or not of the minimum age required for screening. DTC tests were then seen as a way of bypassing the medical screening criteria, as they had requested the test from their doctors but were denied access.

But taking such a test would be pointless if nothing could be done. The chain metaphor has a corollary: the idea that the chain can be manipulated to some extent—that medical interventions can affect genes and that, in keeping with this metaphor, doctors can "cut" the chain, undoing those ties. This would be the main rationale for genetic testing, especially when a genetic disease has already been discovered in the family, as happened in the focus groups with genetic patients.

- I have aniridia, which is a vision problem, the absence of the iris. It's something that started with my father, something hereditary that's pretty much a sure thing—if you have a child and they don't have it, it's because the chain stopped with them, even though another child of yours might still have it. So, genetic tests look for that gene, maybe to eliminate it in the future if I decide to have a child—that was my goal. When I took the test, I was honestly happy because they explained to me how that gene could be eliminated. It's not completely removable, but for the most part, it can be. They give you a 98% probability, and honestly, I was really pleased with that.

- That's why they do the test. The test is done to cut that chain. (TFG1, Genetic patients)

Following this reasoning, genetic tests can also help reconstruct the chain when knowledge of it has been neglected. This was the case in the focus group we conducted with adoptive parents and in other participants who had undergone IVF preimplantation genetic testing privately. They saw the utility of DTC tests for their children as a potential way to recover knowledge of their genetic heritage—something missing, since the biological chain had been broken and they lacked information about health antecedents.

> [Adopted children] don't know if they will be tall or short, fat or thin … They don't know how they will turn out as adults. They don't have a mirror to look into. Because you are their mother or father, but they won't be like you. And one thing my daughter wonders about is, well, "Genetically, how am I?" Like, "Will I have this disease, will I have a certain longevity? How am I genetically? The day I have children, what will I pass on to them?" They do wonder about that. And when they go to the doctor, they cannot tell him anything about their family history. (TFG2, Adoptive parents)

Thus, adoptive parents offered a different metaphor to that of the chain, yet one logically related to it: the image of genetic tests as a "mirror" that can replace direct observation of family health. The idea here is that adoptees lack the ability to anticipate their future adult selves by observing their parents; thus, they do not possess a mirror to reflect their potential future. Unfamiliar with the field of adoption, we initially thought this significant metaphor (DTC tests as a substitute "genetic mirror") was spontaneously expressed by our participants; however, a quick search on the Internet revealed that it is a common metaphor in contemporary psychological therapy surrounding adoption. It is particularly prevalent in the terminology of adoptee associations, having replaced the notion of *genealogical bewilderment* introduced by psychologists in the 1960s to describe potential identity issues faced by children who are adopted or conceived through gamete or embryo donation in assisted reproduction (Humphrey & Humphrey, 1986). The shift to the concept of *genetic mirroring* can thus be seen as an instance of the geneticization of kinship, and also as a notion connected to the metaphor of the chain and its implications for the perceived utility of genetic testing.

Of course, as some participants pointed out, the possibility offered by DTC genetic tests to reconstruct the chain or provide a mirror to understand the connection that binds you to your ancestors can also be dangerous, as it may uncover secrets about the past and reveal discrepancies between the stories your family has told you and the "real" truth about your origins. The term *DNA shock* has been used in the media to describe the surprises about family and ancestry one can experience when taking a genetic test (Johnson, 2023). In our focus groups, this potentially disturbing outcome of private genetic testing was also acknowledged, as DTC tests can make and unmake family ties in unexpected ways. For example, one of our participants, who was not only an adoptive father but also an adoptee himself, expressed the fear he felt when he took a DNA test, hoping to find out whether he was a "stolen child"—he was referring to a scandal revealed recently in the Spanish press in which several hospitals had taken children from poor mothers in the 1960s and sold them to wealthier families.

> Until the issue of stolen children came up in the press, I never considered the possibility that the story my parents told me might not be true. And that was what led me to investigate and get in touch with the [testing company]. The only way to find out if I was adopted, as my parents had told me, or if I was a stolen child, was to take a test, and if someone was looking for me, our files would cross. That's why I did it. Otherwise, I probably wouldn't have. I went to a lab; they took a swab. And that stayed in a database, waiting to see if it matched. I was hoping it wouldn't match because imagine if I found out I was stolen. I think that's why many adoptees wait until their parents have passed away to investigate. (TFG2, Adoptive parents)

Metaphors of Activation: Switches and Lights

The final metaphorical cluster we have identified takes us to a different semantic field: that of "activation," either in relation to the understanding of genetic predisposition or in relation to the tests themselves. In the first sense, participants often described genes as something that can be activated under certain conditions: namely, lifestyle choices and health behaviors. This moves us away from both the merely fatalistic understanding of genetic predisposition expressed in the lottery metaphor, and the view in which inheritance is all-important, as in the chain metaphor. Indeed, it brings us closer to a vision that is more nearly aligned with contemporary scientific discourses on gene expression, given the pervasive use of the concept and imagery of the *switch* in fields such as epigenetics and synthetic biology.

While participants did not explicitly mention gene switches, they repeatedly referred to the idea that genes can be either "active" or "inactive," "on" or "off." The idea of activation and the implied metaphor of the switch marks a shift away from the concept of genetic determinism. It suggests that public perceptions of genetics may be becoming more nuanced, with an increasing recognition of the influence of lifestyle choices and environmental factors on gene expression. They also emphasized that, although there may be a certain random element in that activation, something can be done to influence it, whether through medical supervision, preventive treatments, or lifestyle changes. As one participant put it, "genes do not activate on their own" (TFG5, Health and Nutrition DTCGT users).

> They did a saliva test on my sister, and it showed that the gene for cancer was there, but it wasn't active at that moment. However, it can become active at any time, the gene. So, they started an 8-day treatment to stop the gene from activating. (FG1, General public)
> - It should be said that many genes, cancer for example … which, by the way, I also tested positive for, but I have it as negative. I mean, we all have that gene. I mean, we have to say that we all have that gene. Some are more active than others, some may never activate, and others might suddenly develop it one day, even if there are no family precedents, but it shows up anyway.
> - I'll tell you something, with the test I took, they told me that if I smoke, cancer could activate. And what do you want me to say, I smoke.
> - Do you still smoke?
> - But it comes from family history. I have that gene inactive, but it's there. And maybe I quit smoking now, but in 15 years, that gene that's there, that's inactive, could activate.
> - But you'll have a higher chance of it activating if you smoke. (FG2, General public)

The image of genes as switches inverts the temporal logic of the chain metaphor; while in that metaphorical field the focus of genetic testing was on the past, here it shifts to the future. Knowing where those genetic predispositions come from may be irrelevant, what matters is understanding what kind of behavior might activate those genes in the future. Thus, the utility of genetic tests lies in providing knowledge that is useful for modulating behavior going forward. This is why some participants referred to genetic tests as a "source of light," a common metaphor for knowledge that, in the words of the participant quoted below, was expressed in a specific way: DTC tests are like "high beam" in a car, allowing you to see farther ahead and anticipate at least part of your future.

> I mean, what I believe is that these tests are tools that, well, they kind of activate the high beam in your car, for you to see what genetic possibilities you have and be able to avoid certain things. Like a light that illuminates your path. So you can see that, for example, there are more curves over there, there's a wall over there, and that path is clear … Well, I don't know, like lighting the road ahead, right? (TFG5, Health DTCGT users)

These metaphors of switches, gene activation, and light seem particularly related to positive attitudes toward DTCGT and, indeed, in our fieldwork were more commonly employed by participants who had taken a DTC test, especially those focused on nutrition and sports. Nutrigenomics tests, in particular, were perceived as providing exactly this type of actionable knowledge: detecting which foods you should avoid or what kind of exercise is the most effective for you. This means that such tests do not merely identify "active" or "inactive" genes; they also *activate* the individual, prompting behavioral changes or inspiring a pursuit of further knowledge. As Marres and Stark have noted regarding tests in the contemporary world, testing is often generative (Marres & Stark, 2020); tests lead to more tests—whether through medical monitoring to ensure that a gene linked to a disease remains inactive, or through everyday experimentation, such as trying out different diets.

> The test my niece took … She had three possible results. The negative, which means, "No, you don't have the gene your mom has." The positive,

> which would be, "You have the same gene as your mom, and it's active right now, so we need to take action." And then there was a middle one, which is what my niece got. Not positive or negative. The gene is there, but it's inactive. What they recommended was really close monitoring. Like, since she's still very young, getting an MRI and more tests every two years. (FG1, General public)
>
> After taking the test, I started cutting out certain foods and so on, little by little … Because I wasn't 100% sure it was completely accurate. But thanks to it, I've been trying different foods myself. Like, I think this … For example, milk—it said that I have a predisposition to lactose intolerance. So, I started having lactose-free milk for breakfast, and now I don't feel as heavy in the mornings as I used to. So, now it's about eating with awareness, where you can experiment with what works well and what doesn't for you, based on the test results. (TFG6, Nutrigenomics DTCGT users)

Following the reasoning of this last participant, one final observation seems pertinent: the uncertainty surrounding the reliability of test results does not necessarily diminish their perceived utility. In fact, most participants who had taken a DTC test emphasized that these tests are not diagnostic tools and that, for example, DTC tests indicating a predisposition to celiac disease or certain allergies are distinct from medical tests that confirm an existing intolerance. The key difference lies in how these tests envision a potential future—one that may or may not materialize. Paradoxically, it is precisely this uncertainty, this openness to possibilities, that made DTC tests valuable for many participants.

Conclusions: Mundane Metaphors

As discussed throughout this volume, genomics is becoming mundane. Concepts, practices, and tools that once belonged to the realm of cutting-edge genomic science are now routinely shaping how we construct our identities, self-perceptions, and relationships. However, for genomic knowledge to become truly mundane, it must travel beyond expert circles and find a place in the ways laypeople think and talk about genetics. In this process, metaphors play a crucial role, serving as vehicles for making sense of genetic information and articulating what it means to be "who we are," genetically speaking. While the heuristic use of genetic metaphors by scientists has been widely analyzed, their expansion is not solely the result of expert knowledge transmission. These metaphors travel through media reports on genetic "breakthroughs," through the marketing strategies of DTC genetic testing companies, through books, films, and TV series, doctor-patient interactions, and even everyday conversations with family and friends.

Therefore, the starting point of our analysis was the idea that surveying metaphors connected with genetics can serve as a lens not only for analyzing scientific research programs, theoretical conceptions, and science communication but also as an entry point into lay perceptions and understandings of genetics. Our exploration has focused specifically on some of the metaphors that surround direct-to-consumer genetic testing (DTCGT), aiming to understand how these metaphors may reflect the attitudes and beliefs of the public regarding genetics and genetic predisposition. From the fatalistic imagery of lotteries to the hereditary connotations of chains and the forward-looking symbolism of switches, each metaphorical field highlights a different dimension of genetic testing—whether as a tool for uncovering the past, understanding the present, or shaping the future. These metaphors, far from being purely rhetorical, structure how people make sense of genetic information and, crucially, how they act upon it.

Our findings align with prior research on lay metaphors for genetics (e.g., Condit, 2009; Gronnvoll & Landau, 2010), demonstrating that such public metaphors have a non-linear relationship with those employed by scientists or the media. While newer scientific metaphors—the genetic switches or the scissors of genetic engineering—appear to be gradually entering popular discourse, they coexist with older, more deterministic metaphors, including premodern ones like the Wheel of Fortune; moreover, the latter images sometimes permeate back into scientific discourse. A striking example can be found in a recent and controversial book by behavioral geneticist Kathryn P. Harden, *The Genetic Lottery* (Harden, 2021), which draws on the very lottery metaphor we have explored to discuss the relationship between genetics and social equality.

The three set of metaphors we have analyzed were not evenly distributed among our participants, who varied significantly in age, experience with testing, and even general attitudes toward health and life. Nevertheless, the boundaries between metaphors were not strict. For example, even users who perceived DTC tests as a valuable source of knowledge—a light that illuminates their path toward better health—may simultaneously resort to gambling metaphors to explain why they avoided testing for many conditions. As happens with genetic metaphors used by scientists or the media, everyday understandings of genetics are shaped by overlapping and sometimes contradictory narratives.

Finally, our analysis reinforces the argument that DTC tests are not evaluated by users based solely, or even primarily, on their clinical utility. As Turrini and Prainsack (2016) observed, DTCGT holds different types of value for users, merging personal, health-related, and social motivations. The metaphors our participants used to describe their views and experiences with DTCGT reveal not only the complexity of understandings of genetics in contemporary society, but also the multifaceted utility people may find in these tests.

Acknowledgments This work is part of the research project, "IFGene. Projecting the future of genetic prediction" (PID2020-115899GB-I00, PIs: Mauro Turrini and Rubén Blanco, https://ifgene.eu/), funded by the Spanish Ministry of Science and Innovation (MICIU) through the Agencia Estatal de Investigación (AEI) (MICIU/AEI/10.13039/501100011033), with the support of the European Regional Development Fund (ERDF). We would like to thank the editors of the volume and two anonymous reviewers for their comments on a first draft of the text.

References

Au, L. (2022). Testing the talented child: Direct-to-consumer genetic talent tests in China. *Public Understanding of Science, 31*(2), 195–210. https://doi.org/10.1177/09636625211051964

Black, M. (1955). Metaphor. *Proceedings of the Aristotelian Society, 55*(1), 273–294. https://doi.org/10.1093/aristotelian/55.1.273

Blackman, L. (2016). The New Biologies: Epigenetics, the Microbiome and Immunities. *Body & Society, 22*(4), 3–18. https://doi.org/10.1177/1357034X16662325

Condit, C. M. (2009). Dynamic feelings about metaphors for genes: Implications for research and genetic policy. *Genomics, Society and Policy, 5*(3), 44–58. https://doi.org/10.1186/1746-5354-5-3-44

Covolo, L., Rubinelli, S., Ceretti, E., & Gelatti, U. (2015). Internet-Based Direct-to-Consumer Genetic Testing: A Systematic Review. *Journal of Medical Internet Research, 17*(12), e279. https://doi.org/10.2196/jmir.4378

Curnutte, M. (2017). Regulatory controls for direct-to-consumer genetic tests: A case study on how the FDA exercised its authority. *New Genetics and Society, 36*(3), 209–226. https://doi.org/10.1080/14636778.2017.1354690

Finlay, T. (2017). Testing the NHS: The tensions between personalized and collective medicine produced by personal genomics in the UK. *New Genetics and Society, 36*(3), 227–249. https://doi.org/10.1080/14636778.2017.1351873

Fox Keller, E. (1995). *Refiguring life. Metaphors of twentieth-century biology*. The University of Chicago Press.

Fox Keller, E. (2000). *The century of the gene*. Harvard University Press.

Fox Keller, E. (2020). Cognitive functions of metaphor in the natural sciences. *Interdisciplinary Science Reviews, 45*(3), 249–267. https://doi.org/10.1080/03080188.2020.1794384

Gronnvoll, M., & Landau, J. (2010). From viruses to Russian roulette to dance: A rhetorical critique and creation of genetic metaphors. *Rhetoric Society Quarterly, 40*(1), 46–70. https://doi.org/10.1080/02773940903413415

Harden, K. P. (2021). *The genetic lottery, why DNA matters for social equality*. Princeton University Press.

Hazel, J. W., Hammack-Aviran, C., Brelsford, K. M., Malin, B. A., Beskow, L. M., & Clayton, E. W. (2021). Direct-to-consumer genetic testing: Prospective users' attitudes toward information about ancestry and biological relationships. *PLoS ONE, 16*(11), e0260340. https://doi.org/10.1371/journal.pone.0260340

Hesse, M. (1988). The cognitive claims of metaphor. *The Journal of Speculative Philosophy, 2*(1), 1–16.

Hogarth, S., & Saukko, P. (2017). A market in the making: The past, present and future of direct-to-consumer genomics. *New Genetics and Society, 36*(3), 197–208. https://doi.org/10.1080/14636778.2017.1354692

Humphrey, M., & Humphrey, H. (1986). A fresh look at genealogical bewilderment. *British Journal of Medical Psychology, 59*(2), 133–140. https://doi.org/10.1111/j.2044-8341.1986.tb02677.x

Johnson, R. (2023, October 4). These people took DNA tests – the results changed their lives. *BBC Future*. Retrieved December 28 2024, from https://www.bbc.com/future/article/20231004-these-people-took-dna-tests-the-results-changed-their-lives

Kampourakis, K. (2020). Why does it matter that many biology concepts are metaphors? In K. Kampourakis & T. Uller (Eds.), *Philosophy of science for biologists*. Cambridge University Press.

Kaufman, D. J., Bollinger, J. M., Dvoskin, R. L., & Scott, J. A. (2012). Risky business: Risk perception and the use of medical services among customers of DTC personal genetic testing. *Journal of Genetic Counseling, 21*(3), 413–422. https://doi.org/10.1007/s10897-012-9483-0

Lakoff, G., & Johnson, M. (1980). *Metaphors we live by*. The University of Chicago Press.

Marres, N., & Stark, D. (2020). Put to the test: For a new sociology of testing. *British Journal of Sociology, 71*(3), 423–443. https://doi.org/10.1111/1468-4446.12746

Martin, S. C., Scott, A. M., & Stone, A. M. (2024). Women's metaphors about BRCA gene testing and how they can inform health communication theory and practice. *Health Communication, 39*(3), 603–615. https://doi.org/10.1080/10410236.2023.2178051

Mavroidopoulou, V., Xera, E., & Mollaki, V. (2015). Awareness, attitudes and perspectives of direct-to-consumer genetic testing in Greece: A survey of potential consumers. *Journal of Human Genetics, 60*(9), 515–523. https://doi.org/10.1038/jhg.2015.58

McGuire, A. L., Diaz, C. M., Wang, T., & Hilsenbeck, S. G. (2009). Social networkers' attitudes toward direct-to-consumer personal genome testing. *The American Journal of Bioethics, 9*(6–7), 3–10. https://doi.org/10.1080/15265160902928209

Nelkin, D. (2001). Molecular metaphors: The gene in popular discourse. *Nature Review Genetics, 2*(7), 555–559. https://doi.org/10.1038/35080583

Nerlich, B., & Stelmach, A. (2022). Gene drive communication: Exploring experts' lived experience of metaphor use. *New Genetics and Society, 41*(1), 3–22. https://doi.org/10.1080/14636778.2021.2020633

Oliveri, S., Cincidda, C., Ongaro, G., Cutica, I., Gorini, A., Spinella, F., Fiorentino, F., Baldi, M., & Pravettoni, G. (2022). What people really change after genetic testing (GT) performed in private labs: Results from an Italian study. *European Journal of Human Genetics, 30*(1), 62–72. https://doi.org/10.1038/s41431-021-00879-w

Ongaro, G., Brivio, E., Cincidda, C., Oliveri, S., Spinella, F., Steinberger, D., Cutica, I., Gorini, A., & Pravettoni, G. (2022). Genetic testing users in Italy and Germany: Health orientation, health-related habits, and psychological profile. *Molecular Genetics & Genomic Medicine, 10*(3), e1851. https://doi.org/10.1002/mgg3.1851

Orihuela, P., & Turrini, P. (2022). El imaginario genético en España: un análisis desde el mercado de la genética de consumo, *XIV Congreso Español de Sociología*, Murcia, 29 junio–2 julio 2022

Pauwels, E. (2013). Mind the metaphor. *Nature, 500*(7464), 523–524. https://doi.org/10.1038/500523a

Pigliucci, M. (2010). Genotype–phenotype mapping and the end of the 'genes as blueprint' metaphor. *Philosophical Transactions of the Royal Society B, 365*(1540), 557–566. https://doi.org/10.1098/rstb.2009.0241

Prainsack, B. (2011). Voting with their mice: Personal genome testing and the "participatory turn" in disease research. *Accountability in Research, 18*(3), 132–147. https://doi.org/10.1080/08989621.2011.575032

Reardon, J. (2017). *The postgenomic condition. Ethics, justice and knowledge after the genome*. The University of Chicago Press.

Reynolds, A. S. (2022). *Understanding metaphors in the life sciences*. Cambridge University Press.

Roberts, J. S., Gornick, M. C., Carere, D. A., Uhlmann, W. R., Ruffin, M. T., & Green, R. C. (2017). Direct-to-consumer genetic testing: User motivations, decision making, and perceived utility of results. *Public Health Genomics, 20*(1), 36–45. https://doi.org/10.1159/000455006

Roth, W. D., & Ivemark, B. (2018). Genetic options: The impact of genetic ancestry testing on consumers' racial and ethnic identities. *American Journal of Sociology, 124*(1), 150–184. https://doi.org/10.1086/697487

Ruiz Ruiz, J. (2012). El grupo triangular: reflexiones metodológicas en torno a dos experiencias de investigación. *Empiria. Revista de metodología de ciencias sociales, 24*, 141–162. https://doi.org/10.5944/empiria.24.2012.846

Saukko, P. (2017). Shifting metaphors in direct-to-consumer genetic testing: From genes as information to genes as big data. *New Genetics and Society, 36*(3), 296–313. https://doi.org/10.1080/14636778.2017.1354691

Schaper, M., Wöhlke, S., & Schicktanz, S. (2019). "I would rather have it done by a doctor"—laypeople's perceptions of direct-to-consumer genetic testing (DTC GT) and its ethical implications. *Medicine, Health Care and Philosophy, 22*(1), 31–40. https://doi.org/10.1007/s11019-018-9837-y

Stelmach, A., & Nerlich, B. (2015). Metaphors in search of a target: The curious case of epigenetics. *New Genetics and Society, 34*(2), 196–218. https://doi.org/10.1080/14636778.2015.1034849

Strand, D., & Källén, A. (2021). I am a Viking! DNA, popular culture and the construction of geneticized identity. *New Genetics and Society, 40*(4), 520–540. https://doi.org/10.1080/14636778.2020.1868988

Turrini, M. (2018). Online genomes: Problematizing the disruptiveness of direct-to-consumer genetic tests. *Sociology Compass, 12*(11), e12633. https://doi.org/10.1111/soc4.12633

Turrini, M., & Prainsack, B. (2016). Beyond clinical utility: The multiple values of DTC genetics. *Applied Translational Genomics, 8*, 4–8. https://doi.org/10.1016/j.atg.2016.01.008

Vaage, N. S. (2020). Living machines: Metaphors we live by. *Nanoethics, 14*(1), 57–70. https://doi.org/10.1007/s11569-019-00355-2

Vayena, E., Gourna, E., Streuli, J., Hafen, E., & Prainsack, B. (2012). Experiences of early users of direct-to-consumer genomics in Switzerland: An exploratory study. *Public Health Genomics, 15*(6), 352–362. https://doi.org/10.1159/000343792

Vayena, E., Ineichen, C., Stoupka, E., & Hafen, E. (2014). Playing a part in research? University students' attitudes to direct-to-consumer genomics. *Public Health Genomics, 17*(3), 158–168. https://doi.org/10.1159/000360257

10

Shallow Time Politics: Entangling Ancestors' Qualities and Present-Day Identities by Way of Genomic Ancestry Testing

François Romijn

Introduction

As the twenty-first century progresses, humans are increasingly being exposed to specialist information in molecular biology, especially genetics. First produced in scientists' laboratories, this knowledge is transferred out into society through press releases, popular science books, or commercial markets, as illustrated by the ubiquitous genomic ancestry testing (GAT). Genetics possesses an enduring cultural mystique (Nelkin & Lindee, 2004), with social and political implications that extend way beyond scientific or medical circles. In the general population, DNA is

F. Romijn (✉)
Deakin University, Melbourne, VIC, Australia

Laboratoire d'anthropologie des mondes contemporains, Université libre de Bruxelles, Brussels, Belgium

Department of Human Behaviour, Ecology and Culture, Max Planck Institute for Evolutionary Anthropology, Leipzig, Germany
e-mail: Francois.Romijn@ulb.be

V. Argudo-Portal et al. (eds.), *Mundane Genomics*,
https://doi.org/10.1007/978-981-95-4550-6_10

often perceived as telling us definitive things (i.e., about health issues, origins) that connect with how we understand "who we are" (Hauskeller et al., 2013), while contemporary populations are said to yearn to connect to the past, individual roots, and history. The ways this interest plays out in practice is contextually situated. By virtue of DNA knowledge, users can discursively draw parallels between the experiences endured by their ancestors and their own life or health conditions: when, for instance, the group's historical suffering becomes an explanation for mutations detectable in the present (Gibbon, 2013), or when possessing DNA features associated with certain groups increases a sense of self-confidence (Nordgren & Juengst, 2009). There is something striking, not so much in the observation that the "possession of [genetic] knowledge confers identity" (Kramer, 2015, p. 95), but rather that knowledge of one's ancestors' genetic past "is seen to translate unproblematically to knowledge about the self in the present" (Kramer, 2015, p. 95). Focusing on the still developing field of GAT, this chapter is situated alongside a set of scholarly studies interested in the expansion of new networks of relationships that social actors establish based on DNA knowledge.[1]

To highlight the variety of these networks of relationships, I use the term *entanglements* to reflect the plurality of ways whereby GAT users discursively entangle with DNA knowledge—in specific contexts and temporalities—and establish connections between the self and others. Drawing on an STS ethnographic study of a community of Wisconsin-based descendants of nineteenth-century Belgian immigrants—a rural, mid-western, White community, secure in its identification with a Belgian ancestry—I ask how the group's members imbue their ancestry with meaning, paying particular attention to the use of GAT alongside other practices (e.g., Walloon classes, genealogical research). In recent work, I explored how the group's members negotiate results in relation to their prior self-identification with Belgian ancestry, and illustrated the ways in which DNA knowledge may appear as a significant dimension in constructing ethnic identities. I also addressed how users may re-establish their Belgian ancestry through genomic results, despite the absence of a

[1] These new networks of relationships may unfold in connection with other fields informed by DNA knowledge, such as environmental epigenomics or archaeogenetics.

"Belgian" category derivable from the tests (Romijn, 2022, 2023). In this chapter, I examine another aspect of GAT users' narratives which has received relatively little scholarly attention, studying how the group's members draw on what they perceive as ancestors' qualities to fashion their identity today. The ways members of this community connect their ancestors' past to their own present-day lives, along intimate or more public ways-to-be-in-the-world, and the powers brought forward in this process of *intricating* themselves (a term borrowed from Callon & Rabeharisoa, 1999) with their Belgian ancestors or Belgian ancestry more broadly, are the focus of this paper.

While exploring an American perspective on entanglement from the point of view of a community established in the United States for two centuries, this chapter examines the flexible pathways through which DNA knowledge ends up filtering into everyday discussions and speeches. This sheds light on the mundanity of genomics through the prism of the political. I explore the ways the group's members time travel between past and present by way of DNA knowledge, to assert belonging but also to afford political interventions, overtly or indirectly. Deep or shallow, the depths of time can be exploited in many ways by social actors (Munn, 1992; Van Tienoven, 2019); in this case, the window onto the past is relatively recent or shallow, encompassing events ranging from up to eight generations ago to the interlocutors' contemporary life experience. Private companies selling GAT are providing their customers with insight into more and more "biogeographic" regions, and these results may be exploited for ideological gains. I show how time, genomics, and the politics of identity can be articulated in the daily life of a group. Throughout this chapter, the terms *identity* and *ethnicity* are understood as being discursively (re)produced and (re)negotiated in the context of social interactions (Barth, 1969; Jenkins, 1994), and not as static entities.

In the first section, I discuss the scholarship on social engagement with DNA knowledge, then present my case study and methods. The main section highlights how, by way of DNA knowledge, the group's members entangle their ancestors' factual or imagined past with their present-day lives and sense of identity. I illustrate this trend through three dimensions: (i) the sense of pride and belonging experienced prior to GAT, reflected through the interpretation of DNA results; (ii) the entanglement of the migratory experiences of Belgian ancestors with the perceived current migratory situation; (iii) interpretations of the Covid-19 pandemic health situation in relation to the health issues experienced by Belgian ancestors and their descendants. Throughout these sub-sections, I show how the group members' narratives comprise political views, including the dynamics of racialization through which cultural and/or biological criteria may be used to build either reciprocity or hierarchy between collective social actors (Grosfoguel, 2004; Doytcheva & Gastaut, 2022).

Powers, Depths of Time, and Social Engagement with DNA Knowledge

Recent literature on the interpretations of DNA results converges on highlighting how the often-presumed risk of genetic essentialism reflects an oversimplified view of how users relate to GAT results. Following the seminal work by Alondra Nelson (2008), a number of studies have documented the ways users and non-users make sense of ancestry genomics (see e.g., Tallbear, 2013; Egorova, 2013; Scully et al., 2013; Gibbon, 2013, 2016; Blanchard et al., 2017; Roth & Ivemark, 2018). Social engagement with DNA knowledge often articulates constructivist and essentialist interpretative schemes. Yulia Egorova (2013) shows that the navigation itself between essentialism and constructivism may empower individuals or communities. As she puts it, the gene is both "an immutable determinant of identification," which is irremediably tied to biological determinism (and thus "part of backward-looking socialities") and, concurrently "a site of agency and resignifiability" (pp. 292–295), a tool which has a liberatory potential. Véran puts it in another way, observing that social actors "'prove themselves' on solid substantial elements produced by whichever combined culturalizations of nature and naturalizations of culture" (2012, p. 253). As a further contribution to these efforts to refine the understanding of social and political engagement with DNA knowledge, I wish to offer another perspective.

> A sense of national or regional identity is a complex achievement, which draws upon multiple sets of relations such as kinship, ethnicity, social role, place and ideas of civic citizenship. Many of these are located in the present or the recent past, such as "where I live", "where I was born", "who my parents are". But some stand in a relationship to an imagined "remote past", stretching back over several hundreds of years. For many people this relationship is limited in scope, consisting perhaps of a diffuse image of the past based on images from popular history books, school education, museum visits or broadcast media sources. However, there are occasions when the relationship to the remote past can become suddenly animated. (Scully et al., 2013, p. 922)

In this excerpt, Scully et al. highlight that the confrontation with DNA knowledge is one occasion when social actors discursively tie a recent or distant past to the present in the way they process identities. These discursive practices are at the center of this chapter. In *La Leçon d'humanité de Gino* (1999), Callon and Rabeharisoa used the term *intrications* (in French) to designate how genetics contributes to making visible, objectifying, and amplifying new networks of relationships between human beings. The authors specifically addressed how these *intrications* connect with moral issues, aimed at causing people to behave in a certain way. In the context of their study, they followed a man who may be the healthy carrier of a gene, whose gene may have been transmitted to his children and grandchildren. They explore the rationale for his refusal to hear "the lesson of genetics" (p. 222), which intricates him with others.[2] Following Callon and Rabeharisoa's pioneering efforts, I use the term *entanglement* to designate in a generic sense the plurality of ways social actors intentionally combine in their discourses three aspects: (i) genomic knowledge and identity narratives (i.e., biological information and life history narratives [Nelson, 2008]); (ii) peculiar temporalities that connect past and present, including factual—or imagined—experiences of their ancestors and their own self-identity in the present; and (iii) the establishing of connections between the self and others. In what follows, I draw a

[2] "If he accepts the theory of the healthy carrier and the gene that is transmitted, then [...] he must concern himself with his children and grandchildren, with whom he finds himself inextricably intertwined" (p. 222; freely translated by the author).

non-exhaustive inventory of how the past may unfold in the present by way of DNA knowledge, and highlight the kinds of power that seem to appear throughout these narratives.

In recent decades, critical scholarship has regularly addressed how DNA knowledge can be overtly or indirectly exploited in a political perspective. These narratives occur within global or national contexts, and take many forms. For instance, genomics can be invoked in nationalist discourses because it provides epistemic weight for a community's ancestral claims that helps reinforce the group's present-day sense of belonging (Burton, 2017). Genetic knowledge "continues to be caught up with nation building and the reproduction of different 'national identities'" (Gibbon et al., 2011, p. 10). GAT can be purchased "for reasons related to racial beliefs and ideologies; for White nationalists, GATs may be a means to validate racial purity" (Roth et al., 2020, p. 10). Conversely, advertisements for GAT can be framed as promoting global cosmopolitan values (the unity of "the global human family") or a universal anti-racist view, as in the *Genographic Project* (Nash, 2015). Genomics may appear as a liberal leverage against discrimination (Reardon, 2011), whereby "progressive White people may seek racial mixture, approaching these tests as 'methods for performing racial harmony and assuaging white guilt'" (Roth et al., 2020, p. 10). "In an international context where 'diversity' has become the key paradigm for power redistribution and legitimation" (Véran, 2012, p. 251), "molecular multiculturalism" (Hinterberger, 2012) can be a vector to promote diversity. However, when applied to the exploration of identities ("Know your DNA, know who you are," as a GAT company puts it), "liberal genomics" (Habermas, 2011) also entails risks. This is particularly true when it ignores the multiple normative horizons that support the search for genomic information on origins. As shown by Nash (2004, p. 15), self-declared, anti-racist geneticists may promote DNA knowledge because it contributes to dispelling the myth of pure or distinct population groups, but also often do so without acknowledging the "ways genetics is inevitably implicated in ideas of personhood, nationhood, cultural belonging, identity and community." Similarly, Tallbear (2013) has shown how the non-discriminatory perspective through which Native Americans' genomic research has been promoted has had the effect of re-essentializing Native Americans on a genomic level. In these various illustrations, DNA knowledge is used to promote a common good (Boltanski & Thévenot, 1991) which should prevail, according to their advocates, over any other goals (e.g., a sense of national specificity; views promoting diversity; the unity of the "human family"; to condemn the consequences of colonization on ancestors and descendants' health; to claim social justice; to uphold the importance of scientific research).

Now, I would like to highlight that engagement with DNA knowledge may also follow more intimate paths than formally public claims. In her highly cited ethnographic account of African American and Black British genomic ancestry test takers, Nelson (2008) drew attention to three important areas: site, scale, and subjectification. Following her, scholars have provided further insights into the dynamics of subjectivation at play when genomics supports a narrative of the past that impacts the present, while entangling history, memory, and identity. For users of GAT, the test results may be perceived as revealing "who I am," a statement equating genetics with existential being (Strand & Källén, 2021, p. 535). Folk taxonomies are numerous when it comes to genetics, including in Western cosmologies where "science never eradicated the firm belief that 's/he is a chip off the old block'" (Niewöhner, 2011, p. 290). This is partly due to lay theories on transmission and the long-held appeal to humans of naturalizing a person's character based on biology (e.g., the genes s/he possesses).

A closer look at more intimate engagement with DNA knowledge reveals that the underlying powers at play are diverse and irreducible to a single political or strategic frame or temporality (Beaudevin, 2013; Gibbon, 2016). Being confronted with genomic knowledge causes individuals to produce discourses on ancestry, health conditions, colonization, and migratory history, and also question their relationship to others/the self in sometimes very intimate ways. For some users, discovering that they possess DNA features of certain groups can provide a sense of increased self-confidence or be perceived as granting access to a higher truth. Meanwhile, scholars have showed the inability of some users to make sense of genomic knowledge in a phenomenological way (Nelson, 2008; Gibbon, 2013). They may describe new DNA knowledge as a prior *feeling* (Scully et al., 2013) or perceive DNA as a *substance* that cannot be taken away from a group's members, who carry a valued biological heritage that can be produced "on demand" to substantiate ancestry claims: a substance that empowers or liberates, irrespective of one's social standing, as illustrated by Egorova (2013, p. 299) in her case study of the untouchable Bene Ephraim community in India. For others, ancestral DNA can be the source of another type of empowerment to the extent that it inscribes individuals into "a fantastically present past" (Sommer, 2012). The gene can appear as a mystical object at times providing a bridge to a perceived heroic ancestral past.

Building on these contributions and drawing on an ethnographic STS study of GAT uses, I would like to shift the emphasis to the discursive practices at play when individuals entangle with DNA knowledge, with a view to understanding how exactly they (i) connect and negotiate the past and present; (ii) navigate between intimate and more public ways-to-be-in-the-world; and (iii) how these narratives deploy new kinds of powers located anywhere between the promotion of identities as relational and situational, and essentialist views of identities as mechanisms of racialization and conservative ideas.

Setting the Methodological Stage: A Study of Wisconsin-Based Descendants of Belgian Immigrants

Thousands of Belgians immigrated for economic reasons to Northeast Wisconsin during the nineteenth century, where, by 1860, they formed a population estimated at around 15,000 individuals (Tinkler, 2019). It has been 170 years since the first Belgian settlement, yet there are still significant numbers of descendants of these immigrants actively self-identifying with a Belgian ancestry or present-day ethnicity[3] in Door, Brown, and Kewaunee Counties, located in the Green Bay area. Studying

[3] While being part of the Belgian community and being a descendant of Belgians who migrated to the region appears to be the same thing (Romijn, 2023), further attention should be dedicated to the dynamics of ethnoracial construction in this group as in other communities of later-generation descendants of European immigrants in the post-millennial period.

this Belgian community affords insights and understanding of how members relate to GAT within a group in which long-lasting community affiliations and a common ancestry are in place. This is important in a scholarly context of scarce qualitative research into the engagement with GAT in groups where networks of inter-knowledge and a common ancestry are established (Blanchard et al., 2017). This case study also differs from other qualitative research on the use of GAT, as northeastern rural Wisconsin is mainly populated by White Americans. Overall, the scholarship on social engagement with DNA knowledge, not necessarily GAT, has far more often addressed the practice in communities racialized as non-White (e.g., Nelson, 2008; Abel & Schroeder, 2020; Tallbear, 2013). In this instance, the practice of GAT—alongside other practices related to Belgian ancestry—also offers a perspective on how a group's members process their own settler identity.

I gained initial access to this community through the Belgian Heritage Center (BHC) (Namur, Wisconsin). Since 2010, the BHC has become a cultural center dedicated to the preservation and celebration of Belgian culture in the region. Following initial contacts with BHC's administrators, I had opportunities to meet numerous community members. Since summer 2020,[4] I have conducted ethnographic observation, in-depth interviews with 37 participants (1.5–3 hours long), and three focus groups (gathering 3–5 participants). Participants were aged between 23 and 87, and all self-identify with a Belgian ancestry. All but three of the 45 participants are currently living in the Green Bay area: 15 are members of and/or involved in the Belgian Heritage Center; 22 are women and 23 are men. The interviews and focus groups took place at the BHC (inside or in the vicinity) and in people's homes or backyards, one took place in a pub, and another two at a diner; seven were carried out online. This data collection process allowed me to reach out to active members of the BHC and gradually to more distant members of the Belgian community who are not involved in the BHC.

[4] In 2020, the project was slowed down by the Covid-19 health crisis and fieldwork initially planned to take place in three stages had to be re-scheduled for a single longer-term research stay from August to November.

In the broadest sense, this study has focused on descendants of Belgian immigrants, and perceptions and practices related to their Belgian ancestry and current identity in Northeast Wisconsin.[5] Participants were recruited because they self-identify as Belgian and I explored the practice of GAT in this larger framework; around half of the participants are GAT users. My ethnographic approach also enabled me to explore why people choose *not* to take a DNA test, or why people might see genomic information as (ir)relevant or even as threatening aspects of their identity or social position. In this chapter, the data highlighted are not only excerpts of conversations centered on how participants address DNA results per se (i.e., population categories, percentages), which have been extensively documented in the last two decades, but also other social aspects as the conversation evolves. This sheds light on what may ordinarily affect how users relate to these results.

My analysis is based on a close reading of transcribed, semi-structured interviews and observational accounts. Stemming from participants' work to produce meaning for DNA knowledge, I examine how their narratives ensure a particular agency in their relation to the world. This pragmatic perspective requires ethnographic methods, bringing the researcher close to actors' experiences.

Entangling Belgians Ancestors' Lives with Present-Day Views on Belonging and Politics

Hardships and Achievements of Belgian Ancestors Used to Define Present-Day Identity

The relationship with what the Belgian community was in the past plays a prominent role in how members seem to process their

[5] This study was funded by the B.A.E.F. (2020) and the F.R.S.-FNRS (2021–2024). It has received ethical approval from the Institutional Review Board of the University of Wisconsin–Madison.

self-identification with a Belgian identity today. When I ask Sally,[6] 40 years old and an active member of the BHC, why members of the community mostly seem surprised when I tell them I am interested in the present-day state of the Belgian community, she confidently replies,

> We are not thinking of ourselves as being of as much importance as the ones who came before us, I guess. That they had the harder time, laid the groundwork for all of us. All the things that they did, this was all virgin forests. And they came here to farm… I can't imagine what they thought when they got here. You know. And all the trials they went through, with the Fire and just daily life. The struggles they went through before we got here. We have all these things because of them. So, we don't feel like… a lot people don't think like our stories are as important to us as the ancestors' story.

As illustrated by Sally's comments, members of the group consistently refer to the life and lived experiences of their ancestors when addressing the relationship they have with the category "Belgian," usually connecting with the hardships and achievements of Belgian ancestors as a means to process their present-day identity. In my discussion with Tracey, a businesswoman in her fifties, there is a similar omnipresence of her ancestors' past in the way she processes her identity. Tracey says she feels "thankful for what [her] ancestors accomplished" and "how they had to survive." She goes on to say, "It tells you a little bit of what you are made of. I think it makes you feel kind of proud of that. You know, maybe there is a little of that in me too." Interestingly, GAT users usually make sense of their DNA results in a similar fashion. Tammy's case is enlightening. The 60-something years old claims to have always been particularly aware of her Belgian ancestry and adds that she decided to take a test despite this prior awareness, to "prove it":

> Having proof but not only on paper, but chemically that I am from Belgian descent. Because they had to be tough people, to be strong and smart to be able to come to this rocky territory and make a life. With all the trees here,

[6] All data have been de-identified, the names and other personal information of the participants changed.

> to be able to clear the land, and make a farm; there is a great pride in knowing that I came from stock that was determined and could withstand plagues and natural disaster and difficulty in handling people who were already here. That makes me proud to know that I am part of that. That that blood floods in my veins.

In Tammy's narrative, a power lies in the feeling of knowing herself to be biologically connected to ancestors whom she perceives as determined, resilient, and capable of enduring adversity. The test is about "proving" such an ontology through DNA knowledge ("that blood floods in my veins"). It is about asserting a common membership through "hard facts," which might appear "more resourceful than, say, the situational concept of 'self-affirmation'" (Véran, 2012, p. 251). The feeling of intimate certainty—present beforehand—of being "Belgian" is discursively tied to another logic in which Tammy needs attestable evidence, thus justifying the use of the test. It is also about establishing a common capacity between the ancestors and herself "to survive the Great Fire [a forest fire that destroyed Belgian settlements in 1871] also took a lot of courage and intelligence to be able to survive, they had to survive, or I wouldn't be here." She says she is here today because they were courageous and intelligent, qualities she seems to understand as being passed on from her ancestors' generation to her own.

The oblique reference to the "difficulty in handling people who were already here" should also be noted. The "people" mentioned might be the neighboring settler communities (e.g., Scandinavians, Germans, Irish) established in the region before the Belgian settlements, or it might be a reference to members of First Nations. It is known, moreover, that the *ethnic revival* or the *resurgence of ethnicity* (Jacobson, 2006) among White communities in the United States went with ethnic claims embracing the borders of racial domination in the country (Gans, 1979; Waters, 1990; Alba, 1990). I observe features of this when the pride in, and revalorization[7] of the local Belgian identity is presented together with reference to a feeling of nostalgia for a past, considered bygone, of rural simplicity. A

[7] In a paper in progress, I explore the evolution of perceptions of the discrimination experienced by members of the Belgian ethnic group.

culture of nostalgia emerged in White communities between the 1960s and the 1970s in the United States, in response to, among other things, a context of identity dislocation that accompanied the defense of racial equality (Torkelson & Hartmann, 2020).

Leanne's case epitomizes how the group's members connect past and present. I meet this retired nurse outside the BHC as she completes a tour of the historic campus with visitors. After showing me the oldest building and pointing out her grandmother's portrait hanging there, we sit outside in the shade of an impressive, centuries-old tree, its branches reaching over the graves of the Namur-based Belgian community (Door County). Leanne proudly describes herself as "the historian of the community," adding she is regularly contacted for presentations, the distinctive feature of which is her use of the personal pronoun "I," as she adopts the point of view of her own ancestors: "I become my great great-grandmother and I tell the story through her eyes as I would think." One of her flagship presentations focuses on a traumatic event in the community: the Peshtigo Fire. In 1871, a huge wildfire destroyed part of the area. In the midst of our discussion, she shares long passages of her presentation for the Fire's 149-year anniversary. Remarkably, she navigates from descriptive elements of the 1871 event to a first person narrative of the experience her great-great-grandmother had of the event. Her position evolves in a remarkable way from "I, Leanne's great-great-grandmother" to "I, Leanne engaged in a discussion about Belgian ethnicity." I ask her what disseminating this memory through personified narratives using "I" confers on her:

> It once again emphasizes what they went through. I will tell you, we still have the family farm, not the buildings. We sold the buildings maybe about 10 years ago. And it was really hard for me to do that because I kept thinking: they put their hearts and souls into these buildings, and who am I to sell them? And I kept thinking when I close the door the last time, I thought one of my relatives is gonna come out of the wood and say, "What the heck are..." [laughs]. I and my brother both felt that. It was really eerie.

Leanne feels indebted to her ancestors and especially sensitive to the hardships they endured. She also stresses how she believes her personality arose "from the things that they [ancestors] did because they had to make their decisions. They had to forge forward even when they didn't want to." She establishes a strong connection between her ancestors' characters and her own.

While our conversation evolves toward the stereotypes associated with the Belgians in the region, Alan, a 37-year-old GAT user, tells me that what for him "really encapsulates the Wisconsin Belgian experience is like a sense of flinty, almost medieval grit, as a way to survive in the world." Again, in his own way, Alan provides another illustration of presenting his sense of belonging to the Belgian category as an experience that transcends past and present. Not remote from a culturalist perspective, Alan says this practice epitomizes a psychological trait—shared by members of the community although rarely discussed: "To me, there is this thing, that flintiness, underneath the jokiness, somewhere in us, deep-seated, and I think it comes out sometimes." He then explains that this trait is to him the result of what the ancestors endured; they "had to go through to this tiny little bottleneck of time where things were extremely precarious, like primally brutal, kind of." Hardships would have somehow left their mark: "I think it's still kind of there, still sitting in Belgian people, in like a lot of what we do."

The Migratory Experience of Belgian Ancestors as an Oblique Reference to the Current Issue of Immigration

By way of ancestry genomics, the group's members frequently—overtly or indirectly—connect either the real or imagined migratory experiences of their Belgian ancestors, or their perceptions of their genomic heritage, with the current migratory situation in the United States. GAT users among this group regularly search for signs of their preexisting, valorized Belgian identity (Romijn, 2023). Most users do not specifically express a strong desire to discover something different or considered more exciting; there are, however, cases where GAT users appropriate results in a way that seems to add "color" to an otherwise "bland" White identity. Tammy seems delighted to observe the diversity of her origins as she refers to a

Belgian traditional dish: "It is amazing to think how much of a booyah [a traditional Belgian soup made of diverse vegetables and meat] it becomes [laughs]. I am a pot of booyah. Because I am made up of so much that makes me who I am. And the core is still the hardworking ethical people that I come from. That's still very important to me. It's still part of me today." Tammy points out the varied origins of her ancestors to present her views on her genomic make up as valuably diverse. Showing me the population categories (i.e., "Scandinavian," "French," "Scottish") appearing in her test results, Tammy demonstrates a remarkable ease in concomitantly promoting genetic diversity and stressing the centrality of her "Belgian" heritage.

Some GAT users insist on the degree of "purity" of their "Belgianness," which they expect to appear in the genomic results. There appears to be a form of pride in being the carrier of the genomic stock of Belgian immigrants who decided to settle in this region of Wisconsin. It is also important to identify whether the genetic endowment of the family might include other origins beyond the known recent Belgian ancestry. Some users—sometimes the same ones—also celebrate the diversity of origins their GAT results indicate. Others pick up on the less tangible fact that they are the carrier of genes transmitted through thousands of years and irreducible to Belgium. Some value this observation and others are dissatisfied. They seem to experience a form of estrangement from the self (Ruiz-Junco, 2011) that lives alongside the perspective of realizing one becomes "more and more of everything" over generations. In these narratives, the powers at play seem to range along a continuum that includes the importance of claiming a so-called pure "Belgianness" as well as the celebration of diversity.

Through a unique path, which ties into a liberal conception of migration, Dora establishes a common ancestry with her Turkish daughter-in-law. She notices, after her results are refined (percentages associated with population categories are updated regularly by some GAT companies), that the added "10% Italian"—and an assumed although intricate common ancestry dating back to the time of the Roman Empire through which she had established the connection to her son's wife—is no longer available. This dynamism—which tempers a deterministic understanding of genomics and compels users to consider the probabilistic nature of their test results (Romijn, 2022)—amounts, for Dora, to a loss of a power. With the disappearance of the result, one can glimpse the valued "bit of something nonwhite" or the "cultural cachet" (Roth & Ivemark, 2018, p. 178) that supplemented her Belgian identity. One also realizes the inclination of these tests to promote symbolic identities that are "costless," and would "exist separately from the social and structural consequences with which society has imbued it" (Roth & Ivemark, 2018, p. 178) A bit later, Dora confides that her children and grandchildren "are 'mutts' because they have so many different heritages. My husband and I are someway different. And then of course my son married a girl from Turkey. So, his child is really mixed up you know, heritage. But I think they think it's interesting and they know that I identify with the Belgians."

The current political divisions are particularly notable in the State of Wisconsin, a key state in the 2020 and 2024 presidential elections. In most interviews, participants establish parallels between the experience of their own Belgian ancestors migrating to the United States and today's political situation. As Tracey tells me, "I think that the topic of people who immigrated here, in the past [this to make a distinction with immigration in today's United States], is a safe one," but getting too close to the current immigration policy in the country is something she says I should steer away from. Tracey may not seem to perceive it as such, but mentioning the migratory experience of Belgian ancestors often serves as an indirect means of explicitly positioning oneself on the political spectrum. She continues by saying, "I personally am appalled by the attitude that our government projects toward the immigrants since we are technically all immigrants going back to 18–1900. [...]. I think this country welcomed my family so I don't know why we wouldn't continue welcoming families from whatever."

Along with these narratives, one can glimpse respondents' ideological stance. Janice, for instance, insists on her great-great-grandfather's achievements, telling me he was an active member of the county board and the school board, launched a new church, and at the same time managed a farm. She says her ancestor and other immigrants in the region spoke different languages and therefore had to learn English to communicate. She then adds: "Pardon me if I get a little bit... [she does not finish her sentence; I understand she is warning me with a view to contain the tension in addressing a politically sensitive issue]. Now people don't want to learn English, they like to keep their Spanish or whatever." She then refers to one of her uncles, who lived to be 97, and starts quoting him: "'Grandpa came over, pa came over. He learned English. He learned to work. He set up a farm and worked. Now people come over and expect everything to be handed to them. And they don't wanna speak English.' He said, 'If you come to this country, that's the language you should speak.'" Janice speaks through her uncle's mouth with a view to formulating a political critique, which is based on racialized patterns.

One can see an articulation between, on the one hand, the expression of a valued sense of belonging, sustained by intimate attachments, and, on the other, a political positioning. Explicitly, Janice relies on an "us against them" logic, "in which the ethnic outgroup was expected but deliberately failed to adapt to the dominant in-group culture" (Stevens & Görgöz, 2010, p. 1357): on one side, the members of the group—represented by the Belgian ancestors and implicitly their descendants—who made the effort required for successful integration; on the other side, the members of another group, with fuzzy but also encompassing contours, comprising foreigners who abuse the system. This is a narrative in which ancestral hardships, giving way to successful assimilation, are juxtaposed with oblique references to "problematic" racialized or migrant groups in the United States (e.g., "they like to keep their Spanish," here in reference to Latinx migrants). This narrative depicting a successful integration—or one perceived as such—of white ethnic groups into mainstream American society, serves "as the measuring stick against which other groups' (lack of) successes were evaluated" (Alba, 1990, pp. 315–319, cited by Torkelson & Hartmann, 2010). At the same time, a false equivalence is created between the experiences of migrants and migrant-descent groups that are racialized as White versus non-White, undermining the differential impacts of structural and systemic forms of racism on differently racialized groups: a false equivalence that has been used by European-descent groups to distance themselves from charges of racism, and, more broadly, inherited privileges in the United States (Jacobson, 2006).

In these illustrations, we can see that the migratory experience of Belgian ancestors and the hardships that they would have experienced are summoned into the present with a view either to address the current migratory situation, or to postulate biological or psychological traits inherited from events that occurred in that period or, again, to regret with a certain nostalgia a situation that has changed for recent generations in the Belgian community.

Connecting the Public Health Situation to Ancestors' Health Issues

In 2020, when the first stage of the research took place, the Covid-19 public health crisis was itself strongly politicized, as illustrated by several potential informants refusing an interview if required to wear a mask. In the following excerpt, the conduct of people toward the public health situation may be used to put into perspective how tough people were in the past, compared with today.

> Tracey: Comparing with all the people bitching about wearing a mask, imagine what would these people do if they were set in the 1860s. With all the cholera and typhus, it was really a struggle. We complain all the time. We have things very easy when compared to all the things they had to get through, and even after that period of time. There was the Great Depression in this country. That wiped out people all again. There was that constant, you know, struggle to make a better life. That's what I think of: these were tough people. They couldn't have been afraid of much or they wouldn't have survived.

Tracey tells me how she thinks, every now and then, about how today people are "touchy," "spoiled," or "not able to handle little setbacks" compared to her ancestors. The public health situation has been a trigger for such thinking. "They are times where I think of what our ancestors would say about all you have to do to avoid serious illness is wear a mask and be a little bit careful. And [some people] are just not cooperating. Every now and then something just strikes me as, you know, just contrary to what it took to be productive back in those days." The reference to ancestors is here used to criticize an in-group inclination toward complaining. This critique is expressed from a particular position as Tracey is perceived as a successful businesswoman, highly regarded by members of the community.

These parallels, between ancestors' hardships and the present-day health situation, also unfold explicitly when backed by genetics, as a means through which ancestors' health issues are transmitted to present-day descendants. I discuss the common practice among community members of dedicating time and research constructing narratives about their dead ancestors with Margareth (a respected member of the Belgian community known for her ability to speak Walloon). She tells me it is something important for her. Lately, what interests her most, is the health of her ancestors and how diseases were transmitted from one generation to another. Margareth tells me that she has been matching up information on people whom she knows are connected to the family Dumonts who have developed Celiac disease. She also says it is important to pay attention to the prevalence of diseases in related people, because doctors ignore this aspect most of the time. In a similar vein, Charles, passionate about genealogy and user of several DNA tests, explains he was able to figure out the disease suffered by one of his ancestors suffered by comparing information from different generations.

As genetics is rooted in biology, some group members establish links between past and present, biological health and ancestry, self and others. Members of this community are trying to understand how their ancestry is relevant to their health. Tammy's case is representative of this dynamic. She spontaneously expresses her interest in exploring further links between ancestry, genetics, and health. "First, my dad's mother had uterine cancer and so did I. And her sister had breast cancer and so did I. So, that health factor: am I genetically predisposed to something because of my ancestry? Am I more predisposed because I am of Belgian descent?" She raises many questions ("Where did it begin?" "What caused it?" "Where did it come from?" "Is it just genetic?"), and seems convinced there must be a way to trace the origin of her cancer. Her desire to know more points to an instrumental logic, a way to prepare "ourselves or make lifestyle changes that would help us so that we can live up our lives." "What red flags does that put up for you to pay attention?" "I wanted to know if I was passing that possibility on to my children and then to my grandchildren. [...] If you have it, then you have knowledge. If you have knowledge, you have power. So, you can make decisions." Again, ancestors' views or health conditions are mobilized to address current issues, either at the intersection of intimacy and strategic action ("to be prepared") as for Tammy, or more politically, as in people's conduct toward the public health situation.

Conclusions: The Power of the Dead Ancestors

This chapter contributes to refining understanding of how the exposure to DNA knowledge causes individuals to time travel discursively between their ancestors' life events (especially their perceived hardships and achievements) and their own present-day experiences. Several lessons can be drawn from this study.

In the group studied, these entanglements translate to differing degrees of empowerment and strategic position, which include feeling more resilient toward current societal issues, understanding their ethnic status more fully, and dealing better with disease and pandemics. Commemorating their ancestors serves their present through diverse mandates: "pietistic homages to ancestors; as models of exemplary lives to inspire the living generation; as cautionary tales for the living; as spiritual bridges to the past; as sources of entertainment, not unlike historical fiction" (Lambert, 2006, p. 332). It is also very much about having a story that affirms a desired identity (Nelson, 2008; Romijn, 2023). Although there is relevance in all that, there is more in the powers at play when members of this group establish repeating ties between experiences factually or supposedly lived by their ancestors, ancestral traits, and forms of heritability between ancestors' personality traits and their own today. These narratives entail the acquisition of power by means of the appropriation of virtues attributed to the ancestors. The group's members draw on their ancestors' perceived qualities to fashion their own identities today. As brilliantly demonstrated by Laqueur (2018), the dead and their avatars—ancestors, memory, history—"are individually and collectively, the temporal extension of our present: they are us—our symbolic world—as we imagine them in the long term" (p. 134). This temporal extension of our present is particularly evident in the case studied. The group's members seem to need their deceased ancestors in order to live better today.

These entanglements between the past as experienced by the ancestors, and the present, involve the transmission of some features or traits of character either by way of the cultural (e.g., "a medieval grit" which has left its mark on people's psychology) or—the usual case under discussion here—by way of explicit references and interpretation given to DNA knowledge (e.g., "that blood floods into my veins"). Most participants are well aware of the number of their ancestors associated with Belgium—even though up to eight generations separate them from the migratory experience—claiming to be "100% Belgian," "a full-blooded Belgian," "a pure Belgian," "a 7/8th Belgian." They foster a pride in embodying a biological heritage associated with the category Belgium. The source of pride, and the associated empowering dynamic, seems mostly to reside in the fact of *being the descendant of* Belgian ancestors. The ways they interpret the DNA results often echo these narratives. The appropriation of the ancestors' virtues seems facilitated by dwelling on DNA knowledge, which in a way seems to suspend the inevitable severance between the life and the body (Laqueur, 2018, p. 121) of ancestors. As Tammy made clear, it is important for her to be sure of her claim that their "blood floods in my veins." The group's members want to make known that their ancestors suffered but were also resilient in the face of the adversities they experienced. They express extraordinary interpretations of the past that share features with heroic narratives of heroes they want to resemble. It is not uncommon in the literature on interpretations of GAT results to find users racialized as non-White, asserting historical prejudices, sufferings, or exploitation while reflecting on their DNA. In this case, we see that this view emerges in a White American minority in terms of European national origin.

A second connected lesson of this chapter is to show how the articulation between DNA knowledge and a collective sense of belonging may also support political interventions. It has been long demonstrated that GAT users engage with DNA knowledge through affiliative self-fashioning, interpreting results in the context of their "genealogical aspirations" (Nelson, 2008). We also knew that genomics can be exploited for formal political goals (e.g., Burton, 2017). What was less evident is that DNA knowledge can also be a vector for expressing ordinary political positioning. Genomic information can be integrated into aspects of contemporary society as trivial or banal as vectoring or mediating political views. The relatively recent—or shallow—past of the community, with the first Belgian settlements dating back to the 1850s, provides legitimacy for enunciating and positioning on politics in the present. Here, a genomic technology is engaged and made personally relevant for political positioning. Interestingly, in this case, the use of DNA as a vector for expressing political views is equally located in both directions of the political spectrum. Some participants are committed to the fight against racism as a structure of domination, others may be White nationalists.

Attention must be directed to how these narratives are connected to the actual political climate in the United States. Is the recognition of past and present-day sufferings of racialized, historically discriminated-against communities in the United States producing such narratives among this White American group? Is there a desire to downplay the validity of this recognition? Following Waters (1990, cited by Carlson, 2020, p. 831), is there any correlation between, on the one hand, the efforts of members of this White community to equate their Belgian ethnicity "with the ethnicity of an American minority," whose ancestors would have faced discrimination and, on the other, a reluctance or a lesser likelihood "to accept that minorities [such as Native-Americans; African-Americans; Latinx] faced structural violence"? While further research is needed, I shall bring an element of response by emphasizing that for some group members, the construction of a Belgian identity is discursively produced through borrowing characteristic features of hegemonic Whiteness and racialization. One can observe it in the expression of nostalgia for a bygone past, which comes with the stigmatization of racialized groups (Alba, 1990) and fears connected with "invading" migrants and changing social values: a cornerstone for the construction of White ethnonationalist identity (Ahmed, 2004). The ever-present need to recall the hardships and accomplishments of Belgian ancestors may also be viewed against the background of a renewed pride in being Belgian in this region, where, for decades, Belgians are said to have suffered discrimination and mockery. At the same time, the community's pride in its ancestors can be seen as entangled with tensions and ambivalences regarding its racialization as a White community in the United States. This appears in the fact that racial discrimination is an issue largely confined to the margins, and referenced obliquely[8] (e.g., the reference to the "difficulty in handling people who were already here"; the invitation to stay away from the topic of "politics"; the tendency to promote a primordialist view on Belgian ethnicity; or the great flexibility in interpretating the content of Belgian identity, such as stereotypical phenotypic traits). More recent observations indicate that for some group members the ethnic sense of belonging to the category of Belgian can become more explicitly racialized. For instance, in mid-August 2023, on the occasion of an annual event in the community, a man was wearing a t-shirt proclaiming, "Belgian lives matter." One could see here an explicit use of a sense of "Belgianity" as having features of a specific racialized nature.

Acknowledgments I would like to express my gratitude to participants in this study. I also want to thank the anonymous reviewers of my chapter and editors of this book for their insightful comments.

References

Abel, S., & Schroeder, H. (2020). From country marks to DNA markers. The genomic turn in the reconstruction of African identities. *Current Anthropology, 61*(S22), S198–S209. https://doi.org/10.1086/709550

Ahmed, S. (2004). Affective economies. *Social Text, 22*(2), 117–139.

Alba, R. D. (1990). *Ethnic identity: The transformation of White America*. Yale University Press.

Barth, F. (ed.). (1998 [1969]) *Ethnic groups and boundaries: The social organization of cultural difference*. Waveland Press.

Beaudevin, C. (2013). Old diseases and contemporary crisis. Inherited blood disorders in the Sultanate of Oman. *Anthropology & Medicine, 20*(2), 175–189.

Blanchard, J. W., Tallbull, G., Wolpert, C., Powell, J., Foster, M. W., & Royal, C. (2017). Barriers and strategies related to qualitative research on genetic

[8] In a forthcoming publication, I reflect on my positionality and its influence on how participants produce ethno-racial identities during the research process.

ancestry testing in indigenous communities. *Journal of Empirical Research on Human Research Ethics, 12*(3), 169–179. https://doi.org/10.1177/1556264617704542

Boltanski, L., & Thévenot, L. (1991). *De la justification. Les économies de la grandeur*. Gallimard.

Burton, E. K. (2017). *Genetic nationalism: Scientific communities and ethnic mythmaking in the Middle East*. PhD Thesis. Harvard University.

Callon, M., & Rabeharisoa, V. (1999). La leçon d'humanité de Gino. *Réseaux, 17*(95), 197–233.

Carlson, H. (2020). The route to your roots: New ethnic symbols in the age of the genome. *Nations and Nationalism, 26*(4), 826–844. https://doi.org/10.1111/nana.12642

Doytcheva, M., & Gastaut, Y. (2022). Race, Racismes, Racialisations: Enjeux conceptuels et méthodologiques, perspectives critiques. *Emulations. Revue de sciences sociales, 42*, 7–30.

Egorova, Y. (2013). The substance that empowers? DNA in South Asia. *Contemporary South Asia, 21*(3), 291–303. https://doi.org/10.1080/09584935.2013.826627

Gans, H. J. (1979). Symbolic ethnicity: The future of ethnic groups and cultures in America. *Ethnic and Racial Studies, 2*(1), 1–20. https://doi.org/10.1080/01419870.1979.9993248

Gibbon, S. (2013). Ancestry, temporality, and potentially engaging cancer genetics in Southern Brazil. *Current Anthropology, 54*(S7), S107–S117.

Gibbon, S. (2016). Translating population difference: The use and re-use of genetic ancestry in Brazilian cancer genetics. *Medical Anthropology, 35*(1), 58–72. https://doi.org/10.1080/01459740.2015.1091818

Gibbon, S., Ventura-Santos, R., & Sans, M. (2011). *Racial identities, genetic ancestry, and health in South America: Argentina, Brazil, Colombia and Uruguay*. Springer.

Grosfoguel, R. (2004). Race and ethnicity or racialized ethnicities? Identities within global coloniality. *Ethnicities, 4*(3), 315–336.

Habermas, J. (2011). *L'avenir de la nature humaine. Vers un eugénisme libéral?* Gallimard.

Hauskeller, C., Sturdy, S., & Tutton, R. (2013). Genetics and the sociology of identity. *Sociology, 47*(5), 875–886. https://doi.org/10.1177/0038038513505011

Hinterberger, A. (2012). Categorization, census, and multiculturalism: Molecular politics and the material of nation. In K. Wailoo, A. Nelson, &

C. Lee (Eds.), *Genetics and the unsettled past: The collision of DNA, race, and history* (pp. 204–224). Rutgers University Press.
Jacobson, M. F. (2006). *Roots too: White ethnic revival in post-civil rights America*. Harvard University Press.
Jenkins, R. (1994). Rethinking ethnicity: Identity, categorization and power. *Ethnic and Racial Studies, 17*(2), 197–223.
Kramer, A.-M. (2015). The genomic imaginary: Genealogical heritage and the shaping of bioconvergent identities. *Media Tropes, 5*(1), 80–104.
Lambert, R. D. (2006). Descriptive, narrative, and experiential pathways to symbolic ancestors. *Mortality, 11*(4), 317–335. https://doi.org/10.1080/13576270600945584
Laqueur, T. W. (2018). *Le travail des morts. Une histoire culturelle des dépouilles mortelles*. Gallimard.
Munn, N. D. (1992). The cultural anthropology of time: A critical essay. *Annual Review of Anthropology, 21*, 93–123. https://doi.org/10.1146/annurev.an.21.100192.000521
Nash, C. (2004). Genetic kinship. *Cultural Studies, 18*(1), 1–33. https://doi.org/10.1080/0950238042000181593
Nash, C. (2015). *Genetic geographies: The trouble with ancestry*. Minnesota University Press.
Nelkin, D., & Lindee, M. S. (2004). *The DNA mystique: The gene as a cultural icon*. University of Michigan Press.
Nelson, A. (2008). Bio science: Genetic genealogy testing and the pursuit of African ancestry. *Social Studies of Science., 38*(5), 759–783. https://doi.org/10.1177/0306312708091929
Niewöhner, J. (2011). Epigenetics: Embedded bodies and the molecularisation ofbiography and milieu. *BioSocieties, 6*(3), 279–298. https://doi.org/10.1057/biosoc.2011.4
Nordgren, A., & Juengst, E. T. (2009). Can genomics tell me who I am? Essentialistic rhetoric in direct-to-consumer DNA testing. *New Genetics and Society, 28*(2), 157–172. https://doi.org/10.1080/14636770902901595
Reardon, J. (2011). The democratic anti-racist genome? Technoscience at the limits of liberalism. *Science as Culture, 21*(1), 25–47. https://doi.org/10.1080/09505431.2011.565322
Romijn, F. (2022). We are all cousins. Belgian ancestry and genomic testing in a close-knit community in Northeastern Wisconsin. *New Genetics and Society, 41*(4), 312–333. https://doi.org/10.1080/14636778.2022.2134101

Romijn, F. C. (2023). Negotiating Belgian identity in Wisconsin through ancestry genomics. *Science as Culture, 32*(2), 240–265. https://doi.org/10.1080/09505431.2023.2178401

Roth, W. D., & Ivemark, B. (2018). Genetic options: The impact of genetic ancestry testing on consumers' racial and ethnic identities. *American Journal of Sociology, 124*(1), 150–184. https://doi.org/10.1086/697487

Roth, W. D., Yaylacı, Ş., Jaffe, K., & Richardson, L. (2020). Do genetic ancestry tests increase racial essentialism? Findings from a randomized controlled trial. *PLoS ONE, 15*(1), e0227399. https://doi.org/10.1371/journal.pone.0227399

Ruiz-Junco, N. (2011). Stranger to you and stranger to myself? Theorizing self-estrangement. In H. F. Dahms (Ed.), *The diversity of social theories* (pp. 147–168). Emerald Group Publishing Limited.

Scully, M., King, T., & Brown, S. D. (2013). Remediating Viking origins: Genetic code as archival memory of the remote past. *Sociology, 47*(5), 921–938. https://doi.org/10.1177/0038038513493538

Sommer, M. (2012). Do you have Celtic, Jewish or Germanic roots? Applied Swiss history before and after DNA. In K. Schramm, D. Skinner, & R. Rottenburg (Eds.), *Identity politics and the new genetics: Re/ creating categories of difference and belonging* (pp. 116–140). Berghahn Books.

Stevens, P. A. J., & Görgöz, R. (2010). Exploring the importance of institutional contexts for the development of ethnic stereotypes: A comparison of schools in Belgium and England. *Ethnic and Racial Studies, 33*(8), 1350–1371. https://doi.org/10.1080/01419870903219243

Strand, D., & Källén, A. (2021). I am a Viking! DNA, popular culture and the construction of geneticized identity. *New Genetics and Society, 40*(4), 520–540. https://doi.org/10.1080/14636778.2020.1868988

Tallbear, K. (2013). *Native American DNA*. University of Minnesota Press.

Tinkler, J. L. (2019). *Determinants of ethnic retention as seen through Walloon immigrants to Wisconsin*. PhD Thesis. The University of Texas at Arlington.

Torkelson, J., & Hartmann, D. (2010). White ethnicity in twenty-first-century America: Findings from a new national survey. *Ethnic and Racial Studies, 33*(8), 1310–1331. https://doi.org/10.1080/01419870903434495

Torkelson, J., & Hartmann, D. (2020). The racialization of ethnicity: The new face of white ethnicity in postmillennial America. *Sociology of Race and Ethnicity, 6*(3), 302–318. https://doi.org/10.1177/2332649219892621

Van Tienoven, T. P. (2019). A multitude of natural, social and individual time. *Time & Society, 28*(3), 971–994. https://doi.org/10.1177/0961463x17752554
Véran, J.-F. (2012). Old bones, new powers. *Current Anthropology, 53*, 246–255. https://doi.org/10.1086/662384
Waters, M. C. (1990). *Ethnic options: Choosing identities in America.* University of California Press.

11

What Are We to Each Other? The Limits of DNA in the Search for Origins in Donor Conception

Anaïs Martin

Introduction

Audrey was born in 1980s France. As a teenager, her mother bluntly revealed during an argument that she had been conceived via sperm donation. Audrey immediately wanted to know more about the donor, but strict anonymity rules at the time in France meant the clinic refused to share any information. For years, she let the matter rest, although it resurfaced at various moments of her life, like when she became a mother. It was only 20 years after learning the facts of her conception—about a year before I met her—that she discovered direct-to-consumer (DTC) DNA testing and decided to search for her donor.

She ordered a kit online, sent off her saliva sample, anxious about whether it would arrive safely. A few weeks later, just back from an outing with her partner, she sat at her bedroom desk to read the results posted on the genetic genealogy company's website.

A. Martin (✉)
LISST-Centre d'anthropologie sociale, CNRS, Toulouse, France
e-mail: anais.martin@cnrs.fr

V. Argudo-Portal et al. (eds.), *Mundane Genomics*,
https://doi.org/10.1007/978-981-95-4550-6_11

> So, I see origins that I wasn't expecting at all. [...] I was expecting things like a bit Croatian, a bit Bulgarian, all that, Bosnian, [...] things that I knew. [...] But the test only showed 47% Portuguese ethnic origins. And at that point, I thought, "Well, it must come from there. [...] So, he [the donor] has really strong Portuguese origins, at least very close genealogically." Then suddenly, we move on to the DNA matches [...]. Tiffany ... So, I see that she's not quite a year younger than me, with a 25% DNA match, "aunt" or "half-sister". Given her age, being an aunt would be complicated, so I understood that she was actually my half-sister. [...] then there's also the question of whether she is my half-sister born from a donation? Is she my half-sister, his daughter? At that moment, there is doubt about everything, we know nothing, we are in absolute uncertainty.

Audrey shares several experiences common among other donor-conceived people: being told the circumstances of her conception, realising that legal access to donor information was impossible, moving forward with her life, later discovering new tools that allowed her to circumvent donor anonymity, and eventually uncovering information about her origins and making new connections. In just a few sentences, she shows how making sense of her DNA results required a series of logical operations. Yet she also reveals the ambivalence of the search for origins in donor conception; while she has learned more about the donor and her connections through him than ever before, she stresses the continuing uncertainty of her situation.

DTC DNA-testing emerged in the 2000s and became widely accessible in the 2010s (Darroch & Smith, 2021). These technologies dramatically transformed the landscape for donor-conceived people. For decades, third-party reproduction was kept secret to avoid disrupting family relationships (Frith et al., 2017), and gamete donation was anonymous in most countries (Blyth & Frith, 2009). However, the growing focus on the harmful effects of family secrets in the 1980–90s (Smart, 2009) increasingly led fertility professionals to encourage parents to be open with their donor-conceived child(ren) (Gilman & Nordqvist, 2018). Simultaneously, some donor-conceived people began advocating for access to information about their origins—that is, identifying or non-identifying data about the gamete provider.

Some jurisdictions amended their regulations—such as Sweden in 1984, the Australian state of Victoria in 1998, and the United Kingdom (UK) in 2004—while others, like France, continued to uphold anonymity (Frith, 2001). With the rise of DTC DNA testing, some individuals learned that they were donor-conceived unexpectedly, and others, like Audrey, used the tests to bypass legal barriers (Crawshaw, 2018; Gilman et al., 2024). As a result, many now argue that donor anonymity can no longer be guaranteed (Harper et al., 2016). In this context, more countries, including France in 2021, have reformed their laws to allow donor-conceived people access to information about their origins.

Despite legal reforms, DTC DNA tests remain central to donor-conception experiences. For those born under anonymous systems, clinic records often provide little to no donor information, as records may be destroyed, lost, or legally inaccessible (Haimes, 1990; Klotz, 2014; Martin, 2022). In systems like in the United States (US), parents typically receive donor profiles to select the gametes they want to use. But in countries such as France and historically in the UK, medical professionals (used to) choose donors without disclosing any details to parents. In these cases, the body itself holds the only "trace" of the donor (van den Akker et al., 2015; Courduriès & Gourarier, 2020; Ginzburg, 1980).

At first glance, DNA testing—a "masterkey that unlocks many secrets" (Nelson, 2016, p. 8)—may seem like a straightforward way to bypass the effects of anonymity. However, discussing the independent British genetic database of the Donor Conceived Register (see below), van den Akker and colleagues note that "DNA testing to identify genetic relationships can produce complex results that require scientific and statistical interpretation" (2015, p. 117).

Drawing on an ethnographic study with French and English sperm donor-conceived adults, this chapter explores the ambivalence and uncertainty characterising the search for origins through DTC DNA testing. Focusing on the use of mirror-treeing to identify donors, the categorisation of connections detected through testing, and the negotiations surrounding these categories, it describes the substantial work involved in origin searching. In doing so, the chapter shows how the knowledge produced challenges to the common association between DNA, procreation, and kinship.

This chapter also contributes to discussions about the *mundanity* of genomics by highlighting the culturally, socially, and historically situated framework embedded in genetic genealogy platforms. It underscores the complexities involved in the "lay" use of DNA, nuancing the pervasiveness of genomics in contemporary societies.

Origins, Procreation, and Kinship

Origins have been increasingly recognised in many countries, as well as in international law—first in the context of adoption, and more recently in third-party reproduction. They are conceptualised as an individual right (the right to know), offering a more complete sense of one's identity (Baumann, 2022; Martial et al., 2021). Several studies have highlighted the positive effects of gaining new personal information about one's health, identity, and history (Canzi et al., 2019). However, the questions raised by origins in the social sciences extend beyond the right to information, as they also involve sets of relations, within and beyond donor-conceived families. Research has shown that access to origins may lead to the formation of new and sometimes longstanding relationships with the donor and other individuals conceived with his gametes (Frith et al., 2018; Hertz & Nelson, 2019; Jadva et al., 2010). The unprecedented nature of these connections has been emphasised; they are often described as falling outside existing frames of reference, making those involved pioneers in uncharted territory (Blyth, 2012a, 2012b; Klotz, 2016). As such, origins challenge Western understanding of kinship and identity (Martial et al., 2021), shedding new light on the meanings attributed to procreation and the resulting connections (Courduriès, 2016; Martin, 2022).

Some scholars, particularly in adoption studies, have warned against the essentialising potential of origin searches in shaping understandings of kinship and identity. Kaja Finkler (2001) argues that the focus on origins, especially medical histories, contributes to the medicalisation of kinship, aligning kin relations with biomedical definitions, while French sociologist Sébastien Roux (2015) suggests that discourses on origins naturalise identity, turning "uteruses" and "homelands" (*l'utérus et la patrie*) into ontological anchors rather than relational constructs. From this perspective, origin searching through genetic genealogy participates in the "geneticisation" process identified by Abby Lippman, "by which differences between individuals are reduced to their DNA codes, with most disorders, behaviours and physiological variations defined, at least in part, as genetic in origin" (1991, p. 19). Pascal Ducournau sees DTC DNA testing as a new phase; initiated by individuals rather than by medical professionals or institutions, it supports an "auto-geneticisation" process, wherein genetics complements or replaces "the social marks of identity", such as family histories (2018, p. 139).

Other scholars challenge this view, stressing the relational, temporal, and narrative dimensions of origin searches. Janet Carsten suggests that such quests allow adoptees to reclaim control of an unlived past, although they inevitably confront its partial irretrievability (2000b, p. 694); meanwhile, Agnès Martial (2022) shows that these searches are not conducted in isolation from existing relationships in adoptive families, as they often involve adoptive parents and can support young people's growing autonomy on the path to adulthood.

This debate has often been based on studies examining the outcomes or circumstances of origin searches. Little research has focused on the search process itself. Yet searching for origins involves a range of practices: contacting authorities (Roux, 2017), posting on the Internet and social media (Martial, 2022; Thomson-Sweeny, 2021), seeking help from outsiders (Baden et al., 2021; Modell, 1994), or gathering clothing items, photographs, and official documents (Carsten, 2000b, p. 696; Martial, 2020). These practices unfold over months or years and are central to the searching experience. This chapter therefore examines the practices in which sperm donor-conceived adults engage to search for their origins, with particular attention to their use of DTC DNA testing. How do they use these technologies in their search? What knowledge do they produce, and how do they make sense of DNA in the process?

This chapter draws on anthropological kinship studies—especially those focused on "Euro-American" societies,[1] where procreation is understood as being embodied through transmission (Bamford, 2004; Strathern, 1995). Departing from biomedical definitions, DNA is treated

[1] See a critical definition of "Euro-Americanness" in Edwards (2006).

as a trace of procreation. As such, it is not considered an immutable "given" (Carsten, 2004), but rather as the embodiment of actions and intentions that constitute procreation. Following Edwards and Carsten, I move beyond the biological vs. social divide (Edwards, 2008) to consider the "relatedness" implicated in DTC DNA testing (Carsten, 2000a).

In Euro-American societies, procreation is often seen as a privileged means of creating kinship (Schneider, 1980), based on the principle of exclusive descent: a person is expected to have two parents—no more, no less—who take on all parenting roles and attributes, such as conceiving, giving a name, nurturing, and educating (Goody, 1982; Ouellette, 1998). In donor conception, however, the premise is that one person helps conceive a child without becoming their parent, allowing another person to become the parent of the child they did not conceive (Théry, 2010). Donor conception therefore introduces a "multiparental" situation, where more than two adults may assume parenting roles, even if not all are considered parents (Martial, 2019). Through analysing DTC DNA tests in this context, I explore how the relationship between DNA and kinship is reconfigured in the search for origins in donor conception. I focus on how donor-conceived people produce knowledge about their origins—that is, how they "put in order" (*mettre en ordre*) (Adell-Gombert, 2011) or make sense of the information they gather.

The Study

This chapter draws on an ethnographic study conducted with donor-conceived adults in France and England between March 2015 and December 2019, as part of my Master's dissertation (March 2015–May 2016) and my doctoral thesis (April 2017–December 2019). The study received funding from the School for Advanced Studies in the Social Sciences (Aires culturelles grant), the Origins ANR programme through the French National research agency (no. ANR-18-CE26-001), and the INFODON research project through the French Agency of biomedicine (AOR n° 2018-A02515-50).

Ethics committee approval is not (and was not) mandatory for social science research in France (Derbez, 2023); however, the doctoral study received ethical approval from the internal ethics committee of the Donor Conception Network (UK) in autumn 2017 and from the *Comité de protection des personnes Sud Méditerranée III* (Public University Hospital of Marseille, AP-HM, France) on 1st January 2019. All participants gave their free and informed consent to take part in the study, and each doctoral study participant signed a formal consent form to that effect. All names used here are pseudonyms. Some identifying details, such as professions, were removed or altered, with no impact on the data analysis.

Calls for participation were shared in organisations dedicated to donor conception, on social media via private and public groups, and by word of mouth. Broad inclusion criteria were applied to maximise the chances of recruiting a diverse range of experiences: participants had to be adults and donor-conceived in France or the UK (or residing there at the time). Volunteers were free to contact me to express their interest in participating. I declined only one volunteer due to age, and a few initial contacts did not lead to an interview. Otherwise, the profile of participants reflects that of the individuals who reached out to take part in the study.

The study was conducted through in-depth interviews (Beaud, 1996) with 63 sperm donor-conceived adults between 21 and 72 years of age (with an average age of 37). They were born between the 1940s and the 1990s in France and England, and thus belong to the first generations of donor-conceived people: those mostly born between the creation of the first specialised DC centres (in the 1940s in England and the 1970s in France) and the introduction of the first laws regulating donor conception in each country (1990 in the UK, 1994 in France).

Despite their differences, both countries shared a similar socio-legal landscape during this period: governments were reluctant to legislate on a practice strongly criticised by religious authorities (Betta, 2017; Pfeffer, 1993). The organisation of donor conception was left to the discretion of professionals within DC centres, resulting in varied practices. In France, specialised clinics in public hospitals coexisted with private sector centres until the early 1990s (Novaes, 1994).[2] In the UK, third-party reproduction initially emerged in English private clinics (Frith, 2001), and the

[2] Since 1994, third-party reproduction can only be performed in public or private non-profit medical centres in France.

private sector has remained dominant to this day (Human Fertilisation & Embryology Authority, 2021). Across both countries, DC centres did not seek the same donor profiles (Daniels et al., 1996; Novaes, 1994) and did not always follow the same rules—for example, regarding donor compensation. However, in both France and the UK, DC centres upheld donor anonymity and typically advised recipient parents to keep their use of donor gametes confidential.

Because of the similar socio-legal context in France and England, the participants in this study share a number of characteristics. All were conceived via *anonymous sperm* donation within *heteroparental* families.[3] Moreover, most participants were women (49 out of 63) and held higher-skilled jobs—more than half were employed in "intellectual or scientific" occupations (ESeG), compared to only 26.3% of the British and 18.8% of the French working populations in 2019 (INSEE & Dares, 2020, p. 166–67).

Over half of the participants had searched for their origins or intended to do so, a trend also observed in many studies with donor-conceived people, which may reflect a recruitment bias (Canzi et al., 2019). This chapter is more specifically based on the interviews with 38 participants, mainly from England (23), who used DNA testing, mostly through DTC kits. A few case studies are developed in the text, but the analysis is informed by all the data collected. The next section provides details on how these participants have come to use DNA testing.

Taking a DTC DNA Test

Various motivations can prompt a search for origins: donor-conceived people mention their curiosity about the donor, their wish to find resemblances, to access medical information, or to form new connections (Canzi et al., 2019). The process often coincides with important life transitions and events and may pause or stop at any point (Abel & Pálsson, 2020; van den Akker et al., 2015). These searches follow nonlinear paths

[3] One person was conceived by a solo mother by choice, but for confidentiality reasons, their case is not identified as such.

(Martin, 2022), with DNA testing occurring at different stages and under diverse circumstances.

In my study, some participants inadvertently discovered the circumstances of their conception after taking a "curiosity" test received as a gift, or purchased for genealogical research (Crawshaw, 2018). These tests sometimes revealed unexpected information, such as unfamiliar ethnic origins, prompting them to question relatives. Others were connected with a donor-conceived person—a "same-donor offspring" or a "donor sibling" (Indekeu, 2020)—whose story led them to realise they shared the same circumstances. Other participants deliberately used DNA testing to search for their origins. Those informed of their conception more recently could take a test without delay, unlike those told before the 2000s. The latter had often been told they would never learn more about the donor, and put aside their desire to search, until DNA testing offered new, previously unimaginable possibilities.

These opportunities differ significantly between England and France. In France, DNA testing is restricted to specific situations listed in the Civil Code (art. 16–11), which excludes origin searches.[4] Only licenced medical centres can administer such tests. DTC DNA testing is banned and punishable by a €3,750 fine for users and up to 1 year in prison and a €15,000 fine for unauthorised providers (arts. 226-28 and 226-28-1, French Criminal Code). Nonetheless, French participants had used DTC kits, and no convictions have been reported to date.

In the UK, DTC DNA testing is legal. Since 2008, the Donor Conceived Register (DRC) has also offered a genetic database to individuals who were donor-conceived or who donated before the creation of the national register managed by the Human Fertilisation and Embryology Authority (HFEA) in August 1991. Since 1991, the HFEA national register has provided various levels of donor information, which has included identifying data since April 2005.[5] A few English participants in my

[4] Since September 2022, France has granted donor-conceived people access to their origins upon reaching their majority. A dedicated commission, the CAPADD (*Commission d'Accès des Personnes nées d'une Assistance Médicale à la Procréation aux Données des tiers Donneurs*) was established to handle such requests. This measure was not yet in place when I conducted my study.

[5] All individuals born after April 2005, except in a few cases during the transitional period, are entitled to access the donor's identifying or non-identifying information upon reaching their

study were registered with the DCR or eligible for other information requests, but most actually relied on DTC DNA testing to search for origins.

A myriad of genetic genealogy companies offer DTC DNA kits (Ducournau, 2018; Greely, 2008). Some are designed for specific populations, such as African Ancestry (Nelson, 2016), but the ones most commonly used by donor-conceived people, such as 23andMe, Ancestry, MyHeritage, or FamilyTreeDNA, cater to a broad customer base. These companies offer several types of DNA tests, including Y-chromosome, mitochondrial, and autosomal DNA analysis, along with various complementary services (Greely, 2008). For example, before filing for bankruptcy in March 2025 (Kollewe, 2025), 23andMe was known for its health-related testing, whereas Ancestry provides access to digital records from a range of historical archives. Most donor-conceived people I met chose an autosomal test, which includes information from both the "paternal" and "maternal" lines without distinction.[6]

The kits are purchased online for less than a hundred dollars and delivered by mail within a few days. Samples are collected from saliva without supervision, usually by spitting into a tube or a cheek swab. The sample is then sealed in the provided envelope and returned to the genetic genealogy company by mail. It takes a few weeks to extract the DNA, analyse it, and upload the results to the company's database. There, the DNA is algorithmically compared to other profiles and data. Once the results are ready, users are notified by email and can access them in a personal account on the company's website using a login ID and password.

In the context of origin searches, results typically include two parts. First, a breakdown of the user's "ancestry" or "ethnic origins", generally presented as a geographical distribution on a colour-coded map and accompanying chart. These lists include both geographical regions (e.g. "Iberian Peninsula", "British Isles"), or spatialised cultural groups (e.g. "Ashkenazi Jews"), with each category represented as a percentage. The second part of the results displays the "genetic matches": a list of other

majority, whereas those born between August 1991 and April 2005 have access to the donor's identity only if the donor has previously consented.

[6] Autosomes are the 22 pairs of chromosomes distinct from the one pair of allosomes (also called "sex-chromosomes", X and Y).

users whose DNA profiles are similar enough to suggest a probable common ancestor. These matches are ranked by genealogical proximity and may be accompanied by a name and surname, a photograph, age-related details, a family tree, the estimate level of connection, and the date of the match's last login. The connection is quantified using various metrics (centimorgans, percentages, number of shared segments) and labelled with relational terms such as "close family" or "second to third cousin". I elaborate on this aspect in a later section.

The effectiveness of these websites in identifying matches depends on the size of the company's database and the quality of its algorithm. As databases expand, results become more precise and new matches appear. The donor-conceived people I met would check these platforms regularly, especially during holidays or sales periods when DTC kits are often discounted and given as gifts. People also subscribed to several platforms to maximise their chance of matching. Nevertheless, the matching process was not always straightforward.

Mirror-Treeing

Genetic matches will seldom identify the donors directly, as they are typically not on the databases themselves. Therefore, an investigation needs to be conducted on the results. This investigative work is known as "mirror-treeing". Let us follow the crash-course I was given by an English participant. Jodie was born in the 1980s and had known she was donor-conceived for two years when we first met. The day she was informed, she dived right into the search for the donor, both registering with the DCR and buying a DTC DNA kit. Her search was exceptionally fast: within a year she identified three same-donor offspring, as well as their shared progenitor. During my fieldwork, she unofficially became my DNA-testing advisor, not only because of her acquired expertise but also because of her ability to make the process as plain as possible for a novice like myself.

> It's like acting basically. So, you pretend that you are one of your DNA matches. You pick your best one and you pretend that you are them. So, you start an ancestry tree and you make your home person that person.

Genetic matches first have to be sorted to filter those on the donor's side. Some people ask a maternal relative to take a DNA test, in order to triangulate and set aside the common matches on the maternal line. Others use surnames or ethnic origins to do the same. Once the sorting is done, the closest genetic match on the donor's side should be selected, and a family tree started where the match is "Ego"—or the "home person" as Jodie puts it. From there on, the match's genealogy needs to be built. The closer the match, the better the chance of being able to work on their family tree.

> Then you use Facebook, the Internet, the birth registers, the maiden names, everything you can to get the next line because that line will be anonymous because they're still alive. And that often dictates what mirror trees you can build because if you can't get the parents of that person, you can't move any further.

Some users make their family tree available on genetic genealogy websites, but the identities of living individuals who are not registered in the database are hidden for confidentiality. Identifying the generations above the match may require various tools, such as social media, direct contact with the match, or requesting birth or marriage certificates from official institutions—within local legal limits.[7] As the family tree is built "backwards" through the generations, the platform's algorithm will triangulate it with the genetic database to make suggestions that help the searcher go further.

[7] In England and Wales, anyone can request a copy of a birth, marriage, or death certificate issued after 1837, between 6 and 18 months after the event's registration (Births and Deaths Registration Act 1953, c. 20). In France, such documents are available 75 years after the register's closure or 25 years after the individual's death. Immediate access is granted only to the person's spouse, civil partner, direct ascendants, or descendants—a category that does not include donor-conceived people under French law (*Décret n°2017-890 du 6 mai 2017 relatif à l'état civil* and article L. 213-2 of the French *code du patrimoine*).

> [The website] will search every tree of every one of your DNA links, even if it's someone you share a tiny amount with. And it will throw up the shared lines. And when you see the shared lines, you identify the line that you match that person on. So, it allows you to discount the rest. So, you've got one line then. And you've got the shared ancestor. You know that their fifth-grade grandfather was also your fifth-grade grandfather.

The goal is to get to an ancestor shared with the selected match. The work can then resume in reverse: the shared ancestor becomes "Ego" in the tree, and their genealogy is worked out "downwards" in the hope of identifying the donor's line. The same tools as before are used, but the task is even more tedious, as many possibilities need to be sorted and may result in dead ends. Information needs to be cross-checked: "ethnic" origins, estimated age of the donor, typical donor profile at the time, and so on.

> So, then you start with that fifth-grade grandfather, that's probably 1810, and then you work that person's tree down. So, you find a geographical match or an ethnic match. So, one of the things I had was a 50% Jewish, 50% English tree. So, I was just looking for the Jew that married into the family. [...] And I thought it would have happened post-war. So, I was thinking, well, I'm looking for someone who marries a Jew in the forties and I've got my grand-parents. [...] It didn't work like that at all for me.

Despite her expertise, Jodie herself came to an impasse in her search for the donor. After pursuing several unsuccessful lines, just as she had unknowingly found the right path, it was a new match, the donor's sister that eventually allowed her to complete the process.

As the intricacies of this section illustrate, the process of searching for origins through DTC DNA testing requires lengthy and complex procedures. DNA does not magically unveil origins. As with other DNA tests (van den Akker et al., 2015), results from genetic genealogy websites need to be interpreted. Knowledge about one's origins is produced through techniques requiring a form of expertise—even when a direct match is made in the database.

Categorising Connections

Donor-conceived people do get direct matches on genetic genealogy websites, most often with same-donor offspring who might be searching for their origins or simply using DTC DNA testing recreationally. However, matches will not be specifically labelled "same-donor-offspring" or "donor siblings". The connection might not even appear as a "half-sibling", as Alice explained during our interview. She was born in the 1980s in England. She was told about her donor conception by her mother after finishing university. A decade before we met, she registered on the DCR but did not get a successful match. A few years later, she turned to a DTC DNA testing kit that matched her with two men: "So, when you go in your DNA-matches, it sorts them in order of relationship. [...] It just says 'close family'. That's all it says for both of them".

The labelling system categorising genetic matches draws on the Euro-American kinship vocabulary, which anthropologists have come to know as "Eskimo" terminology (Murdock, 1965):[8] positions are distinguished depending on generation (e.g. parents/grandparents), gender (son/daughter), and whether the connection is one of filiation or collaterality (mother/aunt). Identical terms are used to refer to the father's and mother's siblings (uncle/aunt), as well as their children (cousins).

This description will probably seem obvious to fellow anthropologists and somewhat awkward yet familiar to those who are not versed in classical kinship studies. The reason I present it is that this terminology is not universal: for example, some societies differentiate paternal and maternal cousins, uncles, and aunts; others use the same vocabulary for all members of the parents' generation (Godelier, 2010). By referring to Euro-American kinship vocabulary, genetic genealogy websites use culturally and historically situated categories in labelling genetic matches, despite serving a global user base.

Those categories should not be taken for granted, especially as the actual labels will often combine several kinship terms. Same-donor

---

[8] The term "Eskimo", used in the historical classification of kinship terminologies, originates from the European colonisation of North America. It is now considered derogatory and has been rejected by the Inuit people, whom it was meant to describe. I use it here solely in reference to the historical classification of kinship terminologies, where, to my knowledge, the term has not been updated.

offspring may appear as potential "half-siblings" or "uncles"/"aunts", "nephews"/"nieces", "grandparents", or "grandchildren". They may also be labelled with a broad category, like "close family" in Alice's case. Therefore, for donor-conceived people, the first question arising when they get a match is: what are they to each other?

Alice showed me how she navigated her results, leaning over from a café's brown leather couch so I could follow along on her smartphone's screen.

> So, when you click on them … [The profile of a man appears on screen]. So, that's the guy that I got matched to last year. And then you click on this little bit here and it shows you the centimorgans that we share. And then "What does it mean?" So you go on [that] [She clicks and a window opens.] So, it suggests that he's my uncle. He's possibly my uncle or I'm his aunt. That's the number of centimorgans. But that's just a suggestion. Same to me, he could have been my grandfather with just that amount of DNA, this is potentially what it could be.

The website's labelling system is associated with a unit for measuring genetic linkage, centimorgans (cM) in this instance. Centimorgans represent the probability that two individuals share a common ancestor. The higher the number, the more recent and closer the shared ancestor is likely to be. The unit is derived from a process observed during gametogenesis (Greely, 2008).

Spermatozoa and oocytes are haploid cells, meaning they each carry 23 chromosomes: half the number typically found in an embryo. They are formed through meiosis, a process involving two successive cell divisions of a diploid germ cell, which initially contains 46 chromosomes. Before the first cell division (during prophase I, leading into metaphase I), the chromosomes align in homologous pairs and exchange segments of genetic material, in a process known as genetic crossing-over or recombination (Wetterstrand, 2025b). These cells with recombined chromosomes then proceed through two rounds of divisions, ultimately producing four genetically distinct gametes (Gilchrist, 2025).

In other words, because of the genetic cross-over, each gamete contributes a slightly different version of its original genome to any resulting embryo. As a result, full siblings do not inherit identical genomes, and over successive generations, individuals tend to share fewer and smaller segments of identical DNA. Centimorgans quantify the frequency of genetic recombination (Wetterstrand, 2025a), and serve to measure the probability of connectedness between two individuals (Kling et al., 2021). However, they do not precisely indicate the position of the connection, as Alice highlighted. The average number of centimorgans is very close between a half-sibling, a grandparent, and an uncle or an aunt (1759cM vs 1754cM vs 1741cM according to *DNA Painter*[9]). The measures might even be identical, and other details are necessary to differentiate between them.

Benjamin's profile—the first man with whom Alice was matched—gave little information. No photograph, no birth date. Alice messaged him, hoping he was a relative of the donor. "When I asked him how old he was and he was born two weeks after me, so we were conceived literally at the same time, the same week probably, I was like 'damn…'. For him as well". Given their close age, Alice concluded that Benjamin was donor-conceived and that they shared the same donor, thus referring to him as her "half-brother".

Translating centimorgans into a relational label that makes sense for donor-conceived people is not straightforward. The people I met during the study used online tools to calculate the connection probability or consulted tables that give the minimum, maximum, and average number of centimorgans depending on the type of connection.[10] They also used the knowledge they had gained about genetics, like Jodie, who concluded that Joan and Patrick were her "half-siblings" despite centimorgan measures that could indicate otherwise:

> So, me and Joan we share a very low amount of DNA. Whereas Joan and Patrick share enough to be full siblings. But they're right at the very, very top. And me and Joan are right at the very, very bottom of the bracket. So, if you test through [a DNA testing company], they would put Joan and

[9] "The Shared cM project 4.0 tool v4", https://dnapainter.com/tools/sharedcmv4, accessed on 26-05-2025.

[10] For example, "What are the odds?" on *DNA Painter*, https://dnapainter.com/tools/probability, accessed on 26-05-2025.

> Patrick as full siblings and me and Joan as first cousins. [...] But me and Joan, we're definitely half-siblings because ... we know we are. We share 1300 centimorgans, so that is enough to be half-siblings. [...] Joan could be my first cousin but we share a full X-chromosome. And a full X-chromosome can only be shared by paternal half-sisters.

DNA data need to be interpreted and corroborated by a body of evidence, part of which has nothing to do with the genome. Age indicates the generational level: Benjamin was not likely to be Alice's grandparent or uncle given their two-week age difference. Family history helps differentiate between a donor's son or daughter and same-donor offspring: a long period of time separating the parents' union and the birth of their first child might suggests fertility issues and the resort to donor conception. Resemblances are also a clue to assert a connection.

Melissa was born in the 1980s in the UK. As a student, she did a DTC DNA test along with a friend for recreational purposes. She had forgotten about it until a year before I interviewed her, when she was matched with Gemma, who claimed they were "half-sisters". Melissa's father had just passed away and her partner feared that it was a scam attempt. Melissa was less wary: "I said I didn't think it was [a scam], because of the picture. Just, I looked and I said, 'look, look'. You know ... [...] I knew my parents had had fertility treatments. So ... I knew it was true".

Genetic data do not stand on their own, nor do they substitute for other types of information contradicting the geneticisation hypothesis. DNA tests and genetic genealogy are considered valuable evidence in the search for origins, but they are always complemented by other data that verify them (Freeman & Richards, 2006). Donor-conceived people do not passively accept DTC DNA testing results as irrefutable truth about their origins. These results are put to the test in order to produce knowledge. This proves all the more complex when it comes to qualifying connections in a way accepted by all people involved.

Disputable Results

> Explore your DNA Family! We all have our immediate, biological or adopted families … but did you know you could also share DNA with other 23andMe customers—your DNA Relatives?[11]

Without excluding pre-existing kin relationships, genetic genealogy websites offer to introduce a new, family-sounding type of connection. 23andMe (as cited above) refers to them as "DNA relatives", while AncestryDNA uses the term "family connections",[12] MyHeritage mentions "new parents",[13] and Family Tree DNA's matching tool is called "Family Finder". DTC DNA testing is not intended to replace existing relatives, but the findings are presented as informative about the customer's family.

This seems consistent with what American anthropologist David Schneider declared about the "American cultural conception": "kinship is defined as biogenetic" and "if science discovers new facts about biogenetic relationship, then that is what kinship is and was all along, although it may not have been known at the time" (Schneider, 1980, p. 23). However, the use of DTC DNA testing by donor-conceived origin seekers reveals the limits of this logic. The way the people I met categorised their genetic matches was not always accepted by the individuals concerned. For example, Benjamin, whom Alice considered her "half-brother", rejected her interpretation of their connection. He had never been told that he was donor-conceived and did not want to ask his mother—his father was deceased. Since the genetic genealogy website listed Alice and Benjamin as "close family to first cousins", Benjamin concluded that they must be first cousins, despite Alice's efforts to convince him otherwise.

> I had to try and explain this to him. And a year later, he's still not talking to me. Because he can't accept it. He's in total denial. […] He cannot

[11] "Our service", *23andMe*, https://www.23andme.com/en-int/dna-ancestry/, accessed on 06/06/2019.

[12] "What to expect from your AncestryDNA® test results", *Ancestry*, https://www.ancestry.com/c/dna/what-toexpect-ancestrydna-test, accessed on 06/04/2022.

[13] "ADN: vue d'ensemble", *My Heritage*, https://www.myheritage.fr/dna, accessed on 21/06/2021.

> accept that we're brother and sister. So yes, I'm kind of stuck with that with him and that's been painful for me because I was so excited. He was the first one and I was so excited to have a new sibling.

Benjamin cut all contacts with Alice. Consequently, a few months later, when she received a new match with Trevor, she approached the situation with extra caution.

> I messaged him straight away and just said, "Hello, I don't know if you've checked your results", but obviously I knew this was his first set of results because he had joined three or four weeks before that. And I just said, "We're a really high match. I'd like to talk to you if that's possible". I was really careful. I worded it way more sensitive than I did the first time. I didn't know what I was doing and I jumped in a bit too fast with this other boy.

Alice did not want to replicate the situation with Benjamin. As other participants in the study described, she chose her words carefully to avoid scaring Trevor away and having her interpretation of their connection rejected once again.

> I didn't want it, you know ... It's difficult. I explained I was donor-conceived. So, his obvious reaction was then that his dad must have been a sperm donor. [...] I had a gut feeling. I just thought "no". And he sent me photos of his father and I went "no way".

Trevor had also not known that he was donor conceived. He immediately accepted that Alice was a "half-sister", but he did not ask his mother for an explanation right away. He initially thought his father might be the donor, whereas Alice was convinced otherwise due to their lack of resemblance. The situation remained uncertain for two months, until Trevor's mother confirmed that he was donor conceived. Trevor also matched with Benjamin, which, in Alice's view, corroborated their connection—although Benjamin had not changed his mind.

Nothing guarantees that matched individuals will share the same interpretation of their connection, even if the validity of the match is not contested. The problem lies in how the connection is positioned in "genealogical" terms. Not only are genetic data insufficient to qualify such matches (as showed in the previous section), but the epistemological model employed by the websites to translate them into relationships is not adequate for origin searches in donor conception.

Genetic genealogy websites use kinship as a framework for interpreting the data analysed by the tests. As Catherine Nash notes, kinship provides a "grammar for translating the complexities of new genetics into public culture" (2004, p. 25). While these tests are often perceived as applying genetics as an analytical framework for kinship—the geneticisation hypothesis—Nash argues that, conversely, kinship serves as an analytical grid for genetics and for science more broadly (see also Bamford & Leach, 2012). The rhetoric of affiliation and kinning is prevalent in the discourse of these websites; for example, recurring DNA chains are said to form "clans" or haplogroups, united by a remote ancestor (Nash, 2004, 2015). Thus, while DNA certainly contributes to "deciphering" kinship (Déchaux, 2018, p. 40), kinship, in turn, helps to decipher DNA.

In doing so, genetic genealogy websites do not use just any kinship model. Drawing on the authority of science, they assimilate procreation and kinship using a genealogical framework that is very similar to that which Schneider attributed to American understandings in the 1960s:

> The blood relationship, as it is defined in American kinship, is formulated in concrete, biogenetic terms. Conception follows a single act of sexual intercourse between a man, as genitor, and a woman, as genitrix. At conception, one-half of the biogenetic substance of which the child is made is contributed by the genitrix, and one-half by the genitor. Thus each person has 100 per cent of this material, but 50 per cent comes *from his mother* and 50 per cent *from his father* at the time of his conception, and thereby is his "by birth". (Schneider, 1980, p. 23, emphasis added)

In this Euro-American model, procreation is assumed to produce transgenerational transmissions, creating a "common identity" (Schneider, 1980, p. 25) between *parents* and their descendants. The traces of these transmissions are seen as persisting steadily throughout the individual's life. This understanding of the relation between procreation, kinship, and personhood contrasts with other models in which kinship is processual and based on how individuals act over time (Vilaça, 2002; Weismantel, 1995). In Langkawi, Malaysia, for example, blood is central to the definition of kin relationships, but it is not understood as being exclusively conferred at birth. "It gradually accrues and changes throughout life, as persons participate in relationships" (Carsten, 1995, p. 225), especially through sharing food in the same home: "Blood is always mutable and fluid—as kinship itself" (Carsten, 1995, p. 224).

The discourse of genetic genealogy websites "naturalises" kinship by associating science, procreation, and kinship in a manner typical of the Euro-American kinship model, although this has not always been the case historically (Fine & Martial, 2010). However, the example of Alice and Benjamin shows that this translation of DNA into kinship may be contested. The procreative analytical framework does not perfectly align with the kinship framework in the case of the connections based on origins in donor conception. As Jodie remarked while she was explaining the mirror-treeing process to me, information based on origins is not attached to family histories and memories.

> You have to do some verification. And you have to make sure you're not running down the wrong path. Because it's not your family. You have no other knowledge to check it against. You can't say, "Oh that doesn't feel right because I know that they moved to [city] and they didn't live there." You have to sort of... You have to just sort of verify as you go along the best as you can.

Information about one's origins is not instantly familial, and sometimes not even familiar. When a connection is identified by a genetic genealogy website, nothing predicts what the individuals concerned will make of it, and if or how a relationship will develop. This supports Janet Carsten's assertion that knowledge of procreation does not necessarily equate to knowledge of kinship (2007), or, at least, not straightforwardly so when origins via donor conception are concerned.

Conclusion

> You just need to compare your DNA, its four letters, A-T-C-G, and their variations with the file of the other person. That's what anyone could do with a simple comparative software or by cross-checking the results. That's what [the genetic genealogy websites'] algorithms do.

As Nathalie Jovanovic-Floricourt, a French professional genealogist, stated in the above-cited interview (Bonnet et al., 2021, 00:08:10), DTC DNA testing can initially seem like a trivial modern activity. Kits are purchased online like any product; samples are collected without professional oversight and mailed as ordinary parcels. A few weeks later, results are uploaded to a personal account, much like other online services. With proper tools, it can feel as though anyone can interpret genetic profiles. But as the experiences of donor-conceived people illustrate, this practice is far from trivial.

Identifying a donor often requires time and expertise. The mirror-treeing process can take months and may remain unresolved due to missing information. DNA is not a magical key that reveals identities and connections. Furthermore, while its evidentiary value is acknowledged, it is also tested by origin seekers trying to categorise genetic matches. DNA results alone do not produce knowledge of one's origins. They must be verified and supplemented with other types of information. This echoes what Alondra Nelson (2016) pointed out about some users of African Ancestry who received results conflicting with inherited family histories. Resolving this "genealogical disorientation" (Nelson, 2016, p. 84) involved integrating the odd information into a coherent narrative by selecting what made the most sense to those involved. Similarly, in the context of donor conception, searching for origins through DTC DNA testing means producing knowledge that goes beyond DNA alone. This becomes especially clear in negotiations over how to interpret matches. While genetic genealogy websites frame connections through a model aligning procreation and kinship, donor-conceived people may interpret these connections differently. Their disagreements shed further light on the ambivalence at the heart of DNA-based origin searches.

The use of DTC DNA testing illustrates and nuances the contemporary *mundanity* of genomics. As genetic tools have spread, DNA has moved beyond medicine into non-expert use. On the one hand, the experience of non-experts highlights how genetic knowledge is shaped by cultural, social, and historical contexts. The way genetic genealogy websites translate DNA results into connections relies on Western understandings of procreation and kinship, which can be challenged when origin searches expose gaps between the procreative and the kinship analytical framework. On the other hand, even in the "lay" use of DNA, knowledge production remains complex. The pervasiveness of genomics meets limits, as shown by the rise of new experts assisting with origin searches (Baden et al., 2021). DNA may be widely accessible, but it is not necessarily encountered with equal ease by all.

References

Abel, S., & Pálsson, G. (2020). Dépister l'ancestralité: Machines et techniques généalogiques dans la reconstruction des histoires de famille. *Ethnologie française, 2*(178), 269–284. https://doi.org/10.3917/ethn.202.0269

Adell-Gombert, N. (2011). *Anthropologie des savoirs*. Armand Colin.

Baden, A. L., Harrington, E. S., & White, E. (2021). Searching for contact with birth relatives: Search angels and search facilitators. *Journal of Social Work, 21*(3), 476–496. https://doi.org/10.1177/1468017320911349

Bamford, S. (2004). Conceiving relatedness: Non-substantial relations among the Kamea of Papua New Guinea. *The Journal of the Royal Anthropological Institute, 10*(2), 287–306.

Bamford, S., & Leach, J. (Eds.). (2012). *Kinship and beyond: The genealogical model reconsidered*. Berghahn Books.

Baumann, A. (2022). *Becoming donor-conceived. The transformation of anonymity in gamete donation*. Transcript Verlag.

Beaud, S. (1996). L'usage de l'entretien en sciences sociales. Plaidoyer pour l' "entretien ethnographique". *Politix, 9*(35), 226–257.

Betta, E. (2017). *L'autre genèse. Histoire de la fécondation artificielle* (C. Eloy, Trad.). Hermann.

Blyth, E. (2012a). Discovering the "Facts of Life" following anonymous donor insemination. *International Journal of Law, Policy and the Family, 26*(2), 143–161. https://doi.org/10.1093/lawfam/ebs006

Blyth, E. (2012b). Genes r us? Making sense of genetic and non-genetic relationships following anonymous donor insemination. *Reproductive BioMedicine Online, 24*(7), 719–726. https://doi.org/10.1016/j.rbmo.2012.02.010

Blyth, E., & Frith, L. (2009). Donor-conceived people's access to genetic and biographical history: An analysis of provisions in different jurisdictions permitting disclosure of donor identity. *International Journal of Law, Policy and the Family, 23*(2), 174–191. https://doi.org/10.1093/lawfam/ebp002

Bonnet, P., Denis, C.-M., & Sauvage, B. (Réalisateurs). (2021, juin 17). *Tests ADN: la quête des origines*. France Télévision. https://www.francetvinfo.fr/sante/biologie-genetique/tests-genetiques/video-tests-adn-la-quete-des-origines_4664567.html

Canzi, E., Accordini, M., & Facchin, F. (2019). "Is blood thicker than water?" Donor conceived offspring's subjective experiences of the donor: A systematic narrative review. *Reproductive BioMedicine Online, 38*(5), 797–807. https://doi.org/10.1016/j.rbmo.2018.11.033

Carsten, J. (1995). The substance of kinship in the heat of the hearth: Feeding, personhood, and relatedness among Malays in Pulau Langkawi. *American Anthropologist, 22*(2), 223–241.

Carsten, J. (Ed.). (2000a). *Cultures of relatedness: New approaches to the study of kinship*. Cambridge University Press.

Carsten, J. (2000b). "Knowing where you've come from": Ruptures and continuities of time and Kinship in narratives of adoption reunions. *The Journal of the Royal Anthropological Institute, 6*(4), 687–703.

Carsten, J. (2004). *After kinship*. Cambridge University Press.

Carsten, J. (2007). Constitutive knowledge: Tracing trajectories of information in new contexts of relatedness. *Anthropological Quarterly, 80*(2), 403–426.

Courduriès, J. (2016). Ce que fabrique la gestation pour autrui. *Journal Des Anthropologues, 144–145*, 53–76. https://doi.org/10.4000/jda.6364

Courduriès, J., & Gourarier, M. (2020). Introduction: Tracer l'histoire des sujets. *Ethnologie Française, 2*(178), 257–268. https://doi.org/10.3917/ethn.202.0257

Crawshaw, M. (2018). Direct-to-consumer DNA testing: The fallout for individuals and their families unexpectedly learning of their donor conception

origins. *Human Fertility, 21*(4), 225–228. https://doi.org/10.1080/14647273.2017.1339127
Daniels, K., Curson, R., & Lewis, G. M. (1996). Semen donor recruitment: A study of donors in two clinics. *Human Reproduction, 11*(4), 746–751. https://doi.org/10.1093/oxfordjournals.humrep.a019247
Darroch, F., & Smith, I. (2021). Establishing identity: How direct-to-consumer genetic testing challenges the assumption of donor anonymity. *Family Court Review, 59*(1), 103–120. https://doi.org/10.1111/fcre.12553
Déchaux, J.-H. (2018). Le gène à l'assaut de la parenté ? *Revue des Politiques Sociales et Familiales, 126*(1), 35–47. https://doi.org/10.3406/caf.2018.3261
Derbez, B. (2023). La vertu des sciences sociales à l'épreuve des comités d'éthique de la recherche en santé. *Cahiers Droit, Sciences & Technologies, 16, Article 16.* https://doi.org/10.4000/cdst.6975
Ducournau, P. (2018). *S'entreprendre avec ses gènes. Enquête sur l'auto-génétisation.* Presses Universitaires de Rennes.
Edwards, J. (2006). Reflecting on the "*Euro*" in "*Euro-American*" kinship: Lithuania and the United Kingdom. *Acta Universitatis Klaipedensis, 12*, 129–139.
Edwards, J. (2008). Introduction. The matter in kinship. In J. Edwards & C. Salazar (Eds.), *European kinship in the age of biotechnology* (pp. 1–8). Bergahn Books.
Fine, A., & Martial, A. (2010). Vers une naturalisation de la filiation? *Genèses, 1*(78), 121–134.
Finkler, K. (2001). The kin in the gene: The medicalization of family and kinship in American society. *Current Anthropology, 42*(2), 235–263.
Freeman, T., & Richards, M. (2006). DNA testing and kinship: Paternity, genealogy and the search for the "truth" of our genetic origins. In F. Ebtehaj, B. Lindley, & M. Richards (Eds.), *Kinship Matters* (pp. 67–95). Hart Publishing.
Frith, L. (2001). Gamete donation and anonymity. The ethical and legal debate. *Human Reproduction, 16*(5), 818–824. https://doi.org/10.1093/humrep/16.5.818
Frith, L., Blyth, E., Crawshaw, M., & van den Akker, O. (2017). Secrets and disclosure in donor conception. *Sociology of Health & Illness, 40*(1), 188–203. https://doi.org/10.1111/1467-9566.12633
Frith, L., Blyth, E., Crawshaw, M., & van den Akker, O. (2018). Searching for "relations" using a DNA linking register by adults conceived following sperm

donation. *BioSocieties, 13*(1), 170–189. https://doi.org/10.1057/s41292-017-0063-2

Gilchrist, D. A. (2025, mai 28). *Meiosis*. National Human Genome Research Institute. https://www.genome.gov/genetics-glossary/Meiosis

Gilman, L., & Nordqvist, P. (2018). Organizing openness: How UK policy defines the significance of information and information sharing about gamete donation. *International Journal of Law, Policy and the Family, 32*(3), 316–333. https://doi.org/10.1093/lawfam/eby014

Gilman, L., Redhead, C., Hudson, N., Fox, M., Nordqvist, P., MacCallum, F., Kirkman-Brown, J., & Frith, L. (2024). Direct-to-consumer genetic testing and the changing landscape of gamete donor conception: Key issues for practitioners and stakeholders. *Reproductive BioMedicine Online, 48*(1), 103421. https://doi.org/10.1016/j.rbmo.2023.103421

Ginzburg, C. (1980). Signes, traces, pistes. Racines d'un paradigme de l'indice. *Le Débat, 6*(6), 3–44. https://doi.org/10.3917/deba.006.0003

Godelier, M. (2010). *Métamorphoses de la parenté*. Flammarion.

Goody, E. N. (1982). *Parenthood and social reproduction: Fostering and occupational roles in West Africa*. Cambridge University Press.

Greely, H. T. (2008). Genetic genealogy. Genetics meets the marketplace. In B. A. Koenig, S. Soo-Jin Lee, & S. S. Richardson (Eds.), *Revisiting race in a genomic age* (pp. 215–234). Rutgers University Press.

Haimes, E. (1990). *Family connections: The management of biological origins in the new reproductive technologies* [Ph.D. Dissertation]. University of Newcastle Upon Tyne.

Harper, J. C., Kennett, D., & Reisel, D. (2016). The end of donor anonymity: How genetic testing is likely to drive anonymous gamete donation out of business. *Human Reproduction, 31*(6), 1135–1140. https://doi.org/10.1093/humrep/dew065

Hertz, R., & Nelson, M. K. (2019). *Random families. Genetic strangers, sperm donor siblings, and the creation of new kin*. Oxford University Press.

Human Fertilisation & Embryology Authority. (2021). *Fertility treatment 2019: Trends and figures*. Human Fertilisation and Embryology Authority. https://www.hfea.gov.uk/about-us/publications/research-and-data/fertility-treatment-2019-trends-and-figures/

Indekeu, A. (2020, May 8). *Same donor offspring or donor half-siblings ?* [Communication]. The complicated relationship between genetic relatedness, the family and the State, CFA workshop, Umea University.

INSEE & Dares. (2020). *Emploi, chômage, revenus du travail. Edition 2020.* Insee & Dares. https://www.insee.fr/fr/statistiques/4504425

Jadva, V., Freeman, T., Kramer, W., & Golombok, S. (2010). Experiences of offspring searching for and contacting their donor siblings and donor. *Reproductive BioMedicine Online, 20*(4), 523–532. https://doi.org/10.1016/j.rbmo.2010.01.001

Kling, D., Phillips, C., Kennett, D., & Tillmar, A. (2021). Investigative genetic genealogy: Current methods, knowledge and practice. *Forensic Science International: Genetics, 52.* https://doi.org/10.1016/j.fsigen.2021.102474

Klotz, M. (2014). *(K)information: Gamete donation and kinship knowledge in Germany and Britain.* Campus Verlag.

Klotz, M. (2016). Wayward relations: Novel searches of the donor-conceived for genetic kinship. *Medical Anthropology, 35*(1), 45–57. https://doi.org/10.1080/01459740.2015.1012615

Kollewe, J. 2025. DNA testing firm 23andMe files for bankruptcy as CEO steps down. *The Guardian.* Retrieved May 26, 2025, from https://www.theguardian.com/lifeandstyle/2025/mar/24/dna-testing-firm-23andme-bankruptcy-ceo-anne-wojcicki-data

Lippman, A. (1991). Prenatal genetic testing and screening: Constructing needs and reinforcing inequalities. *American Journal of Law & Medicine, XVII*(1–2), 15–50.

Martial, A. (2019). Multi-parenthood and contemporary family forms in French studies. *Antropologia, 6*(2 N.S), 13–26.

Martial, A. (2020). Les archives des origines. Traces et (dis)continuités narratives dans les dossiers de l'Aide sociale à l'enfance (1995–2015). *Ethnologie Française, 2*(178), 285–298. https://doi.org/10.3917/ethn.202.0285

Martial, A. (2022). Une « parenté pour soi » ? La quête des origines en adoption. *Anthropologie et Sociétés, 46*(2), 101–118.

Martial, A., Côté, I., & Lavoie, K. (2021). De l'adoption à la procréation assistée par autrui: Cadres, pratiques et enjeux entourant la question des origines et de ses récits. *Enfances Familles Générations, 37.* https://journals.openedition.org/efg/11408

Martin, A. (2022). *La parenté d'après le don: Une approche relationnelle du vécu des personnes conçues par don de sperme (France, Angleterre)* [Ph.D. Dissertation]. École des Hautes Études en Sciences Sociales.

Modell, J. (1994). *Kinship with strangers. Adoption and interpretations of kinship in American culture.* University of California Press.

Murdock, G. P. (1965). *Social structure*. The Free Press & Collier-Macmillan Limited.

Nash, C. (2004). Genetic kinship. *Cultural Studies, 18*(1), 1–33. https://doi.org/10.1080/0950238042000181593

Nash, C. (2015). *Genetic geographies: The trouble with ancestry*. University of Minnesota Press.

Nelson, A. (2016). *The social life of DNA. Race, reparations, and reconciliation after the genome*. Beacon Press.

Novaes, S. (1994). *Les passeurs de gamètes*. Presses universitaires de Nancy.

Ouellette, F.-R. (1998). Les usages contemporains de l'adoption. In A. Fine (Éd.), *Adoption: Ethnologie des parentés choisies* (pp. 153–176). Éditions de la Maison des sciences de l'homme. http://books.openedition.org/editionsmsh/621

Pfeffer, N. (1993). *The stork and the syringe: A political history of reproductive medicine*. Polity Press; Blackwell Publishers.

Roux, S. (2015). L'utérus et la patrie: Adoption, origines et politique des identités. *Mouvements, 82*(2), 66–75. https://doi.org/10.3917/mouv.082.0066

Roux, S. (2017). L'État des origines. Histoires adoptives, conflits biographiques et vérités passées. *Genèses, 108*(3), 69–88.

Schneider, D. (1980). *American kinship: A cultural account* (2nd ed.). University of Chicago Press.

Smart, C. (2009). Family secrets: Law and understandings of openness in everyday relationships. *Journal of Social Policy, 38*(4), 551–567. https://doi.org/10.1017/S0047279409003237

Strathern, M. (1995). Displacing knowledge: Technology and the consequences for kinship. In F. D. Ginsburg & R. Rapp (Eds.), *Conceiving the new world order: The global politics of reproduction* (pp. 346–363). University of California Press.

Théry, I. (2010). *Des humains comme les autres. Bioéthique, anonymat et genre du don*. Editions de l'EHESS.

Thomson-Sweeny, J. (2021). Chercher ses origines sur Facebook: Quels liens entre les médias sociaux et la quête des origines en adoption internationale? *Enfances Familles Générations, 37*. https://journals.openedition.org/efg/11419#quotation

van den Akker, O., Crawshaw, M., Blyth, E., & Frith, L. (2015). Expectations and experiences of gamete donors and donor-conceived adults searching for genetic relatives using DNA linking through a voluntary register. *Human Reproduction, 30*(1), 111–121. https://doi.org/10.1093/humrep/deu289

Vilaça, A. (2002). Making kin out of others in Amazonia. *Journal of the Royal Anthropological Institute, 8*(2), 347–365. https://doi.org/10.1111/1467-9655.00007

Weismantel, M. (1995). Making kin: Kinship theory and Zumbagua adoptions. *American Ethnologist, 22*(4), 685–704.

Wetterstrand, K. A. (2025a, May 28). *Centimorgan (cM)*. National Human Genome Research Institute. https://www.genome.gov/genetics-glossary/Centimorgan-cM

Wetterstrand, K. A. (2025b, May 28). *Crossing over*. National Human Genome Research Institute. https://www.genome.gov/genetics-glossary/Crossing-Over

Part IV

Databases and Nations

12

The Daily Life of DNA: Forensic Genetics and the Administration of Death and Hope in Mexico

Vivette García-Deister

Introduction

Over the past two decades, the discussion of human genetics in society has included the biomedical, forensic, and genealogical domains. In analysing how publics interact with genetic knowledge, scholars identified the rise of "new social formations" and "biological citizenship", including individual and collective "genetic identities" (Glasner & Atkinson, 2007; Atkinson et al., 2007; Rose & Novas, 2005). They denounced practices of determinism, stigmatisation, and racialisation mediated by DNA technologies (Koenig et al., 2008; Krimsky & Sloan, 2011). As genetic knowledge and practices leave the laboratories to play an increasingly important role in everyday life (Pollock et al., 2021), the goals and ambitions pursued through genetics have expanded (Nelson, 2016). In the field of forensics, genetics has transformed from a state-controlled

V. García-Deister (✉)
Faculty of Sciences, National Autonomous University of Mexico,
Mexico City, Mexico
e-mail: vivettegarcia@ciencias.unam.mx

V. Argudo-Portal et al. (eds.), *Mundane Genomics*,
https://doi.org/10.1007/978-981-95-4550-6_12

instrument of individual identification into a powerful tool for shaping social narratives and challenging dominant claims to truth and justice. For instance, the use of African ancestry genetic tests has illuminated histories of oppression and systemic inequality in the United States, lending support to anti-racist social movements (Nelson, 2016).

In Latin America, DNA technologies have had, historically, multiple uses and effects. Mothers and grandmothers organised around genetic technologies in the wake of the horrific atrocities of the 1976–1983 dictatorship in Argentina. They found in these "subversive" technologies a path towards recovering their kidnapped grandchildren and identifying the bodies of the victims of enforced disappearance. In this way, DNA technologies became powerful human rights tools that reconfigured social, familiar, and political life (Smith, 2016).

The Mexican context offers a different approximation of what Alondra Nelson calls "the social life of DNA". Since 2012, the year that marked the end of President Felipe Calderón's six-year term and the deadly consequences of his "war on drugs" became palpable, the social and political relevance of forensic genetics has come into sharp relief. The increasing popularisation of genetic analysis for human identification (Canal Once, 2025; CNB, 2025) takes place in a context of systematic violence (ONU-DH Mexico, 2025), where there are 128,715 disappeared persons (RNPDNO, 2025), 72,100 unidentified bodies in forensic services (Red Lupa, 2025), and the discovery of new clandestine graves and human remains is, sadly, part of the daily life of all Mexicans—a phenomenon that has been dubbed "Mexico's forensic crisis" (Movimiento por nuestros desaparecidos en México, 2021).

This peculiar epistemic space, marked by certainty and suspicion, mistrust and reliance, assurance and disbelief—and similarly antagonistic affects—offers a view of how the facts and machines of forensic genetics are being appropriated by different actors to different ends, acquiring a dynamic social and political presence.

What is the role of DNA in the process of human identification? And how does genetic research in everyday forensic practices differ from how DNA is represented in Mexico's political discourse about genetic databases to identify the disappeared? In this chapter, I approach these questions by contrasting the modest place that DNA occupies inside forensic institutions with the over-optimistic representation that portrays genetic technologies as the definitive tool for human identification and, therefore, a solution to Mexico's forensic crisis. This approach allows me to observe the daily social life of DNA in Mexico on three different levels: the social demands for truth and justice that are intertwined with the hopes of families to find their loved ones, the capitalisation of such hopes by political discourse in the form of a constant (unfulfilled) promise of the construction of a national genetic database for human identification, and the commercial interests of biotechnology companies that see the current forensic crisis as an opportunity for profit.

Methodology and Structure of This Chapter

To examine the place of DNA in forensic practices and how it contrasts with the political discourse about the efficacy of genetics for human identification, I combined ethnographic and historical methods with document and discourse analysis. Contrasting the social life of DNA inside and outside the laboratory is an analytical strategy that reveals how the promises of genetic databases function as necropolitical devices to contain social discontent by creating and regulating hope. This approach allowed me to discuss how the administration of hope impacts the families searching for the disappeared in different ways, making justice even more difficult to achieve.

Section 1 of this chapter is an ethnographic vignette from a study I undertook in the summer of 2017. For four weeks, I followed the experts of the Genetics Laboratory of Mexico City's Institute of Forensic Sciences (Incifo) to document their protocols and observe their daily practices. I performed participant observation and conducted ethnographic interviews to understand the place of DNA in this institution. I interviewed the head of the Identification Department, who walked me through the process of opening a file for an unidentified body and the multidisciplinary steps involved in the protocol. Upon arrival, a file is created, and fingerprints are taken from all ten fingers, or as many as are available. A dental chart is completed, a somatological chart is filled out, and both facial and intraoral photographs are taken. Depending on the circumstances, a tissue sample—such as blood, muscle, bone, or cartilage—may also be collected.

When a family member reports a missing person, an ante-mortem interview is conducted according to the International Committee of the Red Cross (ICRC) protocol. The interview lasts 1.5 hours and the questionnaire is 30 pages long, eliciting a broad range of information on the missing person, including personal (habits, hobbies), physical (distinguishing marks such as tattoos or piercings), medical and dental information, as well as information on the circumstances of their disappearance. This information is compared with data collected post-mortem, encompassing findings from the examination of the body, such as detailed pathological, anthropological, and odontological analyses, along with insights into the cause of death. According to the head of the identification unit, the genetics laboratory occasionally contributes to this process, typically to confirm or rule out an identity hypothesis based on the gathered evidence.

During my fieldwork at Incifo, I observed necropsies and spoke to experts in other areas of the Institute, such as dactyloscopy and forensic toxicology. I also attended seminars where techniques and statistics were discussed as part of the training of early-career forensic professionals. The case studies presented in these seminars gave a precise account of the multidisciplinary work involved in human identification and the need to use different methods to restore people's identities.

Observing these practices sparked my curiosity by revealing that DNA is not the central tool in human identification, but rather one of many important techniques. Contrary to its portrayal as the heroic protagonist in public discourse and outside the lab, DNA plays a supporting role within a broader forensic framework. This realisation motivated me to examine the political discourse surrounding genetic databases and to trace the historical narrative of the government's promise to build a DNA-centred sociotechnical infrastructure aimed at "solving" Mexico's forensic crisis (Section 2). Building on this, I analysed official documents and journalistic sources to argue that the promise of a national genetic database operates as a necropolitical device for the administration of death and hope (Section 3). Finally, I reflected on what approaching DNA through the lens of the mundane and examining its social and political life in Mexico reveals about forensic infrastructures, the State's claims to truth and justice, and the cycle of hope and disappointment that is set in motion by an overreliance on DNA.

1. Incifo Genetics Laboratory and the Place of DNA in Human Identification

On 13 July, as happened on other occasions, the head of the genetics laboratory assumed the role of a tour guide for three lawyers and a criminal-court judge. He was already well calibrated in the language he used to describe the different areas of the laboratory: he was descriptive enough for a lay audience to understand the function of each section, but precise enough not to lose scientific rigour. Making the nature and reliability of the processes carried out in the laboratory known to non-specialists is an important part of the work of a forensic expert.

The geneticist explained to the visitors the transformations that a biological sample undergoes in the different areas of his laboratory, and during this explanation he dispelled the doubts that the lawyers had and corrected some of their pre-existing assumptions. For example, the identity of a person does not appear automatically on a computer screen—CSI style—after comparing the genetic profile of an unknown deceased person with that of a family member who is looking for him: "The technology needs the expert to interpret the results", he said. Visitors understood why it makes no sense for a public prosecutor to request a "femicide protocol" for a skeleton whose gender is unknown, or the difficulty of obtaining "sweat DNA" (DNA can only be extracted from tissues containing cells).

The faces of the listeners registered understanding and amazement, which they verbally expressed: "We came to visit you to be certain of what you are working with and what you have". They observed dehydrated and unidentified bone samples in sterile boxes marked with the date and type of bone from which the samples were collected. The objective of "creating a genetic profile of each body that arrives at the facilities" was reiterated by Incifo personnel, and the visitors understood that this is a laboratory where the possibility of identifying a body in the future with the help of forensic genetics is protected, even if it has been deposited with its proper registration in the common grave of the Dolores Civic Cemetery, as happens with those who are not identified and claimed. But visitors were also told that DNA is not a miracle technology and that geneticists do not work in isolation but are linked with all other areas of Incifo. "We are the auxiliaries of the Identification Department", the geneticist insisted.

The characteristics of the Incifo Genetics Laboratory—that is, its accreditation, modern equipment with well differentiated functions to avoid cross-contamination, and in general its good reputation as a reliable laboratory with highly trained human resources—make it susceptible to this type of visit. Incifo's genetics laboratory operates under the jurisdiction of the Court of Justice of Mexico City. While it primarily conducts paternity tests for family law cases, its experts also analyse biological samples related to sexual assault investigations and assist in identifying the human remains in Incifo's custody.

When the Identification Department seeks to confirm an identity hypothesis, genetic profiles are created for both the deceased and potential relatives, then compared for a match. If a family member contacts Incifo with a missing person inquiry, reference samples are collected from the relatives to generate profiles that can be compared to unidentified remains, potentially generating new investigative leads. When the Institute's staff is not fully occupied with these tasks—or when additional qualified personnel, such as scientific interns completing their social service, are available—they work on obtaining genetic profiles from unidentified bodies in cold storage and update the genetic profile database accordingly. They prepare for the possibility that in the future, based on some input of information (a preliminary inquiry, a search request from a family member, a request from another government agency), they may be able to link a genetic profile to a sample and, by matching it with other available information, identify a body.

DNA is an important resource for identification, especially in cases where a complete body is not available, or the remains are decomposed or skeletal. But, as the director of the genetics laboratory emphasises, "[I]n no case that has come to Incifo has a body been identified solely on the basis of DNA". Identification is a multidisciplinary process involving at least three different areas of Incifo: dactyloscopy, odontology, and anthropology. And DNA is not always required to make an identification that satisfies authorities and relatives. Something that characterises Incifo's genetics laboratory is that geneticists see it as "another link" in the chain of processes that are carried out to identify a person.

In the work done in this laboratory, there is methodological rigour, there is an effort to make forensic genetics a tool for justice—for example, by trying different DNA extraction methods to obtain a useful genetic profile from samples that have been chemically altered—but there is no emphasis on DNA as a crucial methodology or on geneticists as chief forensic experts. "In the ten years that I have been working as a genetics expert, a forensic odontologist has never made a mistake" in formulating an identification hypothesis, said one of the experts I interviewed. This moderation of the contribution of genetics to identification, together with the routine, everyday methodical nature of genetic work in identification—along with the understanding that its effectiveness depends on coordination with other disciplines—contrasts sharply with the expectations of first-time visitors to the laboratory, who hope for swift and definitive results.

2. The Promise of Genetic Databases

On 16 November 2009, the Inter-American Court of Human Rights, in its ruling on the case known as *Campo Algodonero vs. Mexico*, for the first time established the obligation of the Mexican State to use DNA technologies in the identification of individuals (Inter-American Court of Human Rights, 2009a). Just a week later, on 23 November, the Court also held the State responsible for the enforced disappearance of Rosendo Radilla during the 1970s *Dirty War*. It ordered the State to locate Mr Radilla or, if deceased, to identify his remains through genetic testing before returning them to his family (Inter-American Court of Human Rights, 2009b). Despite these rulings, Mexico remains far from fulfilling these obligations, while international bodies and civil society organisations continue to press for them with growing urgency in the face of ongoing impunity and violence (ONU-DH, 2025).

The horrific case of Ayotzinapa—where 43 students from a rural teachers college disappeared (Torres, 2022)—the massacres of migrants in San

Fernando, Tamaulipas (Smith & García-Deister, 2017), the discovery of hundreds of bodies hidden in trailers in Jalisco (García-Deister, 2019), and, more recently, the uncovering of mass graves, crematoriums, and personal belongings at what search collectives believe to be a forced recruitment and extermination camp also in Jalisco (Franco, 2025), are all stark—although not isolated—examples of the State's inadequate or inactive forensic apparatus. These cases can hardly be revisited without reference to forensic genetics.

Given the persistence of death and disappearance, the promise of matching the genetic information of unidentified bodies with that of family members in the search for missing persons suggests that the creation of a national bank of genetic profiles is urgent, precise, necessary, and inevitable. In principle, having a data infrastructure for forensic purposes makes sense, but when we analyse the use that government institutions have made of the project to create such a database over several six-year presidential terms, serious concerns emerge. In October 2010, during President Felipe Calderón's administration, authorities reported the existence of a forensic genetic database containing 9758 profiles. A year later, in November 2011, Mexico adopted the FBI's CODIS system. In February 2012, the Chamber of Deputies introduced a draft decree to reform the General Law of the National Public Security System, proposing the creation of a genetic database aimed at locating missing persons and responding to public demands amid the growing crisis of disappearances (Kahwagi Macari, 2012). By August of that year, then-Attorney General Marisela Morales declared forensic genetics a national priority for the justice system, stating that the National Genetic Database—also referred to as the Federal Database of Genetic Profiles—contained 15,618 profiles. This number was later revised to 14,278 (Secretaría de Gobernación, 2012).

In June 2014, Mariana Benítez Tiburcio, Deputy Attorney General for Legal and International Affairs, announced the federal government's plans to develop a national database of genetic profiles as a tool for criminal investigations (*Expansión*, 2014). In March 2016, Eruviel Ávila, Governor of the State of Mexico, proposed establishing a new database—Red Genética Nacional GenMex—comprising genetic information from all 32 federal entities (Jiménez, 2016). In 2018, the Congress of Veracruz proposed a reform to the Constitution to create the "Unique National Database" that would include genetic profiles (Rosas, 2018). Finally, in March 2025, President Claudia Sheinbaum once again pledged to unify and strengthen administrative and forensic records—including DNA databases—to aid in the identification of missing persons (Morán Breña, 2025). This announcement came in response to national outrage over the government's handling of the Izaguirre Ranch discoveries in Jalisco.

Thus, with striking consistency and persistence, the creation or expansion of a national genetic database has been repeatedly presented to the public as the State's response to civil society's demands—framed as an innovative, high-priority initiative that promises to resolve the crisis of the disappeared. This proliferation of official initiatives over nearly two decades underscores the lack of a robust and genuinely useful database. The apparent, uncoordinated duplication or multiplication of these efforts, which have not produced satisfactory results, together with the continued increase in the number of missing persons (more than 120,000), clandestine graves (more than 5000), and unidentified bodies (more than 72,000), is what has been called the "forensic crisis".

In February 2019, one of the first official actions taken by the newly appointed leftist Secretary of Human Rights, Population, and Migration was to publicly acknowledge the country's forensic crisis. Referring to Mexico as a "great clandestine grave" and admitting that the national forensic system was unequipped to address the scale of death and disappearance, Alejandro Encinas introduced his national strategy for combating disappearances and locating missing persons (Secretaría de Gobernación & Subsecretaría de Derechos Humanos, Población y Migración, 2019).

The creation of the National Institute of Forensic Identification (which he later renamed the National Centre for Human Identification), of search and investigation protocols, as well as the creation of a unique technological and computerised information system—that is, a National Forensic Database that would include genetic profiles—were central parts of his declared strategy. The obligation of the Attorney General's Office to create such a bank was already foreseen in the General Law on the Disappearance of Persons, which came into effect in November 2017. "The law has not fulfilled its objective, on the contrary, it has fallen into a deep simulation", Encinas said on that occasion and promised to concretise the initiative of the National Data Bank (*Proceso*, 2019).

However, on 31 December 2021—almost three years after his statements—a district court in Mexico City granted a legal protection against the Attorney General's Office "for the failure to create the National Forensic Database" (*Animal Político*, 2022). The plaintiff, Olimpia Montoya, sister of a young man who disappeared in Guanajuato, claims that her right to truth and justice has been violated. She argues that, like thousands of family members searching for their loved ones, she cannot be sure that the genetic profiles she provided to the Public Prosecutor's Office will be compared with all genetic databases of unidentified deceased persons to find the whereabouts of her brother.

Why do both the promises of the State and the demands of civil society converge so prominently on genetic databases? The main reason is that the crisis of unidentified human remains in Mexico has been conceived as a technological problem, and access to truth and justice depends on whether DNA technologies are used (or not). Thus, on the one hand, the authorities promise genetic databases above all other forensic resources, and on the other, civil society continues to demand them.

The idea of DNA as a technological solution to the forensic crisis is part of a widespread but misleading socio-technical imaginary (Jasanoff & Kim, 2015) about what human identification entails and the role of forensic genetics within that process. This overemphasis on DNA often appeals for its ability to establish a framework of credibility based on the status of forensic genetics as irrefutable scientific evidence. While it is true that DNA can be a very useful resource in certain contexts, especially when there are such large numbers of unidentified bodies, it is by no means a universal solution for all cases, and it would be irresponsible to claim that DNA technologies will identify each and every one of the more than 72,100 bodies currently kept in forensic services.

Human identification is a laborious, time-consuming, and highly contextualised multidisciplinary process that is most often conducted on a case-by-case basis. It does not involve a single line of investigation or method of identification, but many different ones. When these methods converge, full identification is achieved. More importantly, forensic specialists do not always resort to DNA, even when it is available, but only in specific circumstances and often as a tool to corroborate a hypothesis of identity arrived at by other means (such as tattoos, clothing, fingerprints, contextual information, and physical characteristics of the body).

The power of DNA identification—the degree of certainty that a genetic test provides—depends largely on its interaction with other types of evidence (archaeological, cultural, osteological, and others), as well as on the administrative guarantee of the integrity of the chain of custody and the traceability of the samples. Nevertheless, DNA is given a predominant role in the Mexican public and political space. It is repeatedly presented, usually without question, as the most objective, reliable, and rapid method of identification. And this rhetoric is taken up, as we have seen, by public officials or family members in search, but it is also taken up by businessmen. Because of exaggerated expectations of what it can do, DNA, and especially genetic databases, can be used as quotidian tools to control or profit from death.

3. Governing Death: Genetic Databases as Necropolitical Devices

Ariadna Estévez (2020) argues that at least since the six-year term of Felipe Calderón, Mexico has been governed by apparatuses designed to administer suffering. She argues that instead of directly addressing the structural causes of violence and providing legal and moral justice to victims of grave human rights violations, the Mexican government has adopted policies that manage and administer death. Estévez calls this "Mexican necropolitical governmentality". The Cameroonian philosopher Achille Mbembe raised the concept of necropolitics in his 2003 article of the same name to explain the objectives of power relations in the developing world. In these contexts, the actions of both criminal groups and the State point to a logic not centred on the governance of life but on the control and orchestration of death. A notion of Foucauldian genealogy, necropolitics has been expanded by Mbembe himself (2020) and taken up by Estévez and other authors to explain various aspects of violence in Mexico and the relations between citizenship and the State (Smith & García-Deister, 2020; Emerson, 2019).

In particular, Estévez points out that among the legal-administrative instruments that order and systematise the politics of death (i.e., necropolitical instruments) are the National Search Program, the National Centre for Human Identification, the National Registry of Victims, and the Executive Commission for Attention to Victims (CEAV). These institutions, Estévez argues, function as an apparatus for the management of suffering and death through the execution of four strategies: (1) the legalisation of social demands (through the bureaucratisation of claims for justice); (2) the regulation of subjects and their political agency (through the appropriation of discourse and the co-optation of sociopolitical action); (3) the construction of subjectivities (experts, human rights defenders, direct and indirect victims, etc.); and (4) the distribution of resources that commodify and reify justice (payments for reparations for harm to people credited as victims and designed to maintain the status quo).

Genetic databases, as technologies of governmentality, form part of a broader apparatus that institutionalises death. Each time a new official database is announced, or a DNA collection campaign is launched, a necropolitical system is activated—one that manages death and the suffering it leaves behind. Families, often desperate for answers, provide DNA samples multiple times across different states, hoping to increase the chances of identifying their missing loved ones. In return, they receive minimal support from government institutions like the CEAV—if they manage to meet the rigid criteria that define a "victim". In effect, families end up financing and carrying out the very search work the State fails to do, even as the government later claims those efforts as its own.

And so, another necropolitical cycle begins. The structural roots of violence remain untouched, while death persists unchecked: the numbers of disappeared persons, unidentified bodies, and clandestine graves continue to rise. Forensic services are stretched beyond capacity, and the backlog in processing human remains is painfully apparent. Victim assistance programmes are reintroduced—often without adequate funding—and the long-promised National Forensic Database is once again brought to the fore. New DNA collection campaigns are launched, yet the actual rate of successful identification through these technologies remains dishearteningly low.

If genetic databases fail to deliver the rapid, large-scale, and conclusive identification results that were initially promised, yet persist in both political discourse and citizen demands, they are still fulfilling a crucial role—one that is not human identification. As part of the institutional machinery of necropolitical governmentality, the repeated promise of a national genetic database functions not only as a tool for managing death, but also as a mechanism for managing hope. From President Felipe Calderón to President Claudia Sheinbaum, DNA technologies remain a central fixture in a discursive loop: a promise made, unfulfilled, and then reiterated, with both officials and citizens clinging to it as a last beacon of hope.

Profiting from Death: Genetic Databases as Commodities

On 24 July 2020, I received an email from an entrepreneur presenting a project to create a human identification and forensic genetics centre in Mexico. Intrigued, I scanned the document he attached. It was a seven-page "executive summary" of what, on paper, appeared to be the ultimate solution to Mexico's humanitarian crisis, where thousands of people are searching for their missing loved ones while forensic services are overwhelmed with unidentified human remains.

Written in business jargon, the document created a sense of urgency. It spoke of "rapid and efficient" mitigation efforts and provided an "immediate action alternative" to help the federal government locate and identify missing persons. With a private investment of $134 million, it proposed the creation of five regional human identification centres that could be up and running within six months. The centres would handle search, exhumation, recovery, transportation, storage of 5 to 10 thousand bodies per unit, genetic identification and family services. All this through the construction of fully equipped facilities, certified laboratories, validated protocols, and trained personnel. Most importantly, there would be a unified technology and information system: a national database that would be created at the rate of 150 genetic profiles per week, allowing for massive DNA comparisons.

In short, the businessman proposed what he called a "comprehensive solution" to Mexico's forensic crisis. The proposal he sent me (and which, according to his message, he had also made known to authorities at the federal level, with whom he had already "established contact" and to national institutions and international and organisations, with which he was establishing "collaboration links") was formally addressed to the President of Mexico at the time, Andrés Manuel López Obrador.

The proposal also referenced the Extraordinary Mechanism for Forensic Identification (MEIF), a policy initiative launched in March 2020. The MEIF acknowledges the limitations of the State's conventional forensic systems and calls for an "extraordinary intervention" to tackle the growing backlog in human identification. As part of this effort, the Mechanism envisions allocating a dedicated budget and granting special legal powers to a selected team—potentially composed of both public and private actors—to lead and accelerate the identification process.

When I received the businessman's proposal, the coordinating group responsible for developing the MEIF's work plan had not yet been established, and the first Regional Centre for Human Identification linked to the mechanism had yet to be inaugurated (Secretaría de Gobernación, 2020). At that time, the MEIF was perceived as an emerging opportunity for private capital investment. It represented a scenario in which the demand for human identification and remains management services far outpaced the State's existing capacity—effectively, a market niche.

This perception of a market niche was further reinforced on 12 December 2021, when journalists Paula Mónaco and Wendy Pérez published an investigation into the private laboratory Central ADN. The lab markets itself to families searching for their missing loved ones, offering identification through genetic technologies. "The private laboratories do the math", the journalists caution. "[T]here are close to 100,000 missing persons, and each identification requires three genetic tests. At current prices, that's $40 to $50 per reference sample and $400 to $500 to analyse each set of remains. The loot is gigantic … in dollars, it's close to sixty million" (Mónaco & Pérez, 2021).

These dynamics are echoed in recent market analyses positioning Latin America—particularly Mexico—as one of the most promising growth areas for human identification services. This trend gained momentum following former President Andrés Manuel López Obrador's approval of the National Centre for Human Identification (Grand View Research,

2025; GII Research, 2025). Notably, the individual behind the forensic project proposal I received was serving as director of innovation and technology for a group that also had plans to expand into direct-to-consumer genetic services—such as disease susceptibility testing and genealogy analysis—suggesting a strategic interest in both public and private sectors of the genetics market.

Researchers studying the social life of DNA have increasingly shown that the traditional distinctions between biomedical, recreational, and forensic genetics—once based on differing goals and the types of DNA analysed (coding for the former two, non-coding for the latter)—are becoming increasingly untenable, as DNA infrastructures tend to overlap (M'charek & Wade, 2020). Today, genetic samples, data sets, methodologies, and market logics are frequently shared across domains, while regulatory gaps in sample storage and database oversight have made the boundaries between these fields more porous. This convergence is precisely what gives human identification its potential to evolve into a vast and profitable industry. Blood or saliva samples collected for forensic identification purposes can also—if permitted by law, or simply not explicitly prohibited—be used to populate databases for recreational or biomedical purposes.

The framing of Mexico's crisis of unidentified human remains as a primarily technological challenge, alongside the tendency to view access to justice as contingent on the application of DNA technologies, further reinforces the idea of human identification as a business opportunity. This perception is amplified by the growing involvement of the private sector in areas traditionally seen as public responsibilities (Nowak, 2017), helping to consolidate human identification as an emerging market niche.

But conceiving of databases simply as technological products or commodities capable of generating economic returns overlooks a fundamental characteristic of data-centric inquiry: that genetic profiles have no scientific value in and of themselves, only in relation to other data. To be useful, these data must be carefully curated. The data must be selected, formatted, classified, annotated, and integrated to allow comparison with other data that have undergone the same treatment. The scientific usefulness of genetic data for human identification does not lie simply in having a collection of genetic profiles. Instead, their value as evidence emerges when they can be compared with other data in specific investigative contexts (Leonelli, 2016). For this reason, it is essential to develop a standardised system that ensures compatibility and coordination between different databases. Without the ability to compare reference samples with those taken from unidentified individuals, DNA technologies lose their effectiveness.

This need for interoperability becomes even more evident in a case documented by Mónaco and Pérez (2021), when a representative from Central ADN assured a family that the firm would compare genetic material extracted from a bloodstained tooth belonging to a missing person "with the largest identification database in the country, one that uses the Combined DNA Index System (CODIS)". The representative was referring to the database managed by the State Attorney General's Office. In effect, Central ADN was claiming a competitive advantage by suggesting it had access to the official genetic databases of the Mexican State.

At the same time, the lab sought to involve families not only to legitimise its work but also to help secure access to regional forensic service databases. Yet, as previously discussed, identification is not simply a matter of matching profiles across databases. Without standardised systems and reliable comparisons, the promises of DNA technology remain limited—regardless of who makes them.

Conclusion

By examining forensic DNA through the lens of the mundane, this chapter has shown that, whether as a tool of justice or as a business opportunity, DNA does not generate immediate returns, nor does it generate them in isolation. Forensic genetics requires expensive infrastructures involving significant and long-term investments in facilities, and human and material resources. It also requires regulatory and social structures, including institutions and individuals in charge of the development, financing, and oversight of genetic data and—very importantly—their linkage with other types of infrastructures. To be functional, this entire framework must prevail in changing political scenarios and fluctuating budgets. Conceived in this way, forensic infrastructures are long-term, spatio-temporal projects that exceed six-year presidential terms and extend into the future, but are by no means reduced to the existence of a national DNA bank or the possibility of comparing genetic databases.

Relying on DNA's widely accepted status as scientific truth, the Mexican State repeatedly invokes the promise of identification. Yet this promise is rarely fulfilled. Instead, it functions as a necropolitical device that feeds on the hope of searching families, setting in motion a cycle of hope and disappointment that is continuously renewed (see, for example, *El Universal*, 2025). Despite having tens of thousands of genetic profiles, the Mexican State has not been able to clear the backlog in the identification of remains because the simple aggregation of profiles does not guarantee their usefulness. In order to function, the National Forensic Data Bank must be constituted as a platform that makes possible the packaging of data and communication between different sectors (e.g., between the federal and State governments; between the Attorney General's Office and the National Search Commission). This implies creating a technical, legal, and political structure to support and coordinate search and identification efforts.

Beyond generating false promises, the commodification of DNA and its deployment as a necropolitical device obscure structural violence, reducing the complex, lived process of human identification to a positive match in a database. As genetic testing—whether offered by state institutions or private labs—becomes a routine marker of truth or justice, there is a real danger of normalising impunity. It severs the recognition of a life lost from the political and social conditions that enabled that loss. What worries me further is how the digitisation of justice may detach DNA technologies from the lived experiences of violence, grief, and persistence that define the daily reality of searching families. If there is a mundane truth to hold onto here, it is this: forensic genetic databases and sample banks do not exist in abstraction. They are made and remade through the ordinary, often painful, participation of those who give their genetic material, in life and in death.

Acknowledgments I am grateful to Research Assistant Karina Aranda for her support in updating this chapter, and to the forensic experts at Incifo for generously allowing me to observe their daily practices.

References

Atkinson, P., Glasner, P., & Greenslade, H. (2007). *New genetics, new identities*. Taylor & Francis.

Breña, C. M. (2025, March 17). Sheinbaum anuncia reformas para unificar los registros de identificación y forenses sobre personas desaparecidas. *El País*. Retrieved April 14, 2025, from https://elpais.com/mexico/2025-03-17/sheinbaum-anuncia-reformas-para-unificar-los-registros-de-identificacion-y-forenses-sobre-personas-desaparecidas.html

Canal Once. (2024, September 20). Análisis e identificación: Zacnicté Minor [Video]. YouTube. Retrieved April 14, 2025, from https://www.youtube.com/watch?v=pVHtqMqF4cI&list=PLrFkZrRQk9nk1YUJBKfYjdJZ62BJ846I8&index=5

Comisión Nacional de Búsqueda de Personas Desaparecidas (2025, April 4). Si tienes un familiar desaparecido acude a dar tu entrevista y una muestra genética [Post]. Facebook. Retrieved April 14, 2025, from https://www.facebook.com/photo?fbid=981544380833050&set=a.295156532805175%20%E2%80%8C

Demandan a FGR por no crear Banco Nacional de Datos Forenses; debió operar desde 2019. (2022, January 5). *Animal Político*. Retrieved April 14, 2025, from https://animalpolitico.com/sociedad/demandan-fgr-banco-nacional-forense

El Universal. (2025, March 21). Buscaban una esperanza y hallan "circo y simulación" [Video]. YouTube. Retrieved April 14, 2025, from https://www.youtube.com/watch?v=8XiimJibgRI

Emerson, R. (2019). *Necropolitics. Living death in Mexico*. Palgrave Macmillan.

Estévez, A. (2020). Mexican necropolitical governmentality and the management of suffering through human rights technologies. *Critical Criminology, 28*(1), 27–42. https://doi.org/10.1007/s10612-020-09502-8

Evento de presentación del Balance Ciudadano: Desapariciones en México. Siguen sin ser prioridad. (2025, January 24). *Red Lupa*. Retrieved April 14, 2025, from https://imdhd.org/redlupa/eventos/desapariciones-en-mexico-siguen-sin-ser-prioridad/

Franco, D. (2025, March 21). Indolencia y desprecio hacia las familias buscadoras, el sello en la apertura del rancho Izaguirre de Teuchitlán. *Cimacnoticias*. Retrieved April 14, 2025, from https://cimacnoticias.com.mx/2025/03/21/indolencia-y-desprecio-hacia-las-familias-buscadoras-el-sello-en-la-apertura-del-rancho-izaguirre-de-teuchitlan/

García-Deister, V. (2019, August 1). DNA will not solve Mexico's unidentified-body crisis. *Slate*. Retrieved April 14, 2025, from https://slate.com/technology/2019/08/mexico-unidentified-bodies-missing-families-dna-databases.html

GII Research. (2025, March 18). *Human Identification—Market Share Analysis, Industry Trends & Statistics, Growth Forecasts (2025–2030)*. Retrieved April 14, 2025, from https://www.giiresearch.com/report/moi1687234-human-identification-market-share-analysis.html

Glasner, P., & Atkinson, P. (2007). *New genetics, new social formations*. Taylor & Francis.

Inter-American Court of Human Rights. (2009a). *González y otras ("Campo Algodonero") vs. México* [Executive Summary]. Retrieved April 14, 2025, from https://www.sev.gob.mx/unidad-genero/files/2020/02/Resumen-Sentencia-Campo-Algodonero.pdf

Inter-American Court of Human Rights. (2009b). *Caso Radilla Pacheco vs. Estados Unidos Mexicanos Sentencia del 23 de noviembre del 2019* [Judgment Resolution]. Retrieved April 14, 2025, from https://www.corteidh.or.cr/docs/casos/articulos/seriec_209_esp.pdf

Jasanoff, S., & Kim, S. H. (Eds.). (2015). *Dreamscapes of Modernity. Sociotechnical Imaginaries and the Fabrication of Power*. The University of Chicago Press.

Jiménez, H. (2016, March 28). Buscan crear base nacional de ADN. *El Universal*. https://web.archive.org/web/20240102101323/https://www.eluniversal.com.mx/articulo/nacion/politica/2016/03/28/buscan-crear-base-nacional-de-adn/

Kahwagi Macari, J. (2012, February 9). *Decreto que reforma los artículos 5°, 14, 20 y 129 de la Ley General del Sistema Nacional de Seguridad Pública*. Retrieved April 14, 2025, from http://sil.gobernacion.gob.mx/Archivos/Documentos/2012/03/asun_2860865_20120313_1331052107.pdf

Koenig, B., Lee, S., & Richardson, S. (Eds.). (2008). *Revisiting race in a genomic age*. Rutgers University Press.

Krimsky, S., & Sloan, K. (Eds.). (2011). *Race and the genetic revolution: Science, myth, and culture*. Columbia University Press.

La crisis forense en México: más de 52 mil personas fallecidas sin identificar. (2021, August). *Movimiento por nuestros desaparecidos en México*. https://movndmx.org/wp-content/uploads/2021/08/Informe-La-Crisis-Forense-en-Me%CC%81xico.pdf.

La PGR anuncia una base de datos de perfiles genéticos. (2014, June 5). *Expansión*. Retrieved April 14, 2025, from https://expansion.mx/nacional/2014/06/05/la-pgr-anuncia-una-base-de-datos-de-perfiles-geneticos

Latin America Identity Verification Market Size & Outlook, 2030. (2025). *Grand View Research*. Retrieved April 14, 2025, from https://www.grandviewresearch.com/horizon/outlook/identity-verification-market/latin-america

Leonelli, S. (2016). *Data-Centric Biology*. University of Chicago Press.

Mbembe, A. (2020). *Necropolítica*. Melusina.

M'charek, A., & Wade, P. (2020). Doing the individual and the collective in forensic genetics: governance, race and restitution. *BioSocieties, 15*(1), 317–328. https://doi.org/10.1057/s41292-020-00193-6

México, "una gran fosa clandestina": Encinas; presentan plan de búsqueda de desaparecidos. (2019, February 4). *Proceso*. Retrieved April 14, 2025, from https://www.proceso.com.mx/nacional/2019/2/4/mexico-una-gran-fosa-clandestina-encinas-presentan-plan-de-busqueda-de-desaparecidos-219692.html

Mónaco, P. & Pérez, W. S. (2021, December 13). Traficantes de ADN, *Gatopardo*. Retrieved April 14, 2025, from https://gatopardo.com/reportajes/traficantes-de-adn-central-adn-y-la-pgr-desaparecidos/

Nelson, A. (2016). *The social life of DNA : race, reparations, and reconciliation after the genome*. Beacon Press.

Nowak, M. (2017). *Human rights or global capitalism. The limits of privatization*. University of Pennsylvania Press.

ONU-DH Mexico [@ONUDHmexico]. (2025, April 4). Fragmento del discurso de clausura del Presidente del Comité de la ONU contra la #DesapariciónForzada [Post]. X. Retrieved April 14, 2025, from https://x.com/ONUDHmexico/status/1908209966864830524

Personas desaparecidas y no localizadas. (2025, April 12). *RNPDNO*. Retrieved April 14, 2025, from https://versionpublicarnpdno.segob.gob.mx/Dashboard/ContextoGeneral

Pollock, A., M'charek, A., Ehlers, N., Creary, M., & García-Deister, V. (2021). Race and biomedicine beyond the lab: 21st century mobilisations of genetics—Introduction to the special issue. *BioSocieties, 16*, 433–446. https://doi.org/10.1057/s41292-021-00261-5

Rosas, T. (2018, August 15). Preparan Banco Nacional de Datos Genéticos para enfrentar desapariciones. *Excelsior*. Retrieved April 14, 2025, from https://www.excelsior.com.mx/nacional/proponen-base-nacional-de-datos-geneticos-para-enfrentar-desapariciones/1258866

Rose, N., & Novas, C. (2005). Biological citizenship. In A. Ong & S. Collier (Eds.), *Global assemblages: Technology, politics and ethics as anthropological problems* (pp. 439–463). Blackwell.

Secretaría de Gobernación. (2012, April 23). *Diario Oficial de la Federación* (Acuerdo A/078/12). Retrieved April 14, 2025, from https://dof.gob.mx/nota_detalle.php?codigo=5244766&fecha=23/04/2012#gsc.tab=0

Secretaría de Gobernación. (2020, August 26). Con Centro Regional de Identificación Humana, habrá un antes y un después en tareas de búsqueda e identificación de personas desaparecidas [Press release]. Retrieved April 14, 2025, from https://www.gob.mx/segob/prensa/con-centro-regional-de-identificacion-humana-habra-un-antes-y-un-despues-en-tareas-de-busqueda-e-identificacion-de-personas-desaparecidas?idiom=es

Secretaría de Gobernación & Subsecretaría de Derechos Humanos, Población y Migración. (2019). *Plan de implementación de la Ley general en materia de desaparición forzadas de personas, desaparición cometida por particulares y del Sistema Nacional de Búsqueda*. Retrieved April 14, 2025, from https://lopezobrador.org.mx/wp-content/uploads/2019/02/SEGOB_Plan-Busqueda.pdf

Smith, L. (2016). Identifying democracy: Citizenship, DNA, and identity in post-dictatorship Argentina. *Science, Technology & Human Values, 41*(6), 1037–1062. https://doi.org/10.1177/0162243916658708

Smith, L., & García-Deister, V. (2020). Migrant flows and necro-sovereignty: the itineraries of bodies, samples, and data across the US-Mexico borderlands. *BioSocieties, 15*(3), 420–437. https://doi.org/10.1057/s41292-019-00166-4

Torres, M. (2022). Huesos y ADN: autoorganización forense y ADN ciudadano en México. In V. García-Deister (Ed.), *ADN, protagonista inesperado. Promesas y realidades de la investigación genética ante nuestra crisis forense* (pp. 135–158). Siglo XXI.

13

Infrastructures, Power, and Filiation: Genetic Information and the Quest for the "stolen babies" of Spain

Miguel García-Sancho

In October 2023 Antonio Alonso, then director of the Spanish Instituto Nacional de Toxicología y Ciencias Forenses (INTCF), told the influential *El País* newspaper that he did not believe in "a plot to abduct newborns" from hospitals in the country since the instigation of democracy (Ansede, 2023). The INTCF had created a centralised genetic repository in 2012 to address the growing number of complaints about babies having been separated from their biological parents without the latter's full, free consent. According to the allegations, the baby severances started during the dictatorship of General Francisco Franco (1939–1975) and lasted well into the establishment of a democratic regime, with the most recent reported episodes occurring during the 1990s and early 2000s. The most extreme instances, some of them resulting in court cases, involved mothers being deceived about the death of their offspring after

M. García-Sancho (✉)
Science, Technology and Innovation Studies, University of Edinburgh, Edinburgh, Scotland
e-mail: Miguel.GSancho@ed.ac.uk

V. Argudo-Portal et al. (eds.), *Mundane Genomics*,
https://doi.org/10.1007/978-981-95-4550-6_13

birth when the children had actually been given to other families in adoption.

Both grieving families and now-adult adoptees fear having been the victims of such practices when they find gaps or inaccuracies in their records: adoption, death, or burial certificates, among other documents. This has led a substantial number of them to look for their biological relatives and, sometimes, to obtain and share online their genetic profiles. The INTCF repository seeks to place these genetic data—mainly obtained via commercial ancestry companies—on the same platform in order to narrow down the search.

If a match between the DNA sequence markers of two or more users is found, the INTCF technicians can verify the existence and type of family kinship: parental, sibling, or other. However, no matches have been found so far in the INTCF repository, and its user pool is substantially lower than the estimated number of adoptees and families that have contracted a DNA profile. This, along with the results of genetic analyses that the INTCF has conducted at the graves of suspected *stolen babies*—no conclusive evidence of supplanted remains was obtained—led Alonso to be sceptical about a widespread phenomenon; he suggested that a more plausible explanation of paper trail inconsistencies could be irregularities committed at the time of the adoptions (Ansede, 2023).

In 2018, a Spanish court found as fact that, 49 years earlier, a gynaecologist had facilitated the registration of the baby of a single mother as the natural offspring of the adoptive family. After the court sentence, the now-adult infant, Inés Madrigal, found her siblings and aunt via a commercial genetic ancestry company and they told her that the biological mother—by then deceased—had knowingly relinquished the baby for adoption ("DNA Database helps", 2019). Alonso argued that the landscape of residential institutions for unmarried pregnant women during the Franco era may have triggered an enormous number of irregular adoptions like that one but not necessarily the "abduction of newborns". The journalist then asked if those women had "consciously" given up their babies. Alonso replied that the "majority" had probably done so, but "forced by necessity" (Ansede, 2023).

In what follows, I zoom into the institutional architecture of adoption during Franco's regime to query the notion of *conscious relinquishment*, understood as the action of giving up their babies by the single mothers. I show that the *necessity* of those mothers was not merely shaped by their youth, inexperience, or lack of resources; it was the product of an infrastructure designed to remove their agency in decisions over the future of their offspring. This infrastructure was embodied in a tightly knit assemblage of people, organisations, documents, scriptural technologies, and socio-cultural representations that spread from the distant and threatening authorities that governed the country to penetrate the local environments of the single mothers: the parishes of their place of residence, their day-to-day communities, and even their homes and families. The infrastructure that pushed unmarried women to give up their children for adoption was historically situated and changed over time; however, it always reflected a class system and power stratification in which the single mothers were invariably at the bottom.

The chapter begins by unpacking the concept of infrastructure with reference to literature from Science and Technology Studies (STS) (e.g. Star & Ruhleder, 1996). I argue that understanding adoption systems as infrastructures strengthens existing investigations into stolen babies and other practices of coercive infant removal in Spain. Studying the adoption infrastructure that emerged under Francoism provides a better insight into the subaltern role of women and complements recent investigations that have emphasised the particularly oppressive environment that single mothers faced during the dictatorship and the early years of democracy (García del Cid Guerra, 2017). This approach also fruitfully enables scholars to combine the study of infrastructures with the investigation of politics of memory (Marre & Gaggiotti, 2021).

I then look at the role and place of genetics within the adoption infrastructure of Francoist Spain and argue that a eugenic discourse justified the aim of severing any hereditary transmission from the biological families to the children. This required removing the infants from their birth mothers at the earliest opportunity, regardless of the mothers' wishes. I contend that the circulation of people, artefacts, and power throughout the infrastructure coerced the women and stripped them of the agency required for the *conscious relinquishment* of their babies. Although the situation spilled over democracy, in the mid-to-late 1980s the Spanish adoption system was significantly revamped and the biological parents were recognised as consent subjects. A growing consensus emerged that the severed families needed to be repaired.

The chapter finishes by showing that the molecularisation of genetics and the recasting of inheritance from an invisible force to a corporeal DNA sequence provides a unique opportunity to reverse the family severance practices. Yet I argue that for this reversal to be complete, it is essential that the biological mothers and families recover their agency and contribute their experiential knowledge to the public interpretation of the genetic results. Such active participation has not yet fully occurred in the Spanish DNA bank. As I conclude, the obstacles to the involvement of the affected families signal a pervasive problem in the deployment of genetic technologies outside the laboratory, which requires a rebalance of pre-existing hierarchies and power dynamics. In the case of the INTCF repository, mothers and families need to overcome their historical subaltern role and be given the opportunity to uncover the coercive nature of the adoption infrastructure they endured.

Dictatorship, Adoption, and Infrastructure

In a seminal scholarly contribution to STS, Star and Ruhleder considered infrastructure the foundational strata in which social life occurs. From roads to sewage to information networks, infrastructures shape key aspects of the day-to-day running of social ecosystems, including how knowledge is generated, spread, and accepted. Infrastructures involve people, tools, technologies, protocols, and some form of authority—hierarchical or communal—that delimits how they work. Since they are conceived to survive their creators, after their implementation they sink "into an invisible background" where their start and end dates become blurred (Star & Ruhleder, 1996, p. 112).

Infrastructures display a remarkable equilibrium between consistency and change, mutating over time in some aspects while always being recognisable (Larkin, 2013). When existing infrastructures are in place, attempts to create new ones must build on their predecessors or, at least, engage with them, even if the objective is to counter their influence (Edwards, 2002). This is why a complete reversal of an infrastructure is a task that requires changing the habits and frames of mind that have been cultivated over however long the previous *status quo* endured.

In this chapter, I approach the adoption system established during the Spanish Francoist dictatorship as an infrastructure. This enables me to show that, in certain cases, the dictatorial authorities forcibly used the adoptions to sever kinship filiations between the adoptees and biological families they considered undesirable. This was conceived as a means to prevent any hereditary transmission of personality, behaviour, or ideology. The stolen babies and other instances of adoptions in which the biological families were coerced are the result of the work of this infrastructure and its persistence over time.

Existing literature, from journalistic investigations to sociological and legal scholarship, has uncovered the intricate assemblage of organisations, totalitarian ideology, fundamentalist religious values, and gynaecological practices that enabled the separation of infants from their biological families (Arroyo, 2013). It has documented how enacting such severances required the coordinated action of different types of professionals in geographically and institutionally distinct spaces, as well as the mobilisation of technologies to break the official paper trail between the infant's adoptive and birth parents (Duva & Junquera, 2011, see figure on p. 161). The nature of the adoptions and their enabling networks changed over time and, from the late 1950s onwards, shifted from state-led political repression to more decentralised forms of moral punishment for single mothers (Rabidoux et al., 2020). Both networks and adoptions survived the Spanish democratic transition (1975–1978) and current attempts to repair the affected families, including the INTCF databank, have not fully accomplished their objectives (Jimeno Aranguren, 2018). Some scholars have attributed these shortcomings to constraints imposed by the Spanish "infrastructures of memory", which prevent uncovering certain aspects of the past due to the influence that some Francoism enablers still have in the country today (Marre & Gaggiotti, 2021; see also Rubin, 2018).

My approach to the adoption system as an infrastructure focuses on an aspect that the existing literature has only tangentially examined. Along with other historians (e.g. Baelo Álvarez, 2023), I investigate the Francoist adoption legislation and concur that its definition of child abandonment was rather vague. This was shaped by the urgent necessity of housing the thousands of orphans and foundling children stemming from the Civil War that preceded the dictatorship (Vinyes et al., 2003). However, as I show, it also empowered the state to appropriate infants without the explicit consent of their biological families.

Recent investigations have shifted their emphasis from characterising the enabling networks to portraying the insurmountable pressure to give up their offspring for adoption that the single mothers endured (Roig Pruñonosa, 2018). They have uncovered a network of boarding schools that lasted until the early 1980s—well into the era of democracy in Spain—with the objective of re-educating the minds and bodies of women considered morally loose, among them those expecting a baby from an extramarital relationship (García del Cid Guerra, 2017). These institutions have been compared to the Irish laundries where young single mothers were also stripped of their babies (Iglesias Aparicio, 2021). My infrastructural gaze documents the lack of agency and voice of the birth mothers—and other members of the biological families—within the socio-technical assemblages underpinning the adoptions.

I have also conducted oral histories and investigated the personal archives of three women adopted during the last stages of Franco's dictatorship, whom I name Marta, Rosalía, and Piedad.[1] In spite of gaps in their birth and adoption records, they found their biological siblings almost 40 years after their separation. Their lived experiences provide an insider account into the coercive nature of the adoption infrastructure and the long-term effects of the family severance practices emerging from it. These testimonies also portray the concealment of any public documentary trace of filiation as a means to prevent contact with the biological family, something that would have potentially reignited the hereditary flux and revealed the coercion underlying the adoptions.

In what follows, I argue that uncovering this infrastructure and its coercive work is essential to transforming the genetic data of the adoptees and families into restorative narratives. The technical reports in which the

[1] Because of the enormous emotional charge of their stories and the power that some institutions involved in the adoptions still have in Spain today, I decided to pseudonymise my three interlocutors. The oral history with Marta and Piedad was conducted in Valencia on 18th June 2024. Following it, I also spoke to Rosalía—Marta's biological sibling—via WhatsApp on 25th July 2024.

INTCF scientists periodically interpret the results of their repository[2] show that the potential of this DNA bank is founded on the possibility of retrieving what the Francoist infrastructure believed it had severed for good: the hereditary linkage between the adoptees and their biological families. Yet I contend that molecular genetic technologies cannot bring to light the experience of the affected communities and the gruelling constraints under which they operated. These stories are essential to understand the behaviour of the birth families, regardless of them being *conscious* or not of the adoptions.

Francoist Adoptions: From State Repression to Decentralised Practices

In 2004, Marta, one of my oral history subjects, received a life-changing call. She was urged to switch the TV on to a daytime programme where a person almost identical to her was looking for her twin sister. Marta had found out many years before that shortly after her birth in the late 1960s, she had been adopted at a foundling home in North Spain, more than 1000 kilometres away from the town where she subsequently grew up. Her adoptive parents had told her that at the time of the adoption, they had been assured that she had no siblings. The twin sister, Rosalía, had been told otherwise by her adoptive mother who, shortly before taking her from the same foundling home, accidentally found out about the twin sibling. She was, however, informed that Marta had already been adopted and that it was impossible to reverse the process; in her account, she was even threatened with cancellation of Rosalía's adoption if she tried to retrieve her sister (Rosalía, oral history with author, WhatsApp, 25th July 2024). Marta describes watching the image of her twin sister on the TV screen as "getting out" of her "own body" and then "coming in again" (Marta, oral history with author, Valencia, 18th June 2024).

[2] This technical literature will be fully referenced below. The annual reports of the INTCF are available at https://www.mjusticia.gob.es/es/institucional/organismos/instituto-nacional/documentacion/memorias (accessed July 2025). Until after the copy-editing of this chapter, the latest annual report accessible was from 2022.

Marta's adoptive father had told a family relative who was facing difficulties having a baby about the foundling home. The relative and his wife also adopted an orphan girl there, Piedad, two years older than Marta, who also unexpectedly encountered previously unknown siblings in her adulthood. In her case, after several decades of attempts, Piedad's siblings had persuaded an authority in the religious order that ran the foundling home to reveal her contact details so they could explain to her the circumstances of her adoption. Piedad came from a financially humble family and, when she was a baby, had been left in the home by her grandfather following the death of her birth mother. However, he always sustained that this was intended as a temporary relief measure while organising new care logistics, and that when he returned to the home Piedad had already been adopted and his attempts to claim her back were aggressively dismissed. The absence of Piedad caused lasting grief in the biological family, and lifelong efforts to trace her back (Piedad, oral history with author, Valencia, 18th June 2024).

The foundling homes whence Piedad, Marta, and Rosalía were adopted predated Francoism. In her history of midwifery in Spain, Ruiz-Berdún (2022) has argued that the pervasive high levels of mortality at these institutions may not only have been due to the poverty of the foundling children and the harsh living conditions; rather, these figures potentially conceal child trafficking and illegal adoptions. She has documented the existence of these practices since the seventeenth century, and showed that the sale or purchase of infants, as well as the pretence of pregnancy, were legally banned behaviours back then. Informal networks between the foundling homes, the vulnerable mothers or relatives who left the children, and wealthier adoptive families sustained the transfers of the infants, which were sometimes mediated by bribed medical professionals. Midwives would often feature at the heart of these scams and end up as scapegoats, saving more powerful male practitioners from being persecuted (Ruiz-Berdún, 2022, ch. 6 and pp. 307ff).

By the time Franco's dictatorship started in 1939, two important transformations were occurring within the foundling homes. Firstly, the number of orphans living there had substantially increased due to the coup d'état and the vicious Civil War against the prior democratic authorities that General Franco perpetrated before his rise to power (Vinyes et al.,

2003). Secondly, a fledging infrastructure created by the dictatorial regime and fostered by its anti-Communist, ultra-Catholic, and Nationalist ideology allowed the admission into these institutions and subsequent adoption of children that were neither orphans nor abandoned by their families.

At the core of the socio-technical assemblage underpinning this infrastructure was Antonio Vallejo-Nágera. A military physician specialising in psychiatry, he became one of the public faces of Francoist science with a eugenicist discourse that associated Marxism with a number of mental and physical conditions (Novella, 2024, especially ch. 3). Vallejo-Nágera had already been a prominent scientific and political figure before Franco's dictatorship. In 1917, he was appointed military attaché to the Spanish Embassy in Berlin and became fascinated with the degenerationist theories that were then flourishing in Germany. An author who decisively impacted Vallejo-Nágera and whom he translated into Spanish was Emil Kraepelin, who investigated the heritability of several mental disorders and is considered to be the father of psychiatric genetics.

Back in Spain, Vallejo-Nágera became increasingly hostile towards the Parliamentary Republican regime that preceded the dictatorship, and was a key supporter of Franco's coup d'état. During the Civil War (1936–1939), he conducted a number of experiments with prisoners in which he attempted to show a statistical correlation between "Marxist" individuals and certain dangerous "bio-psychic conditions", such as psychosis and alcoholism.[3] Vallejo-Nágera used the term Marxism as a portmanteau that included any sympathising with the wartime enemies or support for the democratic order.

Franco's victory and his self-appointment as Chief of State and Government (1939), led Vallejo-Nágera's doctrine to evolve towards a less genetically deterministic and more Neo-Lamarckian approach. This stemmed from the increasingly strong Catholic orientation of the dictatorship, the inviolability of foetal life and birth within this religious system, and its incompatibilities with Darwinism. Within Franco's regime, Vallejo-Nágera defended the proposition that the strength of hereditary

[3] A selection of Vallejo-Nágera's writings, as well as a broader discussion of his doctrine and the practices it enabled can be found in (Vinyes et al., 2003 especially pp. 233ff).

transmission could be attenuated through school education, family values, and other forms of intervention that countered the power of genetics (Campos & Huertas García-Alejo, 2012). In practice, this involved discouraging infanticide, forced sterilisation, and other strategies that had characterised Nazi eugenics—his previous source of inspiration—and advocating instead the separation of children from parents considered to have *Marxist* inclinations. Rather than preventing socially undesirable people reproducing, Vallejo-Nágera thus suggested removing those babies and raising them in other families.

These Neo-Lamarckian eugenics informed the repressive policies of the Francoist State. A number of decrees issued between the end of the war and the beginning of the dictatorship ensured that the legal custody of children of politically dissenting families could be transferred without consent to organisations aligned with the new authorities. This allowed young infants to be taken by force from female prisons or from the relatives of political prisoners who were raising them. Although the ultimate goal was to have these infants adopted, their first destination was often the foundling homes (Cenarro, 2009; Vinyes et al., 2003).

At the heart of this developing infrastructure was a memory regime. Star and Bowker's scholarship has stressed that a principal function of infrastructures is the creation of an institutional memory through which organisations, from schools or hospitals to more complex political systems, can order the past according to their future aspirations. This memory work involves information systems that preserve—or remember—certain people and facts while discarding—or forgetting—others (Bowker & Star, 1999, ch8 and pp. 40ff; see also Bowker, 2005). In the case of the Francoist adoptions, the past lineage of the children would be deleted or concealed from the records. The future ideal that the regime sought to fulfil with this intervention was that of redemption, an important objective of the dictatorship that consisted in breaking with a perceived past of decline and corruption. Following this severance with atheism, Marxism, and permissive liberal governments, the regime aspired to redeem those who had not been fully corrupted and to build a future of convivial family life in a firmly and morally governed country.

Vallejo-Nágera's infrastructure was the biopolitical means (Cayuela Sánchez, 2014) of achieving this ideal. Removing the children and forgetting their genetic filiation ensured breaking the potential continuity with the past embodied in hereditary transmission from their biological families. Once adopted, the infants would only remember their new homes, and a better, hopeful future could be built under the influence of their compliant adoptive families. In other words, the infrastructure would transform the children into blank slates on which their future (and the future of the nation) could be rewritten.

This infrastructure acquired a life of its own as Francoism unfolded. Vallejo-Nágera passed away in 1960, by which time the regime was attempting to project a less repressive and more liberal-oriented image (De la Torre & García-Zúñiga, 2014). A law issued in 1958 epitomised this new political façade by partially replacing the prior adoption regulation while maintaining some of the underlying practices. The preamble stated that adoption, once a rather "unused" option, was enjoying a "thriving vitality" at that time. It hinted at some terminological confusion and endeavoured to distinguish adoption from other less formal transfers of parental custody that had been historically used before the law, especially for foundling and abandoned children. A specific category, called "full adoption", was established for those types of infants and distinguished from the "less full" adoption that applied to all other adoptees. However, the law did not explicitly supress or restrict the prior, less formal transfers of infant custody connected to unofficial childcare by people other than the biological parents—on a temporary or permanent basis.[4]

For an infant to be eligible for full adoption, the 1958 law required their having lived a minimum of three years in a foundling home without having been recalled by their biological families. Yet the capacity to prove this circumstance lay almost entirely with the foundling home authorities. The law enabled the biological parents to contest the adoptions, but specified that for their challenge to be successful, they needed to prove to

[4] "Ley de 24 de abril de 1958 por la que se modifican varios artículos del Código Civil", *Boletín Oficial del Estado,* número 99, p. 730 (quotes translated by myself). The less formal modalities of transfer of child custody to which the law referred—without defining them—were *acogimiento*, *afiliación*, and *prohijamiento*. According to historians of adoption (Baelo Álvarez, 2023, especially ch. 10), they have been used in Spain since at least the Middle Ages and refer to regimes of childcare outside the parental realm that do not involve the same legal obligations as our current definition of adoption.

the judge in charge of the proceedings their "morality", "good conduct", and "utter lack of culpability" in the abandoned or foundling status of the children. It was explicitly anticipated that these recalls would be "infrequent" and that biological families would not be able to "uproot" fledging adoptees from their new "family and social environment" by "invoking alleged rights whose concomitant duties [they] did not address".[5]

These restrictions, along with Francoism's repressive track record, made it almost impossible for the biological families to reclaim their children. Relatives such as Piedad's grandfather were dismissed or threatened, or simply too scared to emphasise the temporary nature of the foundling situation, note a change of circumstances that enabled them to look after the children, or just declare their change of heart and the wish to take the infant back. Furthermore, the co-existence of the other, less formal modalities of transfer of parental custody enabled the foundling homes to send the children away before the three years, sometimes to distantly geographical areas. In Piedad's personal archive, we can see records documenting that when she was two-and-a-half-years-old, her adoptive parents took her to the city where they lived—one of the most distant locations in Spain from her place of birth—under the modality of *prohijamiento*, which the 1958 law recognised as less formal than adoption. The adoption itself was not sanctioned until almost a year later.[6]

The foundling homes were run by different Catholic orders which, under Franco, had gained increasing power over society and state affairs. This meant that each of these orders could exert almost unlimited control over the populations of a town or neighbourhood via parishes that had the local church at their base. The influence of the parishes spread through an institutional ecosystem that disciplined different aspects of the lives of citizens. Unmarried pregnant women and their children were a principal target of that disciplining. Different spaces, ranging from mother and baby homes to boarding schools or flats managed by pious landladies, enabled these future mothers to spend their pregnancies away from the

[5] "Ley de 24 de abril de 1958 por la que se modifican varios artículos del Código Civil", *Boletín Oficial del Estado,* número 99, pp. 735 and 730 (quotes translated by myself).

[6] Letter to Piedad's adoptive father from the local government (Diputación) of the county where she was born, and "Adopción Plena" (affidavit of Piedad's adoption). I have removed the dates and locations from these documents in order to preserve Piedad's anonymity.

public eye. They fled a judgmental environment and sometimes were forced by their own families (Roig Pruñonosa, 2018).

These young pregnant women could circulate through the different spaces of the parishes with little scrutiny. The pressure they faced to give up their babies for adoption increased over the pregnancy and was overwhelming by the time of the birth. The support and accommodation that the women were offered was often explicitly or implicitly on condition of the future adoption of their child. Some of the nuns who ran the premises worked as nursing assistants in various medical institutions and had contacts with midwives and gynaecologists (Arroyo, 2013; Duva & Junquera, 2011). These professionals could thus discretely monitor the women and attend their labour at the mother and baby homes, shelter flats, or in their own private clinics. After the baby was delivered, it could be easily transferred to a foundling home and treated as an abandoned infant to be given in full adoption.

Once the mothers lost control over their children, the adoptions were brokered by the Diputaciones, institutions that governed the different counties of Spain and were headed by a civil governor appointed by the central Francoist authorities. Among their duties, the Diputaciones oversaw the foundling homes in their constituencies, representing them at the local court that, if all the requirements of the 1958 law were met, would authorise the adoption. The last step was the formalisation of the transfer of the legal custody to the adoptive parents by a notary: a specialised lawyer who would issue an affidavit validating the adoption agreement.

At this point, a key difference emerged between the full adoption of children deemed to be foundlings or abandoned and the rest of the adoptions that the law called "less full". While in the less full adoptions consent from the biological families was sought, in the full adoption modality the biological families were excluded from the agreement and replaced by representatives of the Diputaciones. This meant that no trace of the biological filiation of the children remained in the affidavit. Equally, if the adoptee had been separated from the biological mother at birth, it could have been registered as an offspring of unknown parents. This is the case of Piedad's adoption affidavit and Marta's birth certificate.[7] These records and the scriptural technologies that produced them were part of the information system of the adoption infrastructure and its overall memory regime: forgetting filiation by preventing any traceability between adoptees and biological families (see Fig. 13.1).

In 1970, a new adoption law reduced the abandonment requirement of the infants from three years to six months and made the foundling homes increasingly secondary.[8] By that point, all the actors and institutions promoting the adoptions had attempted to make them look more

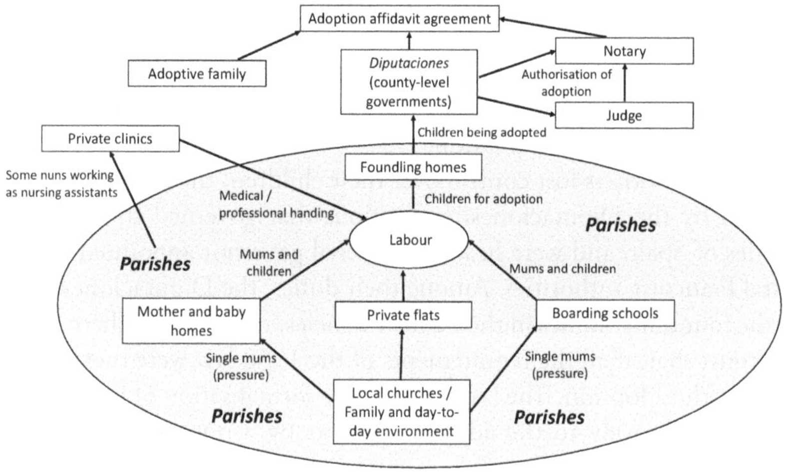

Fig. 13.1 A schematic representation of the Francoist adoption infrastructure and its impact on single mothers and allegedly abandoned children. The biological families lost control over the infants once they were admitted into a foundling home and, until 1987, the rest of the adoption process could be completed in their absence. (Elaborated by author)

[7] Notary affidavit of Piedad's adoption (see note 6) and full birth certificate of Marta. In order to preserve their anonymity, I have removed the dates and locations from these documents, which were produced shortly after Piedad and Marta's births in the late 1960s. Marta's biological parents are recorded "José" and "María". The full birth certificate of Piedad does record the name and surnames of her biological mother, with "José" featuring as her biological father. Yet in Spain, full birth certificates need to be specifically requested at the Civil Registry and Piedad did not have access to one until after meeting her biological siblings.

[8] "Ley 7/1970 de 4 de julio, de modificación del capítulo V del título VII del libro 1 del Código Civil sobre adopción", *Boletín Oficial del Estado,* número 161, pp. 10702–10704.

humane and less repressive. Yet the very infrastructure that Vallejo-Nágera had implemented during and after the Civil War—its values and underlying ideology—persisted and allowed the continuation of the child removal practices. At a purely institutional level, the Francoist imprint was present in the Diputaciones, the arbitrariness of the dictatorial law, and the contempt towards families that did not conform to the ultra-conservative, Catholic pattern. The adoption infrastructure is what stitched all these elements together and camouflaged political persecution as legal procedures.

At a deeper conceptual level, the memory regime on which the infrastructure had been originally founded continued to be endorsed—or, at least, unquestioned—by all the newcomers. The support for this regime was more explicit among some actors than others, like the clinicians and notaries, who may have acted out of fear, indifference, or financial gain. Nevertheless, none of them challenged the principle that the children would be better off with the adoptive families than with the biological single mothers. By action or omission, they contributed to an institutional choreography that enacted the original eugenic vision of the infrastructure: forgetting the hereditary legacy of the children as a means of opening a future of redemption with the adoptive family.

The biological relatives of the adoptees remained passive, powerless actors within the infrastructure. The restriction of their actions stemmed from a sheer lack of agency to consent. The overwhelming and oppressive presence of the parishes conferred on the religious orders the power to supplant or override the decisions of the single mothers on the future of their offspring. This agency theft started in the family or day-to-day local environment of the single mothers and continued all the way to the registration of the children at birth and adoption.

Throughout her scholarship, Star has warned of the shortcomings of merely focusing on the most visible or vocal actors when reconstructing an infrastructure. Subaltern voices tend to inhabit the liminal spaces that the workings of the infrastructure make invisible, both intentionally—as in the Francoist adoption infrastructure—but also unintentionally (Star, 1990). Building on this, Puig de la Bellacasa has called for social scientists to look at and be attuned to the "spaces between" the nodes of a sociotechnical network (Puig de La Bellacasa, 2016, p. 48ff). Feminist scholarship and an emerging historiography of Francoism is starting to achieve this by showing the devastating effects of the dictatorship on women, children, and other vulnerable populations (Cenarro, 2009). With regard to the issue of child removal, this literature emphasises the experience of the single mothers (Roig Pruñonosa, 2018) rather than just reconstructing the enabling infrastructure and thus being co-opted by its concealing effects. By looking at the underlying legislation and the experience of three different children, I have attempted to make the coercive nature of the infrastructure both visible and felt. In what follows, I demonstrate that uncovering this coercion is essential to any attempt to restore the broken ties.

Democracy, Continuity, and Consent

The death of Franco in 1975 and the gradual emergence of a democratic regime led to a growing number of voices denouncing the child trafficking practices behind the adoption system. In 1982, an impactful report published by an investigative magazine showed a photograph of a stillborn foetus stored in the freezer of a private maternity clinic. According to the journalist, it was used to make some mothers believe that their baby had died when it had really been given in adoption (Iglesias, 1982).[9] The clinic had closed shortly before this publication following a police operation to prevent an alleged illicit trade with children that was never brought to court (Duva & Junquera, 2011: 17ff). Despite the use of the foetus not being verified, and the gynaecologist in charge of the clinic claiming that it was awaiting an autopsy, the theft of babies in maternity wards became an increasingly real possibility in the eyes of the public.

In 1987, a new adoption law was passed by the third democratically elected Government of Spain. In its preamble, it specifically stated that a principal motivation of renewing the regulatory landscape was to tackle the "almost absolute lack of control" that "allowed, on occasion, the

[9] The gynaecologist in charge of that clinic was the same one that, in 2018, was accused of having registered a baby as the natural offspring of an adoptive mother (see start of the chapter). The false registration was proven, but he was never convicted, because the court considered that the statute of limitations for the offence had expired.

hideous trafficking of children, denounced by the mass media". One of the ways in which this was addressed was by replacing the vague notions of abandoned and foundling infants of the prior legislation, as well as the informal alternatives to adoption it had upheld, with the specifically defined categories of "neglect" and "fostering". The placement of a child with a foster family was described as a prior and desirable step towards an adoption, one that would be "frequently used", although in some cases it could be skipped in the best interest of the children. This step was reversible and could be undone by the biological parents unless they had been stripped of the legal custody of their offspring. For a child to be considered neglected, legal custody needed to be explicitly and traceably transferred from their biological parents to a state authority. This transfer could be voluntary—and always recorded in the presence of witnesses—or forceful, in which case the authority would need to justify the reasons for the removal of custody. There was thus a detailed and formalised procedure pertaining both to express consent and to stripping biological families of this right in extreme cases.[10]

This new legislation increasingly hindered the workings of the Francoist child removal infrastructure, eliminating components outside state control—such as the notaries—and providing the public administration with the duty of controlling and managing the process. Control was exerted through the newly devolved regional governments that the democratic regime had promoted. Unlike the Diputaciones, these governments were elected by local citizens rather than directly appointed by the central state apparatus. Since their political orientation could be different from that of the central government, their function was one of decentralising and scrutinising power rather than fulfilling a nation-wide ideology.

Another fundamental change introduced by the 1987 law was that of giving a voice to the previously excluded biological families, along with mechanisms for that voice to be heard and have effects in the adoption process. This meant that the process could no longer operate regardless of what the biological parents thought; they could express their opposition

[10] "Ley Orgánica 6/1987 de 11 de noviembre por la que se modifica la sección III del capítulo 4°, título XIII del libro II del Código Penal", *Boletín Oficial del Estado*, número 275, pp. 34158-34159 (quotes translated by myself). I have translated throughout the terms *expósito*, *abandonado*, *acogido* and *desamparado* as *foundling*, *abandoned*, *fostered* and *neglected*, respectively.

to the adoption at different stages, thus exerting their agency in the infrastructure through a regulated consent protocol. The role of the biological mothers was especially important. In voluntary adoptions, they could not confirm their consent earlier than a month after birth and this consent was normally ratified at least once during the process. The infants could still be taken away from their biological parents against their will but, crucially, the authorities saw their role as monitoring and preventing abuses in the process rather than advocating or brokering the adoptions.

As the law stated in its preamble, child trafficking was becoming a growing social concern. From the end of the 1980s, and into the 1990s and early 2000s, media outlets continued to investigate the phenomenon, and TV programmes, like the one where Marta found her twin sister, shared the experiences of affected people (Durán Rodríguez, 2022). This led to an avalanche of cases in which the problem was almost exclusively framed as one of deceiving mothers after birth about the death of their infants and concealing the transfer of the babies' filiation to other families. The underlying adoption process and coercive pre-1987 infrastructure was thus de-emphasised in favour of a narrative that built on the shocking image of the newborn in the freezer.

In spite of the control mechanisms implemented by the democratic authorities, a substantial number of the reports had occurred in public hospitals after the death of Franco, and some of them after the new adoption law. This persistence of repressive Francoist practices under democracy shocked Spanish society and led to the increasing public notoriety of the stolen babies. The persistence can, however, be understood from the viewpoint of the continuity of the infrastructure underpinning and sustaining those practices. As scholars have shown, infrastructures are rarely created *de novo* and need to either adapt to or oppose pre-existing ones (Edwards, 2002). The institutional, medical, and social care assemblage of the new-born Spanish democratic regime had to face decades of ongoing Francoist infrastructures. Some elements of these dictatorial infrastructures were left lingering temporarily or indefinitely, in some cases by conscious decision and, in others, due to oversight, neglect, overcautiousness, or lack of resources. Other preceding infrastructural components were replaced in processes that faced delays and resistance, and in some instances are still ongoing.

In the case of the child removal practices, there was a ten-year gap between the first democratic elections (in 1977) and the new adoption law. The cases did not stop immediately and, even today, Marre and Gaggiotti have argued that the existence of those obscure adoptions has become a "public secret" or a series of "unknown knowns". Building on Rubin, they attribute those ambiguities to enduring "infrastructures of memory", constellations of symbols and entrenched power dynamics that constrain the aspects of the past that can be brought into present-day Spain (Marre & Gaggiotti, 2021; Rubin, 2018). Yet, over the last 15 years, attempts to reverse the separation between the children and their biological parents have been deployed by the democratic Spanish authorities. Rather paradoxically, these attempts present overlaps with the eugenic vision that originally justified the separations: both consider genetic inheritance their point of departure.

What DNA Does Not Show

Marta's youngest son gave his mother a special gift for a recent birthday, submitting a swab of her cheek cells to MyHeritage, a multinational DNA analysis company. This has enabled Marta to compare her genetic profile with those that are publicly shared online by other users of the firm. She hopes that this will lead to DNA sequence matches with other biological siblings. Her dream and that of her sister Rosalía is to reunite with their birth parents and learn about the circumstances of their adoption (Marta and Rosalía, oral histories with author. Valencia, 18th June 2024 and WhatsApp, 25th July 2024, respectively). Like them, thousands of Spaniards have obtained a DNA profile via a private company to find biological relatives.

The genetic repository with which this chapter started is an attempt by the Spanish Government to centralise those multiple and dispersed sources of DNA data (Jimeno Aranguren, 2018, p. 56ff). It is managed by the INTCF (Instituto Nacional de Toxicología y Ciencias Forenses), the public body that channels forensic research across the country. During the early days of the repository, in 2012, the INTCF sponsored a survey in which they found that 2286 people in Spain had requested a commercial DNA test because they suspected or knew they had been affected by baby abductions or coercive adoption processes. By then, some of these individuals were organised into a plethora of national and regional associations that had proliferated since the early 2000s. According to the survey, 72% of all the testing requests came from parents looking for their biological offspring, 15% from now adult infants, and 13% from siblings (INTCF, 2017). Building on this, the INTCF developed its databank as part of an office that assists affected people in their searches, also facilitating access to other public administration records, such as birth and death certificates, adoption files, and documentation about burials (INTCF, 2014).

Since their foundation, the INTCF repository and office have sought to retrieve what the preceding adoption infrastructure severed. Both repository and office are thus part of a new infrastructure deployed to undo the work of its predecessor. As with the Francoist adoption infrastructure, the INTCF bank and office are underpinned by a socio-technical assemblage of people, tools, and institutions coordinated in a choreography and operating according to protocols. The INTCF infrastructure is also connected to a memory regime that seeks to recall what the previous adoption infrastructure intentionally forgot. This is supported by a regulatory framework originally embodied in the Ley de Memoria Histórica (Historical Memory Law), approved by the Spanish Parliament in 2007 and expanded into the Ley de Memoria Democrática in 2022. Both laws have the objective of investigating the crimes of Francoism and offering reparation to victims, tackling the "pact of forgetting" to which the Spanish authorities and society tacitly agreed following the end of the dictatorship. This subtle avoidance of exposing the dictatorial repression is the cornerstone of the entrenched "infrastructures of memory" that have limited political regeneration in Spain (Marre & Gaggiotti, 2021; Rubin, 2018).

The INTCF infrastructure is also founded on an authoritative and promising scientific discourse. By the time the Francoist adoptions started operating in the 1940s and 1950s, genetic inheritance was an abstract property that could be concealed by breaking the flow of documentary information that recorded parent-to-offspring filiations. Yet new genetic techniques in the 1970s and 1980s recast inheritance as shared portions—markers—in the sequence of information of the DNA molecule (García-Sancho, 2015; Lynch *et al.*, 2010). The users of the INTCF service are offered the possibility of storing their markers in a shared repository that automatically detects matches from which family linkages can be reconstructed. Their genetic footprint thus persists regardless of the severance of all the other documentary chains of filiation by the preceding adoption infrastructure.

This genetic retrieval has, however, proved limited. Of the over 2000 genetic profiles that the INTCF estimated are stored in private laboratories, only 663 had made their way to the centralised repository by 2022—or 29%, assuming that the overall number of profiles has remained constant since 2012, the date of the survey (INTCF, 2023, p. 123). No DNA sequence matches enabling family reunions have been found so far (see Table 13.1).[11] The relatively small size and limited representativeness of the sample with which the INTCF works has been attributed to several factors: namely, lack of trust in the state authorities by the affected individuals, and financial and logistic barriers, given that users need to get privately tested before submitting their profiles to the repository (INTCF, 2017, pp. 35–37).

A parallel initiative of the INTCF has been digging up the graves where suspected stolen babies were allegedly buried and genetically analysing the remains, as well as tissue samples from post-mortem biopsies. The results, published in a 2017 report, have been controversial. On the one hand, in 90% of the samples that could be genetically analysed (81 out of 90), the DNA of the bone remains or tissue samples matched those of the parents, suggesting that the buried or biopsied baby was indeed their offspring. On the other hand, it was not possible to extract DNA from 39 bone remains and 2 biopsies, and in 3 exhumations no samples suitable for genetic analysis were found (see Fig. 13.2). The INTCF attributed this shortcoming—affecting almost one third of the 134 exhumations and biopsies on which they worked—to the degradation of the

[11] After the copyediting of this chapter, the INTCF made its 2024 annual report available. In it, the number of profiles in the repository had increased to 717 – 441 of them mothers looking for their biological offspring – and one case of possible half-siblings had been detected, which was being assessed (INTCF, 2025, pp. 124–125). No further news about the outcome of this assessment have emerged.

Table 13.1 Genetic profiles of people knowing or suspecting they have been affected by coercive adoptions or baby removal practices

	Private laboratories (2012 survey estimation)	Public, centralised repository of Instituto Nacional de Toxicología y Ciencias Forenses (2022 data)	DNA matches in public repository
Overall number of genetic profiles/DNA matches	2286	663 (29% of profiles in private laboratories)	0
Parents looking for adoptees (% of profiles)	72%	70.7%	0%
Siblings looking for adoptees (% of profiles)	13%	13%	0%
Adoptees looking for biological family (% of profiles)	15%	16.1%	0%
Other family members (% of profiles)	0%	0.2%	0%

The figures correspond to the centralised repository curated by the Spanish Instituto Nacional de Toxicología y Ciencias Forenses (second column) and a survey that this institution conducted among private DNA testing laboratories (first column). Elaborated by author. Data retrieved from INTCF (2023) "Instituto Nacional de Toxicología y Ciencias Forenses. Memoria 2022". Ministerio de la Presidencia, Justicia y Relaciones con las Cortes, p. 123 and INTCF (2017). "Informe de la actividad pericial y científica desarrollada por los servicios de biología del INTCF con respecto a la prueba del ADN en la investigación de adopciones irregulares y sustracción de recién nacidos en España". In "Comisión Nacional para el uso forense del ADN. Memoria de actividades 2016–2017", Ministerio de Justicia, Secretaría General Técnica, p. 35

biological materials over time. They argued that bone remains, especially those of babies, could undergo "total destruction" under certain burial conditions. According to the report, in two of the three instances where no analysable samples were recovered, the presence of cadaverous fauna, lanugo, surgical clothes, and umbilical cord clamps "were compatible with the prior existence of bone remains at the burial sites", specifically of

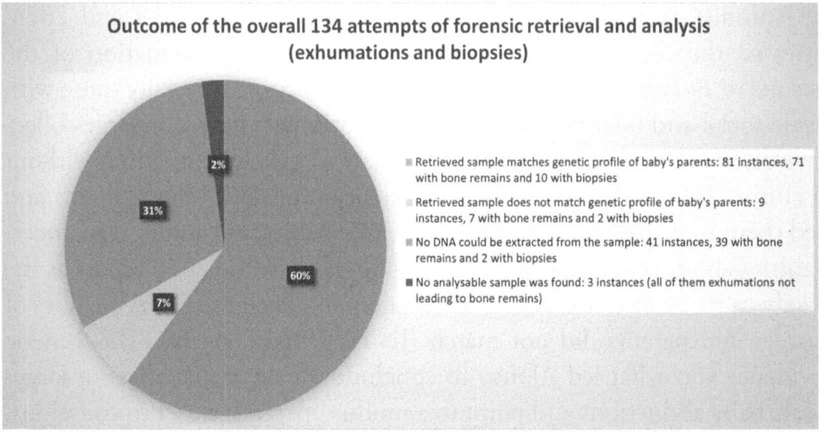

Fig. 13.2 Distribution of the retrieval and analysis of samples from exhumations and post-mortem biopsies of alleged stolen babies according to their genetic results (or lack thereof). The works were conducted by the Spanish Instituto Nacional de Toxicología y Ciencias Forenses (INTCF), and presented in a 2017 report. They were requested by either the Prosecutor's Office or courts investigating over 2000 legal complaints, 25% of which made it to trial. I refer to "attempts" and "instances" of retrieval and analysis rather than "cases", because some legal cases involved several exhumations and examinations of biopsy samples. In two of those cases, the bone remains corresponded to adult rather than baby individuals (these have been excluded from the overall statistics). Graph elaborated by author with percentages rounded to the nearest whole number. Data retrieved from INTCF (2017). "Informe de la actividad pericial y científica desarrollada por los servicios de biología del INTCF con respecto a la prueba del ADN en la investigación de adopciones irregulares y sustracción de recién nacidos en España". In "Comisión Nacional para el uso forense del ADN. Memoria de actividades 2016–2017", Ministerio de Justicia, Secretaría General Técnica (Annex IV, pp. 27–38) and Ansede, M. (2023, October 24). Antonio Alonso, geneticist: "I do not believe that there was a plot to abduct newborns from Spanish public hospitals during democracy". El Pais. https://english.elpais.com/spain/2023-10-24/antonio-alonso-geneticist-i-do-not-believe-that-there-was-a-plot-to-abduct-newborns-from-spanish-public-hospitals-during-democracy.html (Retrieved July 2025)

babies. Yet the report also mentioned two cases that had been excluded from the overall statistics, in which the bone remains corresponded to adult rather than infant individuals (INTCF, 2017, p. 29).[12]

[12] Another limitation of the results is that the overall cases on which the INTCF worked represented only between 6% and 6.5% of the over 2000 legal complaints in Spain involving stolen babies (Ansede, 2023).

Antonio Alonso, director of the INTCF between 2019 and 2024, stressed the difficulties of the exhumations in his explanation of the results to *El País*. Non-christened children in Spain, especially those with little social and financial capital, were traditionally buried in large collective areas in cemeteries rather than in named graves, sometimes without a coffin. This contributed to the degradation of their bone remains and led them to be merged with other deceased babies and adults. The uncertainties about burial sites were also highlighted in the 2017 report as the likely cause of six out of the nine instances in which the DNA analyses of babies and parents did not match (INTCF, 2017, p. 32). This genetic evidence was what led Alonso to conclude on the unlikeliness of large-scale baby abductions and point to a modus operandi closer to conscious, but irregular relinquishment of infants for adoption (Ansede, 2023).

His remarks were strongly challenged by some of the affected communities. Soon after the newspaper interview in which they were published, an association named Plataforma Internacional Te Estamos Buscando issued a statement claiming that there was "no documentary support" proving beyond doubt Alonso's interpretation of the INTCF results. One example was the discussion of the exhumations where no bone remains had been found. For the members of Te Estamos Buscando, the hypothesis of total destruction of the remains was rather surprising, since there were scientific reports of much older burial sites where corpses, including those of children, had appeared "mummified" and presenting both "bone and tissue". Furthermore, in several of the cases of stolen babies investigated over the last 15 years, official records "issued by the public administration" had demonstrated that the infants "were not in the cemeteries". The authors of the statement also contested the allegedly conscious relinquishment of the children, since the institutions that managed the adoptions had exerted "manipulation" and "psychological and physical abuse" over the biological mothers during both labour and pregnancy. They asked to see "signed" consent documents and felt "deeply offended" by the assertions (Foro Internacional Te Estamos Buscando, 2023).

This controversy underscores the ambiguities in the evidentiary power of DNA. Genetic data, in and of itself, can only indicate filiation between individuals and, in the context of the INTCF exhumations, discount baby thefts. When it is not available, as in a substantial number of the examined burial sites, its absence can be variously interpreted. While for the INTCF it is a consequence of the destruction or degradation of the bones, the affected communities view it as suggesting that their offspring may not be there. Similarly, the finding of adult rather than baby remains or a discrepancy between their genetic analysis and that of the parents can be understood as either merged bones in a mass grave or the pretended burial of a baby that was really stolen.

Even a perfect DNA match may be insufficient. In a repository of genetic profiles of living people, such as the one curated by the INTCF, DNA evidence has the capacity to prove family relatedness—with a higher probability for parent to offspring than for other types of kinship—but falls short in answering the questions of adoptees and biological families concerning the degree of coercion and other circumstances surrounding the adoptions. In those scenarios, the experiential knowledge of the people involved and a reconstruction of the infrastructure under which they operated are essential to address the gaps and transform the genetic results into restorative stories.

Conclusion

Alonso's interview hinted at a dramatic shift between Francoism and Spanish democracy, both in terms of institutional architecture and more specific attitudes towards adoption. This perceived rupture informed his interpretation of the genetic data and led him to make the confident assertion that trafficking infants was something confined to the dictatorial past. The contrast between a democratic regime and a dictatorship is outstanding. Yet the deep infrastructures underlying their day-to-day operation cannot change from one day to another.

This chapter has shown that the adoption and family reunion infrastructures deployed by the Spanish democratic authorities have faced, and still face, some challenges in countering their Francoist predecessors. Both democratic and Francoist infrastructures have built their memory regimes on an authoritative (and predominantly male) scientific discourse and the paramount importance of genetic inheritance—either as a tie that needed to be broken during the dictatorship or, subsequently, as the indelible bond that allowed reparation. More importantly, in both cases the communities affected by the child removal practices were assigned, and to some extent continue to be assigned, a rather subaltern role.

Incorporating the experiential knowledge of these communities is essential for the success of any reparation initiative. As this volume thoroughly documents, the day-to-day, mundane deployment of genetic technologies should be accompanied by the participation of the populations affected by the problems that these technologies address. From genetic disease patients to racialised communities or ancestry searchers, these populations are invaluable for the interpretation of the DNA data, and use this DNA data to build an identity that is integral to their wellbeing (Romijn, this volume).

In the case of the Spanish adoptions, when a genetic match is found, the experiences of the biological family are key in uncovering what the Francoist infrastructure attempted to bury into oblivion: the coercive and exploitative nature of child removal. From stolen babies to adoptions that were known but not freely consented to by the birth mothers, the lack of traceability of these practices in the registration records points to a concerted attempt to silence the biological families. Genetic and genomic technologies today have the opportunity of becoming powerful memory tools if deployed in infrastructures that give voice to those at the forgotten end of the paper trail.

Acknowledgements Special thanks to Vincenzo Pavone, Violeta Argudo, and the other editors of the volume, as well as two anonymous referees, for giving me the opportunity to publish this work and providing useful comments. This chapter is based on two talks that I delivered at the 2024 biennial meeting of the European Society for the History of Science and at the Instituto de Filosofía of the Spanish National Research Council (CSIC). I would like to thank Ana Romero, María Jesús Santesmases, Nuria Valverde, and Marta Velasco for inviting me and providing the feedback that has proved crucial in the written version. An advanced PhD student I co-supervise, Cristina Moreno-Lozano, made me aware of the importance of the spaces between infrastructures. Antonio Alonso helpfully clarified and updated the genetic data that I discuss. The three adopted women I approached made time to share emotionally difficult experiences. I dedicate this chapter to them and all the people affected by such a hideous practice. The first of my oral histories took place while I was visiting the

López Piñero Institute in Valencia, and Pedro Ruiz-Castell, José Ramón Bertomeu and Ximo Guillem facilitated the process. Several internal grants from the University of Edinburgh, specifically the Research and Development Fund and the Strategic Research Support Fund, covered the travel and accommodation costs of my investigation.

References

Ansede, M. (2023, October 24). Antonio Alonso, geneticist: "I do not believe that there was a plot to abduct newborns from Spanish public hospitals during democracy". *El Pais*. Retrieved July 2025, from https://english.elpais.com/spain/2023-10-24/antonio-alonso-geneticist-i-do-not-believe-that-there-was-a-plot-to-abduct-newborns-from-spanish-public-hospitals-during-democracy.html

Arroyo, S. (2013). *Los bebés robados de sor María: testimonios de un comercio cruel*. RBA.

Baelo Álvarez, M. (2023). *La adopción en el Código Civil: evolución normativa, doctrinal y sociojurídica desde 1889 hasta la actualidad*. Tirant lo Blanch.

Bowker, G. C. (2005). *Memory practices in the sciences*. MIT Press.

Bowker, G., & Star, S. L. (1999). *Sorting things out: Classification and its consequences*. MIT Press.

Campos, R., & Huertas García-Alejo, R. (2012). Medicina mental y eugenesia: los fundamentos ideológicos de la psiquiatría franquista en la obra de Antonio Vallejo-Nágera. *Historia del Presente, 20*, 11–21. https://doi.org/10.5944/hdp.20.2012.40706

Cayuela Sánchez, S. (2014). *Por la grandeza de la patria: la biopolítica en la España de Franco (1939–1975)*. Fondo de Cultura Economica.

Cenarro, A. (2009). *Los niños del Auxilio Social*. Espasa.

De la Torre, J., & García-Zúñiga, M. (2014). Was it a Spanish miracle? Development plans and regional industrialization, 1950–1975. In C. Grabas & A. Nützenadel (Eds.), *Industrial policy in Europe After 1945: Wealth, Power and Economic Development in the Cold War* (pp. 162–183). Palgrave Macmillan. https://doi.org/10.1057/9781137329905_8

DNA database helps one of Spain's "stolen babies" find family. (2019, July 11). *The Guardian*. Retrieved July 2025, from https://www.theguardian.com/world/2019/jul/11/dna-database-helps-one-of-spains-stolen-babies-find-family-franco

Durán Rodríguez, J. (2022, December 14). Paco Lobatón: "Es un déficit democrático enorme que no se haya aprobado la ley de bebés robados". *El Salto*. Retrieved July 2025, from https://www.elsaltodiario.com/personas-desaparecidas/entrevista-paco-lobaton-programa-television-quien-sabe-donde-bebes-robados

Duva, J., & Junquera, N. (2011). *Vidas robadas*. Aguilar.

Edwards, P. N. (2002). Infrastructure and modernity: Force, time, and social organization in the history of sociotechnical systems. In A. Freenberg, P. Brey, & T. J. Misa (Eds.), *Modernity and technology* (pp. 185–226). MIT Press. https://doi.org/10.7551/mitpress/4729.003.0011

Foro Internacional Te Estamos Buscando. (2023). Contestación de la plataforma Foro Internacional Te Estamos Buscando al artículo publicado en *El País* el pasado 24/10/2023. Retrieved July 2025, from https://sosbebesrobados.es/blog/2023/10/25/comujivafijjjkk/

García del Cid Guerra, C. (2017). *Las desterradas hijas de Eva*. Anantes.

García-Sancho, M. (2015). Genetic information in the age of DNA sequencing. *Information & Culture, 50*, 110–142. https://doi.org/10.1353/lac.2015.0000

Iglesias, M. A. (1982). Niños muertos en la nevera. *Interviú, 299*, 6–10.

Iglesias Aparicio, P. (2021). *Políticas de represión y punición de las mujeres: las lavanderías de la Magdalena de Irlanda y el Patronato de Protección a la Mujer de España*. Círculo Rojo.

INTCF. (2014). Balance del Servicio de Información a afectados por la posible sustracción de recién nacidos. Ministerio de Justicia, Subdirección General de Documentación y Publicaciones. Retrieved July 2025, from https://www.mjusticia.gob.es/es/Ciudadano/Victimas/Documents/1292427063522-Balance_del_Servicio_de_Informacion_a_afectados_por_la_posible_sustraccion_de_recien_nacidos.PDF

INTCF. (2017). Informe de la actividad pericial y científica desarrollada por los servicios de biología del INTCF con respecto a la prueba del ADN en la investigación de adopciones irregulares y sustracción de recién nacidos en España. In Comisión Nacional para el uso forense del ADN. Memoria de actividades 2016– 2017. Ministerio de Justicia, Secretaría General Técnica (Annex IV, pp. 27–38). Retrieved July 2025, from https://www.mjusticia.gob.es/va/ElMinisterio/OrganismosMinisterio/Documents/1292428944758-Memoria_de_la_Comision_Nacional_para_el_Uso_Forense_del_ADN._Actividades_20162017.PDF

INTCF. (2023). Instituto Nacional de Toxicología y Ciencias Forenses. Memoria 2022. Ministerio de la Presidencia, Justicia y Relaciones con las Cortes.

Retrieved July 2025, from https://www.mjusticia.gob.es/es/AreaTematica/DocumentacionPublicaciones/InstListDownload/Memoria%20INTCF%202022%20web.pdf

INTCF. (2025). Instituto Nacional de Toxicología y Ciencias Forenses. Memoria 2024. Ministerio de la Presidencia, Justicia y Relaciones con las Cortes. Retrieved January 2026, from https://www.mjusticia.gob.es/es/ElMinisterio/OrganismosMinisterio/Documents/Memoria%20INTCF%202024%20%28PDF%2067%2c6%20MB%29.pdf

Jimeno Aranguren, R. (2018). Genetic data and the right to identity in cases of enforced disappearances of children in Argentina, Spain and the Republic of Ireland: A comparative study of historical and legal evolution. *Revista de Derecho y Genoma Humano, 48*, 52–73.

Larkin, B. (2013). The politics and poetics of infrastructure. *Annual Review of Anthropology, 42*, 327–343. https://doi.org/10.1146/annurev-anthro-092412-155522

Lynch, M., Cole, S. A., McNally, R., & Jordan, K. (2010). *Truth machine: The contentious history of DNA fingerprinting*. University of Chicago Press.

Marre, D., & Gaggiotti, H. (2021). Irregular adoptions and infrastructures of memory in Spain: Remnant practices from the Franco Regime. *Childhood, 28*, 570–584. https://doi.org/10.1177/09075682211061448

Novella, E. (2024). *Las políticas de la locura: psiquiatría y sociedad en la España de Franco*. Universitat de València.

Puig de La Bellacasa, M. (2016). Ecological thinking, material spirituality, and the poetics of infrastructure. In G. Bowker, S. Timmermans, A. E. Clarke, & E. Balka (Eds.), *Boundary objects and beyond: Working with Leigh Star* (pp. 47–68). MIT Press. https://doi.org/10.7551/mitpress/10113.003.0005

Rabidoux, G., Lencina, M., & Vila Torres, E. (2020). *Stolen babies of Spain: The book*. ValMar Books.

Roig Pruñonosa, N. (2018). *No llores que vas a ser feliz: el tráfico de bebés en España: de la represión al negocio (1938–1996)*. Ático de los Libros.

Rubin, J. S. (2018). How Francisco Franco governs from beyond the grave: An infrastructural approach to memory politics in contemporary Spain. *American Ethnologist, 45*, 214–227. https://doi.org/10.1111/amet.12633

Ruiz-Berdún, D. (2022). *Historia de las matronas en España*. Guadalmazán.

Star, S. L. (1990). Power, technology and the phenomenology of conventions: On being allergic to onions. *The Sociological Review, 38*, 26–56. https://doi.org/10.1111/j.1467-954X.1990.tb03347.x

Star, S. L., & Ruhleder, K. (1996). Steps towards an ecology of infrastructure: Design and access for large information spaces. *Information Systems Research, 7*, 111–134. https://doi.org/10.1287/isre.7.1.111

Vinyes, R., Argengou, M., & Belis, R. (2003). *Los niños perdidos del franquismo*. Plaza & Janés.

14

Forensic Genetics and Mass Incarceration: Widening the Net of the DNA Databases in Brazil

Ricardo Urquizas Campello

Introduction

In November 2015, during a press conference dedicated to the launch of a government report on the situation of the Brazilian prison system, the then Minister of Justice, Eduardo Cardozo, caused considerable uproar among journalists and political forums by referring to the country's prisons as "medieval dungeons." Listing the vast repertoire of violence and human rights violations committed against prisoners, his speech echoed a previous statement in which the minister had said he would prefer death to being held in a correctional facility in Brazil: "We have a medieval prison system. From the bottom of my heart, if I had to serve many years in one of our prisons, I'd rather die." Cardozo's words were met with criticism, as they came from the head of the national executive branch in charge of managing the prison system. It was as if the state of calamity in

R. U. Campello (✉)
Institute of Social Medicine, Rio de Janeiro State University, Rio de Janeiro, Brazil
e-mail: ricardo.campello@uerj.br

V. Argudo-Portal et al. (eds.), *Mundane Genomics*,
https://doi.org/10.1007/978-981-95-4550-6_14

the Brazilian prison archipelago was independent of his political will and the work carried out by his ministry. It was as if the inhumane conditions of survival in the correctional facilities scattered across the country were not also his own responsibility.

Apart from the minister's cynicism, highlighted by the press and political opponents, his characterization of Brazilian prisons sought to draw the attention of the judiciary and civil society to the absolute precariousness of the country's criminal justice system. In the year following Cardozo's statement, Brazil would reach the position of the country with the third highest incarceration rate in the world in absolute terms—behind only the United States and China. Overcrowded institutions, deteriorated units, lack of medical care, insufficient legal assistance, aggression and torture committed by prison guards, and the strengthening of criminal organizations formed from inside prisons—the so-called *facções criminais*—transform the penal system into an incubator of cruelty whose management capabilities are already beyond the control of government authorities (Dias, 2011; Biondi, 2014; Godoi, 2017; Mallart, 2021). In criminal justice matters, the beginning of the twenty-first century ushered in the era of "dungeons" in Brazil.

As might be inferred, the militarized structure of the country's crime control policies is pointed out by analysts as one of the core factors in the set of arbitrary elements that feed back into mass incarceration (Schlittler, 2020; Sinhoretto, 2021; Felitte, 2023). From the decades of civil-military dictatorship (1964–1985) to the present day, the Brazilian public security system has been made up of two main police organizations: the Military Police, responsible for repressive and ostensive policing functions, and the Civil Police, in charge of the judicial and investigative activities of the security forces. In Brazil, the management of the police and prisons is the administrative responsibility of the federal states, and the situation varies according to states. In general, however, research on the subject highlights the predominant role played by the Military Police in the production not only of security but also of justice, since the vast majority of prison sentences handed down by the judiciary are the result of flagrant offenses, rather than investigative procedures (Schlittler, 2020; Sinhoretto, 2021). Moreover, the word of military police officers is generally taken as the only testimony and main source of evidence by magistrates (Jesus,

2020). This means that the country's prison demographics is predominantly made up of people imprisoned based on the subjective and often abusive criteria of police testimony, to the detriment of criminal investigations based on material evidence.

Among the set of probative elements resulting from arrests *in flagrante* presented by police officers to judges and prosecutors, "informal confessions" obtained through threats, physical violence, or psychological pressure often acquire the status of legal truth, without being properly investigated through judicial procedures based on the gathering of evidence, the adversarial principle, and the constitutional right to a full defense (Rigon & Jesus, 2019). The use of torture, which runs throughout the history of the Brazilian police forces (Felitte, 2023)—whether as a method of penalizing enslaved people in the colonial period, or as a tactic for coercing political enemies of the dictatorial regime—has been repositioned and updated in the democratic period as a widespread technique for producing criminal evidence (Rigon & Jesus, 2019).

It is against this backdrop that genetic identification technologies for criminal investigations have been implemented in Brazil, with the stated aim of improving the judicial functions of the police and overcoming coercive criminal investigation procedures. The introduction of forensic DNA databases into the country's legal framework took place in 2012, with the approval of Federal Law No. 12.654, supported by discursive practices emphasizing the need for the "modernization and scientificization" of the criminal justice system. In 2013, the Ministry of Justice and Public Security (MJSP) established the Integrated Network of Genetic Profile Databases (RIBPG) and the National Genetic Profiles Database (BNPG), with the aim of maintaining, sharing, and comparing DNA samples nationwide. Each federal state became responsible for implementing its own genetic databases and for sharing the collected profiles with the BNPG (Richter, 2016; Campello, 2021). Since 2018, the MJSP has been significantly expanding the registration of genetic samples through a series of programmatic actions supported by the advocacy of efficient crime-solving methods, made available by technoscientific innovations at the service of justice and public security (Brasil, 2024). Between 2018 and 2024, the number of genetic profiles stored in the BNPG grew at a rate of over 1700%, surpassing a total of 239,400 stored samples.

Nevertheless, little is known about the operational functioning of the BNPG and the social and political consequences of the process of *geneticization* of crime control strategies in the Brazilian penal scenario. Given that research dedicated to the emergence of new technologies and scientific knowledge in the field of crime control remains largely confined to Western European and US contexts, this chapter aims to contribute to the decentralization of this field of study, taking the Brazilian case as the subject of analysis.

The chapter is organized as follows: First, I briefly present the emergence of forensic DNA databases within a global context of penal escalation and the development of new punitive control technologies, highlighting the racial and social components of this process, as well as its colonial backgrounds. Next, after describing the empirical *corpus* and the methodological approach on which this research is based, I address the contemporary expansion of Brazil's DNA database, driven by the institutional strategy of collecting genetic material in the country's prison units. I then summarize the arguments put forth by the operators of forensic DNA technologies related to the need to attribute greater objectivity to criminal investigation procedures, confronting their rationale with the socio-racial biases that characterize the net-widening of genetic databanks. Finally, I focus on the biological sampling procedures carried out in São Paulo's prisons, emphasizing the institutional illegalities and abuses that underpin the genetic identification of prisoners. Despite the official statements that forensic DNA databases are designed to overcome the traditionally abusive and discriminatory nature of judicial procedures in Brazil, I demonstrate the specific ways in which their use articulates with the historical inequalities and institutional violence that guide and qualify the ordinary functioning of the country's criminal justice system.

Colonial Traces

Over the past three decades, genetic profiling technologies have had substantial impacts on crime control strategies and the development of forensic sciences worldwide (Machado & Granja, 2022; Toom et al., 2023; Vailly, 2024). The examination of biological traces found at crime scenes and their cross-referencing with the DNA profiles of individuals considered suspects are presented as infallible methods for obtaining evidence within criminal prosecution (Jasanoff, 2006; Lynch et al., 2008). In light of their ability to produce the truth about crime based on the biological identification of the criminal, DNA tests have become crucial elements in criminal justice and public security systems (Heineman et al., 2012; Machado & Granja, 2022; Vailly, 2024).

On the international level, researchers focused on recent innovations in forensic genetics emphasize the necessity of deeper analytical investigation and public scrutiny of the social, political, and bioethical dimensions of their developments, particularly those concerning the potential for abusive uses of DNA databases (Wienroth et al., 2021; Vailly, 2024), access to sensitive information through new "genetic surveillance" techniques (Machado & Granja, 2022), and the (re-)emergence of the biological stigmatization of specific ethnic and racial groups (Duster, 2006; M'Charek, 2008; Vailly, 2024).

In general, the specialized literature links the emergence of forensic DNA databases to the set of changes in crime control practices that took place in the last quarter of the twentieth century in some of the core countries of global capitalism (Cole, 2001; Lynch, 2003; Lazer, 2004; Hindmarsh & Prainsack, 2010). Guided by efficiency and productivity parameters, the criminal justice systems, particularly in the United States, the UK, and France, began to rely on an extensive network of technological mechanisms aimed at controlling popular illegalisms in metropolitan contexts. Alongside GPS tracking devices for offenders, urban video surveillance circuits, and predictive policing systems, genetic profiling techniques became part of the techno-managerial arsenal directed at fighting crime, the main effects of which consisted of enhancing police surveillance capabilities and the demographic explosion of prison systems (Marx, 1988; Lyon, 2001; Wacquant, 2009; Fassin, 2017).

The restructuring of the technical and institutional apparatus for dealing with crime was underpinned by a political enterprise based on the penal and police administration of poverty and the reaffirmation of ethno-racial hierarchies (Davis, 2003; Wacquant, 2009; Alexander, 2010). In the United States, the rise of civil rights movements in the northern ghettos and the end of the segregationist regime in the southern states were met with a political response marked by decades of punitive and securitarian practices, amalgamated with the intensive incarceration of the black population and the resolute criminalization of urban marginality under the rubric of law and order (Davis, 2003; Wacquant, 2009; Alexander, 2010). In Western Europe, the intensification of the migratory influx of impoverished people from post-colonial African and Asian territories has become the object of a vast technological apparatus for identification and surveillance within national borders, parallel to the widespread construction of detention centers for illegal immigrants, turning migration policy into a kind of simulacrum of criminal policy (Aas, 2011; Woude et al., 2017; Fassin, 2017; Durão, 2023). On both shores of the North Atlantic, the activities of justice and police agencies at the turn of the century would have as a common denominator the mass absorption and obstinate capture of specific population segments belonging to the impoverished classes and ethnic minorities.

In Brazil, the importation of information technology and biotechnological systems for security and punishment from the United States and Europe is linked to endogenous statements related to the fight against organized crime, everyday violence and the "culture of impunity" (Cardoso, 2014; Duarte & Ceia, 2022; Campello, 2023). From the second half of the 2000s, a series of programs aimed at the "technological modernization" of crime control policies began to be implemented in the country, with the overarching justification being the increased efficiency of the criminal justice system and the strengthening of public security. In 2005, the first Command and Control Center (CCC) was inaugurated in the city of Rio de Janeiro, with the aim of integrating the Military Police's CCTV system (Cardoso, 2014). In 2008, digitized biometric identification mechanisms began to be implemented in the country as a method of civil and criminal registration (Doneda & Kanashiro, 2012). In 2010, the electronic monitoring of convicted individuals was authorized at the federal level, through the use of satellite-tracked ankle bracelets (Campello, 2023). That same year, the Federal Police Department began implementing the Combined DNA Index System (CODIS) software through an agreement with the US Federal Bureau of Investigation (FBI), with the aim of starting the process of setting up forensic laboratories for the storage and analysis of genetic samples across the federal states (Richter, 2016; Campello, 2021).

In the Brazilian context, the racial and social selectivity of penal and police institutions is also evident, as shown by the relative incarceration rates and police targeting of black and poor populations. Situated in the discursive framework of combating urban crime and the activities of criminal factions, the operational patterns of police departments and criminal justice courts maintain a distinctive direction in their practices, with clearly defined racial and class targets (Alves, 2018; Schlittler, 2020; Sinhoretto, 2021). Over recent years, researchers focused on the social and political impacts of new crime control technologies in Brazil have drawn attention to the ways in which the adaptation of technoscientific and modernization paradigms from the Global North intersect with the reactivation of the historical racism and social inequalities that structure punitive and securitarian strategies in the country (Silva, 2020; Duarte & Ceia, 2022; Campello, 2023).

This chapter contributes to this research agenda by analyzing the current process of consolidation and expansion of the National Genetic Profiles Database. What political strategies have been adopted so far to widen DNA databases in Brazil? How does the technical and scientific rationality underlying genetic profiling technologies fit in with the biased criminalization and penalization practices prevalent in the country? What are the social implications of the procedures for collecting and storing genetic material in the Brazilian carceral context? These are the main questions guiding this study.

From a theoretical and epistemological standpoint, this study draws on dialogues with postcolonial approaches situated at the crossroads of the Anthropology of the State, the Sociology of Punishment, and Science and Technology Studies. In *Biometric State*, Keith Breckenridge (2014) examines the political origins and consequences of compulsory biometric identification systems, tracing their genesis to the bureaucratic mechanisms of population control and criminal justice in colonial orders, in which countries such as South Africa and India were taken as laboratories for their implementation and development under the British "carceral empire" (Breckenridge, 2014, 2021). Used as identification methods based on the fingerprints of individuals from marginalized racial and social groups, biometric registration technologies evolved in the post-colonial period as central elements of the South African apartheid bureaucracy and the Indian caste system, spreading throughout the transatlantic world in the second half of the twentieth century.

From the 1980s onward, genetic profiling techniques began to promise levels of accuracy unmatched by their fingerprint predecessors, while being imbued with similar functionalities (Cole, 2001). More recently, the profusion of DNA databases and the development of new uses and technical capabilities for forensic genetics have spurred a series of studies dedicated to their political configurations and their effects on social stratification in the postcolonial world. Tunisian anthropologist Amade M'Charek et al. (2020) investigate the implications of forensic DNA phenotyping technologies and ancestry testing, drawing attention to emerging forms of racialization based on the intersection between genetic determination of geographical origins and the molecular analysis of individual genetic markers. In Ghana, Aaron Amankwaa and Judith Addo (2023) examine the governance of the country's DNA database, highlighting legal and political issues related to the absence of specific and thorough legislation for the protection of private data and for assuring the reliability of the public institutions in charge. In Colombia, Olarte-Sierra and Bermúdez (2023) focus on the genetic identification of victims of armed conflicts, pointing out the prominent place occupied by forensic geneticists in the production of legal-scientific truth based on genetic analysis. They also highlight the need to take into account that the knowledge mobilized by these professionals is always partial and situated, as it is inserted within a specific social and institutional context.

The current intensification of debates surrounding the uses, discourses, and performances of forensic genetics beyond the United States and Western Europe has led to a repositioning of the analytical gaze toward the impacts of technosciences aimed at crime control in contexts marked by the persistence of authoritarian and militarized models of security systems, high rates of violent deaths, and exacerbated levels of inequality in legal and penal treatment. This research seeks to provide a useful contribution to this set of discussions.

Materials and Methods

This chapter results from a broader study carried out between 2021 and 2024 on the development of forensic genetics in Brazil. The methodological strategies employed here sought to identify the discourses and rationalities underlying the implementation of DNA databases for criminal prosecution purposes in Brazil, the legal-political-scientific complex that constitutes the BNPG, and the ways in which biological sampling procedures are carried out in Brazilian prisons.

The empirical material consists mainly of interviews and official documents relating to the activities of the RIBPG. A total of 35 criminal experts and forensic geneticists from technical-scientific police departments in 7 Brazilian states were interviewed. Their reports were gathered either through their participation in focus groups, or through in-depth individual interviews. In addition, I had access to 17 interviews with prisoners subjected to DNA collection procedures, conducted inside prisons by the inspection team of the São Paulo State Public Defender, the body that provides legal assistance to prisoners who cannot afford private lawyers. Prisoner testimonies were supplemented by inspection reports drawn up by the same entity, which systematize information on incarceration conditions in the State of São Paulo. To preserve the confidentiality of the reports presented here, the names of the professionals interviewed have been omitted and the prisoners who took part in the research have been anonymized by the use of pseudonyms.

The documentary analysis is based on the laws that regulate genetic identification policy for criminal prosecution purposes in Brazil; the normative resolutions drawn up by the Management Committee of the Integrated Network of Genetic Profile Databases (CG-RIBPG), the entity responsible for coordinating the activities of the forensic genetics laboratories that supply the BNPG; as well as 21 reports produced by the MJSP, presenting statistical data on the number of DNA samples stored in the national database and their different categories. In addition, I surveyed and systematized information produced by the São Paulo State Department of Penitentiary Administration on the social and ethno-racial profile of prisoners genetically registered within the São Paulo State Genetic Profile Database.

In general, the research techniques sought to provide pertinent material for analyzing the heterogeneous character of forensic DNA databases in Brazil, understood as an assemblage (Abrahamsen & Williams, 2011) made up of practices, discourses, legal-normative documents, as well as judicial, police, and scientific institutions, whose arrangements and configurations entail specific social and political effects. The empirical data and methodological strategies deployed in this study highlight the ways in which the penal management of genetic content intersects with the arbitrary and socio-racially selective features of Brazilian security and criminal justice agencies, to the detriment of its official purpose of "democratizing" crime control policies through science and technology.

Captive Bodies

In May 2012, the approval of Federal Law No. 12.654 amended the Penal Execution Law (Law No. 7.210/1984) and the Criminal Identification Law (Law No. 12.037/2009), mandating the collection of genetic material from convicted or investigated individuals in Brazil. The following year, Decree No. 7950 established the BNPG within the Ministry of Justice and Public Security (MJSP). The BNPG's purpose is to store and compare DNA profile data supplied by state, district, and federal forensic genetics institutes. In the federal states, genetic storage and analysis are carried out in official forensic genetics laboratories, using state DNA profile databases managed by official experts. Currently, the largest of these in absolute terms is the São Paulo State Genetic Profile Bank (BPG-SP), with approximately 28,000 samples stored at the end of 2024.

According to reports produced by the RIBPG Management Committee, the main categories of DNA samples collected and stored for criminal investigation purposes are those of convicted individuals and investigated individuals, and biological traces found at crime scenes or on crime victims (Brasil, 2024). The aim is to identify potential matches between the different categories of profiles, thereby assisting in the process of producing criminal evidence.

From 2019 onward, the Brazilian Government began to promote the expansion of genetic databases throughout the country. It was the first year of Jair Bolsonaro's term as president, and strengthening the criminal policy of genetic identification was one of the flagship programs of his newly constituted Ministry of Justice and Public Security. In order to promote the growth of DNA profile storage, the strategy adopted by Bolsonaro's new ministry was to initiate this expansion through a rapid and widespread process of biological sample collection within the prison system. Given the continental dimensions of the Brazilian prison population, a systematic DNA collection campaign in prisons would allow a large volume of genetic samples to be obtained more quickly and efficiently than the gradual, court-ordered collection of DNA profiles from investigated individuals apprehended by the police. This was also the strategy recommended by criminal experts and entrepreneurs from the international biotechnology industry, who were committed to widening Brazil's forensic DNA bases (Richter & Louzada, 2023). Prisoners' bodies would therefore become the primary source of supply for the technoscientific complex formed by genetic profiling databases, guided by the assumption that such individuals are likely to commit future crimes, as pointed out by one of the criminal experts interviewed in this study:

> Once the person is registered, once the bank is well established, they are less likely to reoffend. And since it's not possible to put the entire population in the bank, you must prioritize a certain group in terms of public policy. The group that seems to make the most sense is the group of people who have already been convicted, because of the percentage of recidivism. (Geneticist 3, March 2022)

As he states, prior conviction justifies prospective suspicion and legitimizes DNA collection, thus creating a sort of porosity between conviction and suspicion (Vailly, 2024). The result was a rapid growth in the number of DNA samples stored in the BNPG, with a significant predominance of genetic profiles of convicted people. During the first years of its implementation, the volume of samples registered in the national database increased at a steady and moderate pace, from 2584 profiles stored in November 2014 to a total of 13,197 profiles in May 2018. As of 2019, when DNA collection campaigns intensified in prisons, the total number of samples registered in the BNPG jumped to over 220,000 by 2024, as represented in the graph below (Fig. 14.1):

Of the total number of samples stored in the national database by November 2023, 74.8% came from convicts, 14.5% consisted of samples from traces collected at crime scenes or victims, and 1.8% from people under investigation (Brasil, 2024, p. 28) (Fig. 14.2).

Just like the strategy adopted in countries such as the UK and the United States (Lynch et al., 2008; Krimsky & Simoncelli, 2011), Brazil's DNA identification policy has focused mainly on convicted prisoners, supported by discourses that emphasize the need to reduce the risk of criminal recidivism through the technical and political modernization of the judicial apparatus.

When faced with occasional queries from jurists and civil society organizations about the legal, social, and bioethical implications of the compulsory collection of genetic content from large groups of incarcerated people (Schiocchett, 2021), criminal experts equate their activities to the

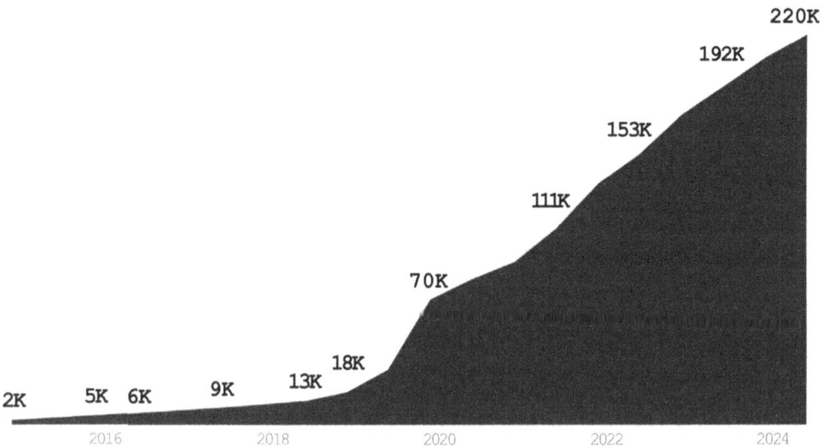

Fig. 14.1 Growth in the number of DNA profiles Stored in the BNPG. Source: Ministry of Justice and Public Security

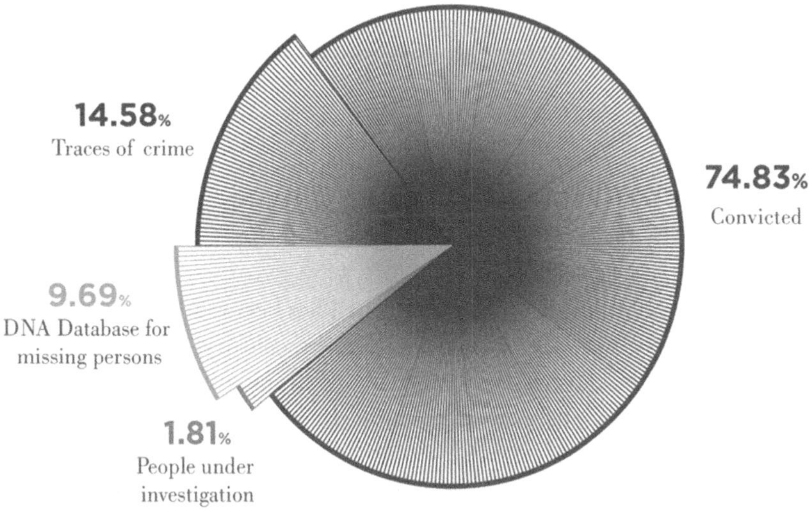

Fig. 14.2 Distribution of DNA profiles in the BNPG, by category. Source: Ministry of Justice and Public Security, 2024

traditional fingerprint identification procedures: "What's the difference?" asks the former manager of the BNPG. "Nobody refuses to donate their fingerprint" (Geneticist 3, March 2022). The expert relies on the argument that forensic genetics in Brazil is restricted to the analysis of the sequence of DNA markers classified as non-coding, which allegedly do not provide information relating to physical characteristics or pathological predispositions. As discussed by Krikorian (2021) in regard to the debates surrounding the French DNA database, this is the scientific statement that allows the depoliticization of forensic genetics, enabling its public acceptance and trivialization. The incorporation of DNA databases for criminal identification purposes into the contemporary universe of *mundane genomics* is enabled by the scientific category of "non-coding DNA." This equates forensic genetics with the old techniques of papilloscopic identification, thus justifying the massive practices of biological collection inside prisons, legitimized by the presumption of dangerousness of their captive population.

Briefly stepping back toward the criminological matrices of forensic identification practices and their epistemic connections with the current development of police DNA databases in Brazil, the next section investigates the institutional reiteration of criminalization procedures based on biophysical traits.

From Phenotypic Suspicion to Genetic Incrimination

The biological identification of the criminal has been pursued by criminal sciences since the second half of the nineteenth century, based on theories developed by the Positivist School of criminology, with Dr. Cesare Lombroso as one of its main exponents. In general terms, Lombrosian criminal anthropology draws its theoretical formulations from anthropometric data, assuming that social behaviors are biologically determined and that the physical and hereditary features of certain individuals indicate their atavistic propensity for crime. In Brazil, positivist criminological thought found widespread acceptance in legal and scientific circles at the beginning of the twentieth century, underpinning legal and institutional reforms in penal policy (Alvarez, 2003, 2014). As the century progressed, discourses grounded in the biological characterization of the criminal came under extensive criticism for lending scientific legitimacy to discriminatory policies and justifying unlimited punitive intervention, ostensibly for corrective purposes (Foucault, 1975; Baratta, 1999; Alvarez, 2014).

Today, forensic sciences are restoring the biological correlation between crime and the criminal by comparing the genetic data of convicted individuals and suspects with those found at crime scenes. As pointed out by Duster (2006), despite the widespread assertion among forensic geneticists that "correlation" does not mean "propensity," the social and institutional contexts within which these correlations are established require further investigation and rigorous analysis if we are to avoid the mistakes of the past. Although contemporary DNA identification procedures using forensic databases do not imply the alleged bio-determined detection of criminal behavior, biased patterns of police activity driven by the targeting of "suspect phenotypes" inevitably imply a racialization of DNA databases, as will be demonstrably shown below. Anchored in a techno-scientific pragmatism in the service of justice and security (Machado, 2011), forensic DNA testing technologies raise a set of questions regarding the convergence between the epistemological reactivation of producing truth about crime based on biological evidence, and the reshaping of racial and phenotypic selectivity that characterizes criminal justice and security agencies in Brazil.

As defined by Brazilian forensic geneticists and criminal experts, the aim of the BNPG is to provide "greater objectivity" to criminal investigation procedures developed in the country, enabling the overcoming of abusive and extrajudicial practices of evidence production, which are centered on police testimony (Jesus, 2020) and the reiteration of torture as an evidentiary technique for extracting confessions (Felitte, 2023). In the words of the former administrator of the national database, "If we don't have proper identification methods, how will crimes be solved? Through torture, which leads to false confessions and the conviction of innocents" (Geneticist 3, March 2022). Similarly, the administrator of the forensic genetics laboratory in the State of São Paulo notes that "DNA testing comes as a way to exonerate people who are being unjustly accused" (Geneticist 4, March 2022). The epistemic qualification of criminal evidence could be one of the fundamental conditions for eliminating abusive evidence production techniques that lead to judicial errors and wrongful convictions.

However, the option adopted by the MJSP to develop the BNPG based on samples from convicted individuals has little or nothing to contribute to the goals of preventing wrongful convictions and exonerating unjustly sentenced individuals, since the practices of storing DNA from convicted individuals do not involve a post-conviction process of sentence review based on prisoners' samples. This is what one of the geneticists who participated in the DNA collection campaigns in the prison facilities of Rio de Janeiro comments, with some discomfort:

> In fact, what bothers me is that the role of the state is to map this person without giving them any kind of feedback. You arrive at the prison to collect the sample and you say almost nothing. You're supposed to collect this because the law says so. And then there have been cases of prisoners saying:

> "Oh, fine, but is this going to be used, for example, to clear me?"—because there are people who have been convicted and are innocent, we know that from experience. And no, it won't exonerate you. The aim is not to clear you. The aim is that if you commit another crime, you'll have your data, your genetic profile in the database. The logic of the DNA database is structured to incriminate as many people as possible. The logic is oriented in this direction: incriminate, imprison. (Geneticist 5, February 2022)

According to him, the constitution of a vast genetic database aimed at producing future evidence against those who have already had contact with the criminal justice system demonstrates its punitive character, by no means favoring the alleged purpose of exonerating innocents—quite the opposite. Inserted into a context of mass incarceration, forensic technologies for genetic identification act as yet another mechanism of the penalization procedures around which the strategies for conflict management in Brazilian society are centered, relying on prison confinement often based on discretionary and imprecise evidentiary methods.

Furthermore, as extensively shown in studies carried out in the United States and the UK (Duster, 2006; Krimsky & Simoncelli, 2011; Skinner, 2013), the strategy of promoting DNA bases from convicts has led to a predominance of genetic samples from black people belonging to disadvantaged social classes, since the prison demographic is largely composed of this specific population, as a result of police and penal racial profiling procedures. According to data provided by the São Paulo State Secretariat of Penitentiary Administration regarding the BPG-SP, out of a total of 8169 convicted or investigated people registered in the São Paulo DNA database by November 2022, 54.3% were black, while only 40.9% of the total population of the State of São Paulo was made up of self-declared black people (Fig. 14.3).

Likewise, taken here as social indicators and corresponding to the profile of Brazilian prison demographics, the education levels of the population genetically registered in the BPG-SP indicate a large majority of people belonging to the marginalized social classes within the database. Of the total number of convicted or investigated people aged 25 or over who had their genetic profile entered into BPG-SP by November 2022, 78.46% had not completed high school, while in the total population of

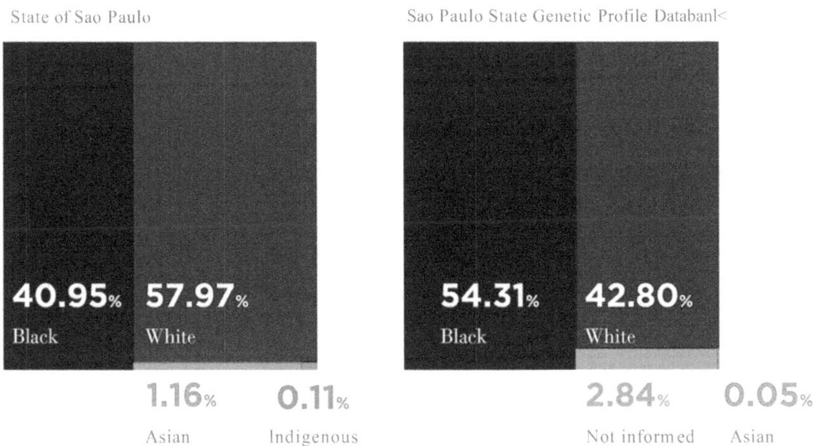

Fig. 14.3 Race 2022. Source: FundaçãoSEADE; Secretaria de Administração Penitenciária do Estado de São Paulo

the State of São Paulo, only 37.2% of people aged 25 or over had not completed high school in 2022 (Fig. 14.4).

It should be noted that research carried out in Brazil on the performance of the public security and criminal justice agencies demonstrates that black people belonging to disadvantaged social classes are more likely to be approached by the police and convicted by the courts of justice. Sinhoretto (2021) and Schlittler (2020) show that black people living on the urban periphery are more likely to be arrested *in flagrante*, as they are more exposed to the rigors of police surveillance. Adorno (1996), Vargas (2008), and Ribeiro (2010) found that although whites and blacks commit crimes in equal proportions, black defendants tend to receive stricter criminal treatment, as they face greater obstacles in accessing a full defense and procedural rights in the courts of justice. The mechanics constituted by the sequential pair *police arrest-judicial conviction* shape the demographic profile of the Brazilian prison population, which is mostly black and poor.

Thus, connected to the social and ethno-racial filtering mechanisms of police and judicial procedures, forensic DNA technologies act as a feedback vector for the biased features of punitive and securitarian practices. If the BNPG is fostered through the collection of genetic material

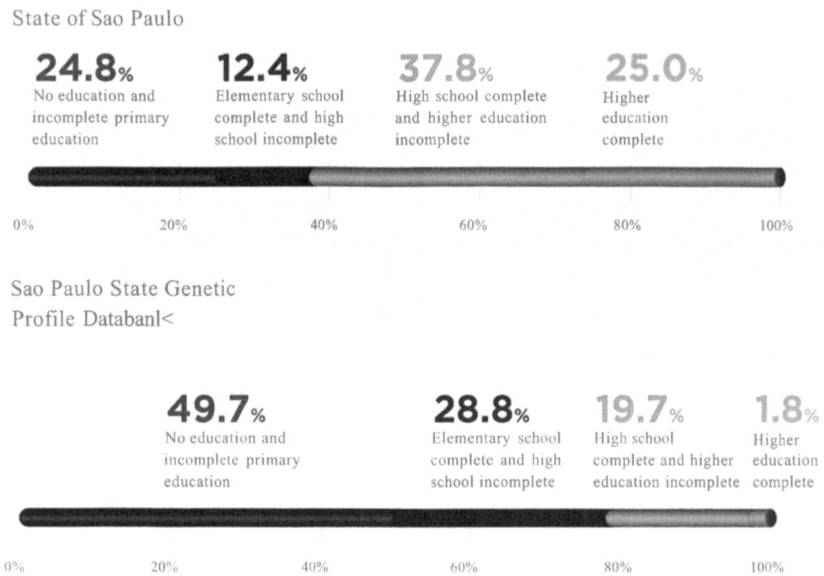

Fig. 14.4 Education 2022. Source: IBGE; SAP SP

conducted in prison units, the socioracial composition of DNA records will predominantly consist of profiles of black people belonging to the underprivileged classes, since the Brazilian prison population is mostly made up of this specific group. To this extent, genetically based incriminations are institutionally prone to produce evidence against black and poor people who have already had contact with the criminal justice system, creating a vicious circle of police arrest, judicial conviction, and *anticipatory suspicion of criminal recidivism.*

At the interface with the framework analyzed so far, the next section focuses on the discretionary features of DNA sampling procedures in São Paulo's prisons.

Prisons Above the Law

Article 9-A of the Penal Execution Law and Resolution No. 10 of the MJSP establish the legal and normative parameters for the collection of genetic material from individuals convicted or investigated by the Brazilian criminal justice system. These are guidelines that describe the basic procedures to be followed by the Technical-Scientific Police teams with regard to the DNA sampling activities carried out in prisons, police stations, or forensic institutes, depending on the specific case. In terms of convicted individuals, the legislation stipulates that the procedure should be carried out "upon entry into the prison system" or "during the serving of the sentence." The law also establishes that the collections must be carried out by "an official expert" (§ 7, Art. 9°-A, Law 7.210)—an agent of the Technical-Scientific Police duly trained to carry out the task. Based on this legal framework, the CG-RIBPG drew up the rules that define the methods by which samples should be obtained, with the aim of regulating and standardizing the procedures to be carried out in the different states of the federation.

According to the guidelines set forth in Resolution No. 10 of the MJSP, the methodology to be used involves the collection of oral mucosa cells, which is considered the most appropriate technique by national and international forensic genetics protocols (UNESCO, 2004; SENASP, 2013), as it is a painless procedure deemed minimally invasive. To this end, in addition to providing training for experts, the RIBPG Management Committee developed a set of instructions for sample collection in correctional facilities, using appropriate kits that include buccal swabs attached to so-called "FTA cards," which are used for storing and transporting saliva samples, aluminum-lined envelopes, and personal protective equipment such as masks, gloves, and lab coats. The stages of the collection process are detailed in the Standard Operating Procedure of the National Secretariat for Public Security, which specifies biosafety precautions and the fulfillment of legal and normative criteria. After the procedures are carried out within the prison units, the collected material must be sent in a sealed envelope by the forensic agents to the genetics laboratories linked to the Technical-Scientific Police, where the sample analysis and registration processes are carried out in the CODIS software.

The importance of the technical detailing and procedural definitions of the collections lies in the need to preserve the chain of custody of genetic evidence, a fundamental condition for ensuring the conservation, identity, and traceability of the material to be used as judicial proof. As the scientific substrate of criminal evidence, biological samples constitute the material foundation of legal truth based on forensic evidence, thus requiring a set of technical and epistemic controls for their production and handling. Furthermore, the establishment of protocols aims to mitigate the intrusive aspects of the collection procedures by employing appropriate methodologies. While it is true that the insertion of a collection device into the mouths of incarcerated individuals for the extraction of their saliva, with the purpose of possible incrimination, already presupposes a form of bodily intrusion, this would tend to be minimized when performed by health professionals or specialized agents.

Nonetheless, cases of non-compliance with DNA sampling protocols in prisons are not uncommon, attributing an abusive or even illegal character to the collection procedures. This is indicated, for example, by the following record:

> Iaras, interior of the state of São Paulo. Orlando Brando Filinto Penitentiary, April 2023. Alessandro, an inmate in a closed regime, informs the Public Defender's inspection team that his DNA was collected in 2022. The procedure was part of the nationwide genetic sample collection campaign for convicted individuals, coordinated by the Management Committee of the Integrated Network of Genetic Profile Databases, as provided by Article 9-A of the Penal Execution Law. However, contrary to the provisions of the law, which requires the collection of biological material using an "appropriate and painless" technique—as well as Resolution No. 10 of the MJSP, which establishes the collection of oral mucosa cells as the standard operational procedure—Alessandro reports that the sample was taken through a blood draw. Furthermore, Alessandro reports that the collection was carried out by the prison guards themselves, in violation of Paragraph 7 of Article 9-A of the Penal Execution Law, which stipulates that samples must be collected by an "official expert". (Public Defender's Office of the State of São Paulo, Internal Inspection Report, April 2023)

Prison officers carrying syringes would have extracted the blood of prisoners for the purpose of inclusion in the DNA database. Defined as an "invasive method" of obtaining genetic information by national regulations governing collections and by international Human Rights and Bioethics treaties, such as the UNESCO International Declaration on Human Genetic Data (2004), venous puncture procedures for blood sample extraction have been reported to the Public Defender's Office by inmates in Iaras and other prison facilities in São Paulo. This is further aggravated by the fact that the samples were taken by prison guards without any specific preparation or training, leading to the risk of nerve injuries and bacterial infections, particularly when considering the precarious sanitary conditions in Brazilian prison institutions (Godoi, 2017; Mallart, 2021). The potential compromise of the chain of custody of the DNA samples handled by agents without any technical qualification for performing the task should also be noted.

The events recorded in Iaras have other parallels. In addition to the information about blood collections and the handling of samples by prison guards, there are also reports of fraudulent tactics used by prison guards to obtain genetic material from inmates, as indicated in this second record:

> Vila Independência, East Zone of the city of São Paulo. Provisional Detention Center, May 2023. Edson, a convicted prisoner held in a provisional detention center due to overcrowding and lack of vacancies in the state's penitentiaries, informs the Public Defender's Office inspection team that he had his biological material collected in early 2022, through a blood draw using a syringe. According to him, the prison staff had informed him that they would draw his blood to carry out a serological test for Covid-19. After the procedure, Edson would have been notified by the prison guards that it was, in fact, a biological sample collection for insertion of his DNA in the Genetic Profiles Database of the State of São Paulo. (Public Defender's Office of the State of São Paulo, Internal Inspection Report, May 2023)

According to the Public Defender's Office, the incident with Edson was not an isolated case. In 2019, Roberto had his saliva collected by agents of the Technical-Scientific Police while he was incarcerated at the

Serra Azul Penitentiary, in the western region of the State of São Paulo. According to him, just before the procedure, "they passed around a list saying it was health care." The collection was carried out in one of the prison's cells, under the pretext of conducting medical exams. The lab coats of the forensic agents seemed to confirm the false explanation provided by the prison staff. Roberto also mentioned that, in 2022, he was subjected to a second DNA collection procedure after being transferred to the Itaí Penitentiary, another facility within the vast São Paulo carceral network. He was given no information about the need for a second collection. The prison officers allegedly told him that he "had to do it," disregarding legal provisions that require the "donor" to be informed of the legal grounds for the procedure. Without access to a defense counsel, Roberto consented.

The information gathered by the Public Defender's Office on DNA collection procedures in São Paulo prisons also indicates extrajudicial consequences resulting from prisoners' refusal to provide their genetic material. In the Florida Paulista Penitentiary, for example, an anonymous complaint issued in September 2022 indicated that the refusal of one of the prisoners to provide a sample of his saliva had led to the suspension of visits to all prisoners in the unit for more than a month, by way of "collective punishment"—an illegal procedure, prohibited by the Penal Execution Law, based on the constitutional determination that criminal liability cannot extend beyond the person responsible for the fault or infraction.

In the same facility, another complaint issued in April 2023 reports disciplinary isolation procedures, for periods longer than the 30 days stipulated by law as the time limit for maintaining isolation, for prisoners who refused to provide their genetic material. In the Brazilian prison system, disciplinary sectors are intended for prisoners who violate sentence rules, or who come into conflict with the prison administration. In general, these are spaces made up of small, poorly lit cells without adequate air circulation, in which the people sent to them are kept on a meager diet, under water rationing and deprived of sunlight for days or weeks. In ethnographic research carried out in São Paulo prison units, Mallart (2021) defines isolation sectors as spaces "made up of suffocating, damp and dark corridors, with around 12 cells, where men and women subjected to so-called disciplinary sanctions remain, in theory, for up to 30 days" (p. 224). In the case reported, three prisoners who refused to have their saliva samples taken were said to have been held for more than four months in spaces like these.

The series of complaints documented by the Public Defender's Office signals the intersections between the power dynamics in Brazilian prisons and the genetic identification policy for criminal purposes developed in the country, centered on the storage of convicted individuals' biological samples for net-widening the DNA databases. In particular, the accounts given by prisoners describe some of the techniques used by prison authorities to meet the targets set by the MJSP regarding the strengthening of state and national genetic bases. Embedded in the mechanics of the Brazilian prison system, forensic genetics seems to be tightly linked to the arbitrary and illegal treatment of prisoners. Despite the determination of normative criteria and legislative parameters regulating genetic sampling procedures, the development of forensic DNA technologies is ultimately conditioned by the ways in which their direct operators perform, establishing feedback relationships between judicial techno-scientific identification practices and the discretionary power of "prisons above the law" (Godoi, 2020).

Final Remark

After its development in the UK-US axis in the latter part of the twentieth century, followed by rapid expansion to Western Europe (Lynch et al., 2008; Machado & Granja, 2022), the use of genetics by the criminal justice system has been adopted throughout the world, driving a broad research agenda concentrated mainly in the so-called Global North. Brazil, therefore, offers a privileged setting for shifting the gaze and studying how the recent process of consolidating and expanding DNA databases for criminal investigations operates in an increasingly threatened democracy, characterized by serious shortcomings in the penal system. As I have demonstrated, it is these very shortcomings that fuel the development of forensic DNA bases in Brazil, enabling their inclusion into the widening scope of *mundane genomics*.

Intensive investment in "modernizing" criminal justice through the use of DNA technologies, combined with deepening racial disparities and the perpetuation of abusive punitive practices, indicate an interweaving whose specific forms this chapter has addressed. As research carried out in other national contexts has shown, racially and socially targeted criminalization procedures materially configure biased genetic technologies, revitalizing the links between gene, phenotype, and race that guide the activities of the Brazilian penal system. Furthermore, this study highlights arbitrary and even illegal procedures committed by officials inside prisons with the aim of promoting forensic DNA databases. Rather than overcoming old practices through the introduction of science- and technology-based criminal investigations, in accordance with the professionals' stated aim, the new technologies of genetic identification are articulated with the reiteration of institutional racism and carceral coercitivity that qualifies the Brazilian criminal justice system—one composed of a vast network of penitentiary institutions ironically likened to "medieval dungeons" by the ministers responsible for them.

In that vein, the analysis of the recent implementation of forensic genetic databases in the Brazilian criminal justice system indicates the ways in which some of the long-standing forms of violence and abuse practiced by law enforcement agents are deepened and intensified by importing new punitive and security technologies from the United States and European contexts. Far from being antagonistic, authoritarian institutions and techno-scientific penal measures reinforce each other in a carceral scenario still strongly following its colonial blueprint.

Acknowledgments I would like to thank Joëlle Vailly and Susana Durão for their friendly supervisory work during the collection and systematization of the information presented in this chapter. I also thank the Vladimir Herzog Institute (Instituto Vladimir Herzog) and the Prison Situation Center of the São Paulo State Public Defender's Office (Núcleo de Situação Carcerária da Defensoria Pública do Estado de São Paulo Paulo) for the partnerships that enabled this work to be carried out. Special thanks to the participants in this research who contributed their testimonies.

References

Aas, K. F. (2011). "Crimmigration" bodies and bona fide travelers: Surveillance, citizenship and global governance. *Theoretical Criminology, 15*(3), 331–346. https://doi.org/10.1177/1362480610396643

Abrahamsen, R., & Williams, M. C. (2011). Security privatization and global security assemblages. *The Brown Journal of World Affairs, 18*(1), 171–180.

Adorno, S. (1996). Racismo, criminalidade violenta e justiça penal: Réus brancos e negros em perspectiva comparativa. *Estudos Históricos, 18*(2), 1–22.

Alexander, M. (2010). *The new Jim Crow: Mass incarceration in the age of colorblindness.* The New Press.

Alvarez, M. C. (2003). *Bacharéis, criminologistas e juristas: Saber jurídico e Nova Escola Penal no Brasil.* IBCCRIM.

Alvarez, M. C. (2014). Teorias clássicas e positivistas. In R. S. Lima, J. L. Ratton, & R. G. Azevedo (Orgs.), *Crime, polícia e justiça no Brasil* (p. 59). Contexto.

Alves, J. A. (2018). *The anti-black city: Police terror and black urban life in Brazil.* University of Minnesota Press.

Amankwaa, A. O., & Addo, J. A. (2023). Forensic DNA analysis and database governance in Ghana. In V. Toom, M. Wienroth, & A. M'charek (Eds.), *Law, practice and politics of forensic DNA profiling* (pp. 56–71). Routledge.

Baratta, A. (1999). *Criminologia crítica e crítica do direito penal: Introdução à sociologia do direito penal.* Revan.

Biondi, K. (2014). *Etnografia no movimento: Território, hierarquia e lei no PCC* (Tese de doutorado, Universidade Federal de São Carlos, São Paulo, SP).

Brasil, Ministério da Justiça e Segurança Pública. (2024). *XXI Relatório da Rede Integrada de Perfis Genéticos (RIBPG).* MJSP.

Breckenridge, K. (2014). *Biometric state: The global politics of identification and surveillance in South Africa, 1850 to the present.* Cambridge University Press.

Breckenridge, K. (2021). Recasting the technologies of the carceral empire: India, South Africa, and the political paradoxes of post-colonial citizenship. Keynote lecture, *Behind Walls, Beyond Discipline: Science, Technology & the Carceral State, Webinar Series*, University of Michigan.

Campello, R. U. (2021). Biocárcere: Cartografia política dos bancos forenses de DNA no Brasil. *Anais do 45° Encontro Anual da ANPOCS, GT 45—Violência, punição e controle social: Perspectivas de pesquisa e de análise.*

Campello, R. U. (2023). *Short circuit: Electronic monitoring and the crisis of the Brazilian prison system.* Springer Nature.

Cardoso, B. d. V. (2014). *Todos os olhos: Videovigilâncias, voyeurismos e (re) produção imagética.* Editora UFRJ.

Cole, S. A. (2001). *Suspect identities: A history of fingerprinting and criminal identification*. Harvard University Press.

Davis, A. Y. (2003). *Are prisons obsolete?* Seven Stories Press.

Dias, C. N. (2011). *Da pulverização ao monopólio da violência: Expansão e consolidação do Primeiro Comando da Capital (PCC) no sistema carcerário paulista* (Tese de doutorado, Universidade de São Paulo, São Paulo, SP).

Doneda, D., & Kanashiro, M. (2012). O novo sistema brasileiro de identificação: Traços exclusivos de uma transformação geral. *Politics*.

Duarte, D. E., & Ceia, E. M. (2022). *Tecnologia, segurança e direitos: Os usos e riscos de sistemas de reconhecimento facial no Brasil*. Konrad Adenauer Stiftung.

Durão, S. (2023). The Portuguese police and colonial sedimentations. In J. Beek, T. Bierschenk, A. Kolloch, & B. Meyer (Eds.), *Policing race, ethnicity and culture* (pp. 47–64). Manchester University Press.

Duster, T. (2006). The molecular reinscription of race: Unanticipated issues in biotechnology and forensic science. *Patterns of Prejudice, 40*(4), 325–345. https://doi.org/10.1080/00313220601020065

Fassin, D. (2017). *Punir: Une passion contemporaine*. Seuil.

Felitte, A. (2023). *A história da polícia no Brasil: Estado de exceção permanente?* Autonomia Literária.

Foucault, M. (1975). *Surveiller et punir: Naissance de la prison*. Gallimard.

Godoi, R. (2017). *Fluxos em cadeia: As prisões em São Paulo na virada dos tempos*. Boitempo.

Godoi, R. (2020). A prisão fora e acima da lei. *Tempo Social, 31*(3), 141–160. https://doi.org/10.1590/ts.2020.313.141160

Heineman, T., Lemke, T., & Prainsack, B. (2012). Risky profiles: Societal dimensions of forensic uses of DNA profiling technologies. *New Genetics and Society, 31*(3), 279–297. https://doi.org/10.1080/14636778.2012.711898

Hindmarsh, R., & Prainsack, B. (Eds.). (2010). *Genetic suspects: Global governance of forensic DNA profiling and databasing*. Cambridge University Press.

Jasanoff, S. (2006). Just evidence: The limits of science in the legal process. *Journal of Law, Medicine and Ethics, 34*(2), 256–265. https://doi.org/10.1111/j.1748-720X.2006.00022.x

Jesus, M. G. (2020). Verdade policial como verdade jurídica: Narrativas do tráfico de drogas no sistema de justiça. *Revista Brasileira de Ciências Sociais, 35*(102), 1–15. https://doi.org/10.1590/0125-14052020037

Krikorian, G. (2021). Usages de l'AND par la police. Les conditions épistémiques de l'absence d'um débat public. In J. Vailly (Ed.), *Sur la trace des suspects. L'incorporation de la prevue et de l'indice à l'ère de la génétique* (pp. 109–136). Éditions de la Maison des sciences de l'homme.

Krimsky, S., & Simoncelli, T. (2011). *Genetic justice: DNA data banks, criminal investigations and civil liberties*. Columbia University Press.

Lazer, D. (Ed.). (2004). *DNA and the criminal justice system: The technology of justice*. MIT Press.

Lynch, M. (2003). God's signature: DNA profiling, the new gold standard in forensic science. *Endeavour, 27*(2), 93–97. https://doi.org/10.1016/S0160-9327(03)00110-1

Lynch, M., Cole, S., Jordan, K., & Reardon, J. (2008). *Truth machine: The contentious history of DNA fingerprinting*. University of Chicago Press.

Lyon, D. (2001). *Surveillance society: Monitoring everyday life*. Open University Press.

M'Charek, A. (2008). Silent witness, articulate collective: DNA evidence and the inference of visible traits. *Bioethics, 22*(9), 508–515. https://doi.org/10.1111/j.1467-8519.2008.00672.x

M'Charek, A., Toom, V., & Jong, L. (2020). The problem with race in forensic identification. *Science, Technology & Human Values, 45*(5), 804–828. https://doi.org/10.1177/0162243919898909

Machado, H. (2011). Construtores da bio(in)segurança na base de dados de perfis de ADN. *Etnográfica, 15*(1), 153–166. https://doi.org/10.4000/etnografica.3847

Machado, H., & Granja, R. (2022). *Genetic surveillance and crime control: Social, cultural and political perspectives*. Routledge.

Mallart, F. (2021). *Findas linhas: Circulações e confinamentos pelos subterrâneos de São Paulo*. Etnográfica Press.

Marx, G. (1988). *Undercover: Police surveillance in America*. University of California Press.

Olarte-Sierra, M. F., & Bermúdez, J. E. C. (2023). Technolegal worlds in an armed conflict: The forensic making of victims in Colombia. In V. Toom, M. Wienroth, & A. M'charek (Eds.), *Law, practice and politics of forensic DNA profiling* (pp. 23–37). Routledge.

Ribeiro, L. (2010). A produção decisória do sistema de justiça criminal para o crime de homicídio: Análise dos dados do estado de São Paulo entre 1991 e 1998. *Dados, 53*(1), 1–25. https://doi.org/10.1590/1678-458320204303

Richter, V. S. (2016). *Identificação genética e crime: A introdução dos bancos de DNA no Brasil* (Tese de doutorado, Universidade Federal do Rio Grande do Sul, Porto Alegre, RS).

Richter, V. S., & Louzada, L. (2023). Travelling promises: Forensic DNA databases in Brazil's technolegal setting. In V. Toom, M. Wienroth, & A. M'charek (Eds.), *Law, practice and politics of forensic DNA profiling* (pp. 38–55). Routledge.

Rigon, B. S., & Jesus, M. G. (2019). Testemunho policial como prova no processo penal brasileiro: Uma análise crítica da presunção de veracidade das narrativas policiais nos processos de tráfico de drogas. *Revista Brasileira de Ciências Criminais, 27*(162), 85–119. https://doi.org/10.1590/1678-4583201927162

Schiocchett, T. (2021). Bancos de perfis genéticos para fins de persecução penal. In *Série Pensando o Direito* (Vol. 43). Ministério da Justiça.

Schlittler, M. C. (2020). "Matar muito, prender mal": Desigualdade racial como efeito do policiamento ostensivo militarizado em SP. *Autografia.*

SENASP. (2013). *Procedimento operacional padrão—Perícia criminal.* Ministério da Justiça.

Silva, T. (2020). *Comunidades, algoritmos e ativismos: Olhares afrodiaspóricos.* LiteraRUA.

Sinhoretto, J. (2021). *Policiamento ostensivo e relações raciais: Estudo comparado sobre formas de controle do crime.* Autografia.

Skinner, D. (2013). "The NDNAD has no ability in itself to be discriminatory": Ethnicity and the governance of the UK national DNA database. *Sociology, 47*(5), 976–992. https://doi.org/10.1177/0038038513476657

Toom, V., Wienroth, M., & M'charek, A. (Eds.). (2023). *Law, practice and politics of forensic DNA profiling.* Routledge.

UNESCO. (2004). *Déclaration internationale sur les données génétiques humaines de l'UNESCO.* United Nations Educational, Scientific and Cultural Organization.

Vailly, J. (2024). *Genetics and the politics of security: A social science perspective.* Routledge.

Vargas, J. (2008). Análise comparada do fluxo do sistema de justiça para o crime de estupro. *Dados, 50*(4), 567–587. https://doi.org/10.1590/S0011-52582008000400003

Wacquant, L. (2009). *Punishing the poor: The neoliberal government of social insecurity.* Duke University Press.

Wienroth, M., Granja, R., Lipphardt, V., Amoako, E. N., & McCartney, C. (2021). Ethics as lived experience: Anticipatory capacity and ethical decision-making in forensic genetics. *Genes, 12*(5), 611–623. https://doi.org/10.3390/genes12050611

Woude, M. V., Barker, V., & Leun, J. V. (2017). Crimmigration in Europe. *European Journal of Criminology, 14*(1), 3–6. https://doi.org/10.1177/1477370816682264

Part V

Imaginaries and (In)equities

15

Mundane Heroics: A Narrative Analysis of Gene Therapy in the Media

Eva Hilberg, Aleksandra Stelmach, Tineke Kleinhout-Vliek, and Paul Martin

Introduction

After almost fifty years of development, gene therapy is finally becoming a reality (Dunbar et al., 2018). It is, however, a reality that is still uncertain in its actual content and derives much of its present appeal from the

This research was made possible by a Wellcome Investigator award (219875/Z/19/Z) '*Orphan drugs: high prices, access to medicines and the transformation of biopharmaceutical innovation*' (Principal Investigator: Paul Martin); and Research Stimulation Funding awarded by the Sociological Faculty, University of Sheffield (Principal Investigator: Eva Hilberg).

E. Hilberg (✉) • P. Martin
School of Sociological Studies, Politics and International Relations, University of Sheffield, Sheffield, UK
e-mail: e.hilberg@sheffield.ac.uk; paul.martin@sheffield.ac.uk

A. Stelmach
King's College London, London, UK

Department of Management, University of Exeter Business School, Exeter, UK
e-mail: a.stelmach@exeter.ac.uk

V. Argudo-Portal et al. (eds.), *Mundane Genomics*,
https://doi.org/10.1007/978-981-95-4550-6_15

anticipated results of treatments that have to still prove their long-term efficacy. Gene therapies, or the concept of "genetic modification by exogenous DNA" (Dunbar et al., 2018, p. 1), are usually presented as one-dose treatments, and are hoped to outlast other medicines, but their effects are often still conjecture, and not as clear-cut as the hyperbolic coverage in the press makes them out to be. The present moment marks the transition from hotly anticipated futuristic treatments into mundane, present-day, medical care, where different standards of proof, reporting, and reliability come into play. Here, points of reference shift from those that are hoped for to those that can be proven and relied upon in an everyday setting. In the case of gene therapy, this shift is of an even greater magnitude, as medicine is undergoing a paradigm shift towards a new form of treatment that "uses a gene(s) to treat, prevent or cure a disease or medical disorder" (National Human Genome Research Institute, 2025). Gene therapy holds vast promise for a number of previously untreatable and often rare genetic conditions, and addresses the body in an entirely new way, promising to undo the errors of genetic variation in specialised cases, and even holding out hope for new treatment regimens geared at large-scale chronic health issues, such as diabetes or hypertension. These hotly anticipated future gains are being discussed in the media in a variety of ways and, as our sample shows, optimistic coverage outweighs more critical voices at this juncture.

The discovery of the genetic code was a heroic story from its beginning in the early 1960s, eventually promising the dawn of a "new era of genetic medicine" (The White House, 2000a). In 2000, a seminal White House press conference announced "the completion of the first survey of the entire human genome" and distilled into one event the exorbitant scale of expectations connected to this new sector of science and economy, one that anticipated anything from a cure for cancer to "learning the language in which God created life" (The White House, 2000b). Twenty-five years

T. Kleinhout-Vliek
University College Dublin, Dublin, Ireland
e-mail: tineke.kleinhout-vliek@ucd.ie

later, new treatments that modify genes to cure disease (such as Zolgensma, which treats spinal muscular atrophy) at last seem to bring about the therapeutic sea change that the genetic revolution has promised all along (see Fig. 15.1). But the journey from discovery to actual availability of gene therapies has been, and still is, far from smooth, hampered by failed drug trials, at least one direct, high-profile fatality, and dashed economic expectations. A new wave of excitement has defined this latest stage in the turn towards genetic medicine, and once again captured the imagination of the wider public, but this time arguably with a changed emphasis:

> Back in 2000 Bill Clinton announced the first draft of the human genome, or as he called it, a map. [...] In the decades since, many have entered a trough of disillusionment about the promise of genetics. In his speech Clinton promised us personalised medicine, personalised disease risk, precise diagnostics, molecular medicine. Where are they? But, quietly, science pulled itself from that trough. Not so much by doing new things but by doing old things better. Molecular scissors called CRISPR came along that could edit DNA. We could edit DNA before, but they made it easy. [...] We know it because, like jet engines and computers before, genetics is becoming mundane. (*The Times* 31/05/2021)

Now, reports of new gene therapies regularly feature in the media, focusing especially on previously incurable rare genetic conditions. But media coverage often also goes much further than that, discussing future scenarios of widespread genetic treatments and a wholesale overhaul of medical approaches akin to the "new era of genetic medicine" set out in 2000. Public media narratives are a significant place in which imaginaries are presented (Petersen, 2001), just as they were in the first wave of hype surrounding the deciphering of the genetic code. But media narratives are more than just records of opinion, they also shape future expectations of gene therapies in a structural sense. This chapter argues that a closer look at a common narrative of the emergence of gene therapies can show the ways in which these presentations direct public attention and lead to structurally determined omissions in debates. For instance, if gene therapies are truly meant to become a widespread treatment option, then health services, insurers, and patients need to be able to afford these

	Foundations
1953	Watson and Crick describe the structure of DNA, with prior work by Franklin.
1961	Crick, Brenner, Barnett, and Watts-Tobin first demonstrate the three bases of DNA code for one amino acid.
1972	Friedmann and Roblin publish "Gene therapy for human genetic disease?" in *Science*, proposing gene therapy.
14.9.1989	NIH approves the first human gene therapy trial for ADA-SCID.
	1990s: First Human Therapies and a Major Setback
14.9.1990	First gene therapy in humans: A four-year-old girl with ADA-SCID receives genetically modified T-cells.
17.9.1999	Jesse Gelsinger, 18, dies in a gene therapy trial at the University of Pennsylvania for OTC deficiency. FDA and NIH increase oversight of gene therapy trials following Gelsinger's death.
	2000s: Recovery and Technological Refinement
April 2000	French researchers report successful treatment of X-linked SCID with retroviral vectors.
26.6.2000	Bill Clinton and Tony Blair announce completion of rough draft of the human genome at the White House.
2002–2003	Some SCID patients develop leukemia due to unintended mutations as a side effect of gene therapy.
14.4.2003	Human Genome Project completes the first sequence of the human genome.
2003–2007	Lentiviral vectors introduced to improve safety and reduce side effects such as unintended mutation.

Fig. 15.1 Timeline of key developments in gene therapy

2010s: Breakthrough Approvals

November 2012	First approved gene therapy in Europe: *Glybera* approved in the EU for lipoprotein lipase deficiency (LPLD).
2012-2013	Charpentier and Doudna develop CRISPR/Cas9 "genetic scissors" to perform targeted gene editing.
30.8.2017	First approved gene therapy in the USA: *Kymriah* approved by FDA for B-cell precursor acute lymphoblastic leukemia.
19.12.2017	*Luxturna* approved by FDA for RPE65-mediated retinal dystrophy.
18.10.2017	*Yescarta* approved for relapsed/refractory large B-cell lymphoma.
26.11.2018	Jiankui announces gene editing of human embryos and subsequently receives a three-year jail sentence.
24.5.2019	*Zolgensma* approved by FDA for spinal muscular atrophy.

2020s: Precision Gene Editing and Expanded Approvals

December 2020	*Libmeldy* approved by the European Medicines Agency for metachromatic leukodystrophy (MLD).
2020–2023	CRISPR-based therapies show high efficacy in clinical trials for sickle cell disease and beta thalassemia.
16.11.2023	*Casgevy* becomes the first CRISPR-based therapy approved (UK).
8.12.2023	*Casgevy* approved by FDA for sickle cell disease and beta thalassemia.
15.5.2025	First patient treated with personalized CRISPR gene editing therapy at Children's Hospital of Philadelphia, USA.

Fig. 15.1 (continued)

medicines to ensure equitable access, and not only in select cases or at an entirely unsustainable rate of personal payments. At the same time, calculations of affordability show that even relatively short windows of efficacy can be sustainable in the case of relatively rare conditions, but these findings may not translate to larger numbers of patients. These issues are not often addressed in popular discourses on gene therapies, which more often discuss the promise of gene therapy in more glowing yet general terms. These oversights are not directly intentional; rather, they are a result of narrative structures which simply make it easier to present positive, hopeful factors than persistent and complex challenges.

Heroic stories operate according to their own logic. In many ways, the modern bioscience sector has relied on the anticipation of great things from its inception (Cooper, 2008; Kay, 1993; Fortun, 2008). But these expectations also have considerable real-world consequences in the shape of promissory regulatory regimes that promote change (Hogarth & Martin, 2021), resulting in a wider financial "fallout" across the sector (Whitacre, 2024); thus, hype can become something of a self-fulfilling destiny in which the promises of a new technology actually bring about an entirely new sector of the bioeconomy (Pavone & Goven, 2017). This chapter contributes to the study of these cycles of expectation by looking more closely at heroic promises on the level of metaphors and narrative arcs in media coverage of gene therapies. It argues that narrative structures affect more than the enjoyment of a particular article; they also work to direct attention towards some elements of a story at the expense of others. This is, on some level, a simple effect of the selection and presentation of facts but, in the case of medical storylines, a preference for stories reporting heroic revolutions can skew the representation of likely gains and issues towards curable conditions, successful interventions, and expectations of future potential. Issues connected with costs and access to treatments become folded into this configuration, which always promises a moment of resolution after any existing obstacles are overcome. This leaves several structural challenges, such as the burden of payment for highly priced treatments and their still uncertain outcomes, severely underappreciated, and thus mostly omitted from public debate. Understanding the mechanisms at work in the representation of the promise of gene therapy can be a first step towards a more balanced appraisal of medical gains (such as the first personalised CRISPR gene editing therapy in 2025, see timeline) and possible limits to this revolution.

The chapter begins with a closer look at the ways in which the operation of metaphors and practices of deliberate framing have influenced the reporting of advances in the field of genetics and genomics. It then introduces our data set of media items discussing gene therapies between 01/01/2020 and 30/04/2023. This period captures the growing excitement around a number of new gene therapies, and represents a new, more intensive stage in their coverage. The chapter then draws out different examples of heroic narratives presented during this time, and contrasts them with more critical storylines. The marked contrast between these different approaches then shows the shortcomings of heroic narratives, especially with regard to discussion of the high costs of these new therapies. While such relatively mundane factors are mentioned in heroic narratives, they often do not play more than a minor role in a report of a miraculous treatment of an otherwise untreatable condition. They represent, however, the challenge that is at the heart of a larger scale turn towards gene therapy, and need to be discussed more prominently in order to identify and address crucial issues surrounding this long-anticipated transition towards "a new era of genetic medicine" (The White House, 2000a).

Heroic Narratives: Why Do They Matter?

The coverage of scientific research in the media is never an entirely neutral undertaking, something particularly noted in the context of genetic medicine. Petersen argues that "at the interface between genetic researchers and the public, [media stories] are likely to play an important role in shaping public perceptions of genetics and its value and applications, by selectively presenting some subthemes and not others" (Petersen, 2001, p. 1256). This process of selection most often takes the shape of editorial control and the need to simplify complex scientific contexts for publication in public media outlets (Priest, 1994); however, the "popularisation" of scientific knowledge is not the only role of the media, whose publications can also have a role to play in the structuring of assumptions and expectations of scientists (for the complex interplay of imagery and science, see van Dijck, 1998). Hilgartner (2017) points out these interactions and the feedback process between science and media, while Petersen et al. (2005) follow the influence of science fiction imagery and metaphor on contemporary public debate, finding that some literary devices "encourage or invite certain preferred formulations of medical genetics, thereby raising questions about their influence in shaping public deliberation about scientific developments and their consequences for society" (Petersen et al., 2005, p. 338). These formulations could be anything from a metaphor to a more deliberate act of framing, to name but two possible literary devices. Framing refers to "the selective presentation of some themes, facts, and claims but not others" (Petersen, 2001, p. 1258)—by means of imagery, metaphors, and rhetorical devices, for instance. The following discussion enlarges on the efforts of these scholars by adding another dimension, that of the narrative arc, either on the level of metaphor or of the story as a whole.

Metaphors are often explored as a central vehicles of influence. They can be divided into small-scale *metaphorical expressions*, *conceptual metaphors* such as the blueprint or code, and *grand metaphors*, such as the "book of life", that seek to tell us something about the human condition or what it is meant to be (Stelmach & Nerlich, 2015, pp. 198, 201). Previous research surrounding the emergence of genetic science showed that "public understanding of science is, at least in part, a struggle over metaphors" (Nerlich et al., 2002, p. 465). Looking at "spirals of meaning" with regard to the book of life metaphor, Nerlich et al. warn scientists employing this (and similar) imagery, of "semantic traps" that can alter meaning surreptitiously, leading to a shift of emphasis away from scientific specificity towards more hyperbolic expectations of magic powers. They can also introduce other frames of reference with influence on meaning: "scientific descriptions rely heavily on imagery and metaphors circulating in the broader culture, and reflect social biases and assumptions" (Petersen, 2001, p. 1257). In Nerlich's account, metaphors accompanying the revelation of the Human Genome Project to the public range from notions such as blueprint or instruction manual to more complex elements, such as the moon landing, the printing press, Galileo's "mapping of the heavens", and even Aldous Huxley's novel *Brave New World.* While some of these metaphors stop at being a particular turn of phrase, or a specific image, others invoke an entire story—in the case of *Brave New World,* quite directly.[1] It is argued here that these types of references operate in a complex way, surreptitiously managing the expectations of outcomes contained in the reference to a particular narrative arc.

Narratives are more than sequences of events; there is a structure that ties them together, that makes sense of the sequence and holds the attention of the audience. Narrative arcs are writerly devices that carefully manage expectations and play on a reader's affective responses, such as curiosity and the desire for a revelation. They direct attention and determine which thoughts may be invoked as part of the story, making it difficult to include ideas or points that simply do not "fit the script".[2] Narrative structure is an essential part of a story, and any number of books will tell you that "[a]ll stories, from Aristotle through all the constellations of civilisation, embody the same dramatic principles"—these being an arc consisting of a set-up, then a confrontation, followed by a resolution (Field, 2005, pp. 3, 21). Aristotle calls the turning points in this arc "reversals", and they denote a moment of change of fortune (often in an inciting incident); a story without reversals would be a "simple" story that introduces changes of fortune without a reversal (Aristotle, 2020, p. 14). In both cases (complex or simple), a sequence of changes in fortune makes up the plot of the story in imitation of actual life, not in a faithful retelling. There are, of course, also "experimental" ways of telling stories (Jeffreys, 2019, p. 59), but the arc of the story remains important (arguably is even more so), down to the character arcs of each participant.

[1] In the case of *Brave New World*, the story arc moves from an entirely technicised world towards a return to nature, and thus makes quite a clear critical point in opposition to the more enthusiastic coverage. While it works as a metaphor of an undesirable world of technological abstraction, there is more invoked in this reference—as the arc heads towards either a complete surrender of all human physical processes to machines, or a complete repudiation in the wilderness.

[2] Narrative arcs also operate on the level of metaphor. For instance, calling a research programme a "Moonshot" introduces not only the metaphor for a virtually impossible task, but also the underlying expectation of eventual success—at huge (astronomical!) expense. A "moon landing" introduces a slightly different slant, as this skips ahead on the narrative arc towards its resolution. The moon landing *is* the moment of success, making this a much more simplistic metaphor; however, it also opens a hidden door towards another arc: the eventual need for the return journey. This (potentially unintentionally) leaves the reader to ask where the journey will go from here.

Popular media aim to tell us something that is noteworthy, that is surprising or entertaining, while also imparting information, and often employ similar storytelling techniques in the process.

Hero stories are a particularly successful type of narrative arc, especially in medical storytelling. Arthur Frank's *Wounded Storyteller* (2013) distinguishes several different forms that such a story can take: the restitution narrative, the chaos narrative, and the quest narrative. The restitution narrative most commonly emphasises an eventual return to health, thanks to medical intervention, while in quest stories "[i]llness is the occasion of a journey" towards an eventual goal that may not be immediately obvious (Frank, 2013, p. 115). The chaos narrative does not imagine an improvement, but rather recounts incidents without structure in a more immediate account of the nature of illness. These more "realistic" chaos narratives "are hard to hear. […] since in life as in story, one event is expected to lead to another" (Frank, 2013, p. 97). In his opinion, patients' stories most closely resemble a quest narrative, which "affords the ill their most distinctive voice […]. Yet only a few quest stories are published" (Frank, 2013, p. 115). This leaves the restitution narrative, which may be most familiar from a wide range of popular medical dramas. Here, "over the decades TV storytellers have actually made quite narrow choices regarding key aspects of the physicians' professional lives" such as favouring "acute illnesses that can be cured" (Turow, 2010, p. 2). The focus on cures directs attention to particular types of conditions, and also, as we argue at the end of this chapter, towards issues that can be solved. More complex deliberations that balance different factors in potentially uncomfortable ways are occluded in these storylines (or narrative arcs) by the expectation of a surprising turnaround, such as, for instance, a magical cure. The examples in this chapter show how problematic an unhappy ending is in a storyline that expects success, for instance, and also explores the effects of a reliance on fairy tale imagery for the discussion of an issue that has very real economic and ethical implications.

Method

In this study, we undertook a thematic narrative analysis to explore stories about gene therapies published in major UK and US newspapers. Riessman defines a narrative as a particular sequence and consequence in which "events are selected, organised, connected, and evaluated as meaningful for a particular audience", and where narratives are told by storytellers who "interpret the world and experience in it", "sometimes [creating] moral tales—how the world should be" (Riessman, 2005, p. 1). The aim of the analysis is to identify common thematic elements across analysed stories, such as the types of events, actions, actors, or issues they report, which helps determine the types of narratives that are being produced in a given context. Unlike in other qualitative approaches, the aim of the analysis is not to break down texts into coding segments focussed on issues, but rather to attempt to "keep the story intact" (Riessman, 2008, p. 111) and evaluate its overall direction. We asked: What are the main narratives of gene therapies in the media stories? How are these narratives structured thematically in terms of key events, actions, actors, and issues? How do these narratives reflect and affect the ways in which this technology is understood and imagined in the public realm? The findings in this chapter are part of a larger project that addresses these questions in more detail (Hilberg et al., 2026), but here we focus on one prevalent narrative archetype: the story of the heroics of biomedicine. It shows how this type of story tends to report some issues, while others are jarring and usually left out. This is not to say that other types of coverage do not exist; however, the chapter looks more closely at the narrative dynamics of the most common type of story routinely encountered in media coverage of gene therapies.

We used the media database Factiva to search for articles containing the term *gene therapy* in the United Kingdom and the United States, two countries with a significant number of research institutions and biotech companies pioneering this innovation. We decided to search top sources in these countries, such as newspapers with the reputation of being agenda-setters and/or with the largest readership. We also made sure that each newspaper represented a diverse readership and presented different perspectives on this technology. We excluded financial newspapers, such as the *Wall Street Journal* and the *Financial Times*, to avoid over-representing financial news in the sample. In order to produce a focused sample, we chose three top newspapers in each country with the largest volume of published articles on gene therapies and with the largest readership. In the United Kingdom, we selected *The Times*, the *Daily Mail*, and *The Independent*, and in the United States, *The New York Times*, the *Boston Globe*, and *USA Today*. The search in Factiva with these criteria returned a sample of 447 articles (194 from the UK and 253 from the US press). We excluded duplicates as well as articles where gene therapies were mentioned but received only cursory treatment, including only those articles where gene therapies were the main topic and/or covered in depth. The final sample consisted of 176 articles (69 articles from the UK press and 107 from the US press). The US sample is bigger than the UK one; this is consistent with our general findings on media coverage across the United States, the United Kingdom, and Europe. The authors read the sample twice and met regularly online to analyse the texts and develop and discuss codes.

The following sections set out some of the data on the most prevalent theme, that of heroic narratives. Several examples of storylines demonstrate the types of heroic stories most often reproduced in the media, and the directions in which these narrative arcs guide the reader's imagination. This is then contrasted with what is not covered in these heroic stories, arguing that narrative structures can work to create expectations and guide attention in particular directions.

The Heroic Tales of Biomedicine

The most common heroic story about gene therapies is that of an almost magical cure to an otherwise incurable condition: an increasingly mundane and expected form of heroism. These narratives are often reported in association with the emergence of a new medicine, or in cases of rare conditions. One such case involved the family of a young British boy, Edward Willis-Hall, whose story was closely followed by multiple news outlets in the United Kingdom. As one article recounted:

> Little Edward has the genetic condition spinal muscular atrophy (SMA), in which the lack of a protein called SMN, vital for muscle development and movement, results in progressively weaker muscles. Until three years ago there was no treatment and babies affected by Type 1—the most severe form—gradually lost the use of their muscles before dying, usually before age two as a result of breathing problems caused by weak chest muscles. Yet there was a glimmer of hope. The consultant who gave them the heartbreaking diagnosis said a one-off jab could halt Edward's rapid deterioration. The catch—it costs £1.7 million. Zolgensma is thought to be the most expensive drug in the world. [...] Zolgensma is not yet available in the UK on the NHS or privately. It is being appraised by the National Institute for Health and Care Excellence (NICE), with a decision expected by March. It is prescribed in other countries including the U.S., Germany, Brazil and Japan. So the only way for Edward to get the jab is for his parents to raise the money for treatment abroad. Since last month they have received more than £120,000 in donations through a crowdfunding site—a fantastic sum, but still a fraction of what's needed. "£1.7 million is a lot of money but it is a one-off treatment and research has shown that children who have been given it are surviving and thriving five years on," says Megan [Edward's mother]. "We would love to see Edward walk and talk and this offers the best chance. Wouldn't all parents want the same?" (*Daily Mail*, 15/12/2020)

The relentless campaign mounted by Edward's parents was successful and during the following year, their son received the treatment as part of NHS (publicly funded) healthcare. The news was greeted in the press which reported,

> The drug is expected to give Willis-Hall, from Colchester, Essex, a better chance of sitting, crawling and walking. [...] Edward's parents Megan Willis, 30, and John Hall, 37, said they were delighted when the drug was approved for use by the NHS. [...] "We're so grateful to have this opportunity," Willis said. (*The Times*, 13/08/2021)

More coverage followed a few months later when little Edward took his first assisted steps after receiving gene therapy. His mother, as the media reported,

> watched with tears in her eyes as her son was able to move with gaiters to support his legs. […] The NHS said that babies born with severe type 1 SMA had a life expectancy of two years but that Zolgensma […] had helped babies breathe without a ventilator, sit up on their own and crawl and walk after a single infusion treatment. […] His mother said that he had lost his lethargy and become a normal, playful baby. "This drug has given Edward his life back. […] It's going to take a long time for Edward to catch up but he is slowly getting there." She hopes that her son could be part of a new generation of babies with SMA to reach adulthood. "I thought his time with us was numbered but now I don't think that at all," she added. (*The Times*, 01/01/2022)

This story combines several different heroic narratives, from the effects of the cure for a previously incurable condition, to a triumph of medical development, and also the more procedural issues of Health Technology Assessment (HTA) approval and costs per dose. It shows that one news item can contain more than one narrative, and also how these are combined into a larger arc that goes from the discovery of a new medicine, via doubt about access to the treatment, to a resolution in which the treatment actually helps a child live a better life. This is an archetype of medical storytelling that is familiar from many books and films, and it never fails to engage an audience. Its goal is a restitution of health, which ultimately validates "the modernist expectation that for every suffering there is a remedy" (Frank, 2013, p. 80). This hope "crowds out any other stories" and "reflects one of the best impulses in modernity: the heroism of applied science as self-overcoming" (Frank, 2013, pp. 83, 92).

This emphasis on applied science features prominently in stories of heroic medical breakthrough, further exemplified by a success story centred on gene therapies for various eye conditions, hailed by the press as a revolution and an example of treatments that would soon reach the masses. The focus of such coverage was on patients whose sight, against the odds and hope, was improved or restored. One such story, entitled "Doctors saved my sight by injecting a gene into my eye", recounted the experience of a British patient:

> Earlier this year, life suddenly looked so much better for Lee Morris. After struggling with worsening sight all his life, within days of a pioneering operation, he could see all the details he had previously missed. [...] The dramatic change happened after Lee became one of the first people in the UK to have ocular gene therapy, a new treatment, which it's hoped could benefit hundreds of thousands of Britons. In simple terms, it involves an abnormal gene being replaced by a functioning one. [...] For those like Lee for whom there has been no other treatment option, it is a huge step forward. "All of my life, I've been told there was no treatment. [...] So when my consultant told me about a new therapy, I knew it could be life changing", says Lee. [...] Lee is revelling in the improvement to his vision. He says: "I'm so grateful that gene therapy has given me, and others with similar conditions, hope when for so long there wasn't any." (*Daily Mail*, 06/10/2020)

This story sets out a virtually biblical miracle—restoring sight to those who otherwise were left entirely without hope—and this by means of a simple injection. Of course, the content of the injection is far from simple, but some of the galvanising appeal of gene therapies rests on their promise of working miracles in a single dose, restoring health entirely instead of following longwinded chronic treatment protocols.[3] It contains the most compelling reversal of fortune, going from no hope to hope, and leaves the audience with the expectation that this type of life change could be possible for a greater number of patients in the future. In many ways, these early gene therapies are only the beginning of this medical revolution, which is a heroic story in its own right.

These heroics of medicine celebrate the achievements of the past few years and are building expectations for the road ahead: "When you step back, the scientific advancements in gene and cell therapies achieved in the past 10 years are simply astounding. Certain types of rare, inherited diseases can be cured by replacing malfunctioning genes with ones that are healthy and function normally" (*The Boston Globe*, 01/01/2020).

[3] This, of course, is a somewhat unrealistic expectation, but one often repeated in media coverage of new gene therapies. Yet, Edward's experience of Zolgensma (see above) shows that results are not often as clear-cut as that. Instead, this real-life example contextualises the hyperbole of many descriptions by stating that "Zolgensma [...] had helped babies breathe without a ventilator, sit up on their own and crawl and walk after a single infusion treatment."

Rare genetic conditions are often at the heart of these narratives, as gene therapies have been particularly effective in this field so far. They are also "big business", giving rise to an entirely new branch of the bioeconomy made up of often very highly priced treatments (Zolgensma, as the story above pointed out, is listed at £1.7 million per dose). The price of these treatments is, however, not regularly problematised in accounts of this emergence:

> Gene therapy [...]—a scientific vision for decades—is finally becoming a more common reality in the United States. Dozens of disorders are now being treated, though most still only in clinical trials [...]. Within the next few years, experts say, gene therapies could soon be available for conditions never effectively treatable before, such as sickle cell disease, Huntington's, ALS, Parkinson's, some forms of heart disease and a host of very rare diseases. "The exponential growth phase" of gene therapy has arrived. (*USA Today*, 27/04/2021).

There is nothing like talk of an "exponential growth phase" to capture the imagination of investors in these new medical companies focused on developing gene therapies. It is also part of a medical fairy tale in which a range of untreatable degenerative genetic conditions finally may be overcome by means of medical advancement. The narrative arc stretches from the present, detailing clinical trials, to the near future, when all these conditions will be undergoing change and lots of money will be made, presenting this as a near certainty. The price of the treatment is not being seen as a problematic issue in this context; indeed, it could much rather be considered a selling point.

Other points of reference do not work as economic fairy tales, but are even more overt in their fairy tale imagery. In one case, hyperbolic coverage about the potential of gene therapies, for instance, promised "to reset the biological clock in mice", prompting hopes that scientists "were edging closer to unearthing a fountain of youth" and reversing ageing in humans (*The Boston Globe* 12/01/2023). This is representative of straightforward metaphors of fundamental biological processes that are otherwise considered irreversible—such as the biological clock—and, like the fountain of youth, are plucked wholly from the realm of fairy tales and legend. In the context of the present study, even the throwaway reference to the fountain of youth contains a narrative arc, stretching from the well-known elusiveness of this legendary spring to its imminent discovery, thus ending centuries of quest. The likely benefits of finding it are widely understood to be immeasurable, rendering worthwhile the labour and expense involved in the search. While this is clearly not the most precise and factual reporting of the promise of gene therapies, it nonetheless introduces powerful and evocative imagery that directs attention in particular directions, towards a heightened state of expectation.

The first real experiences of gene therapy are, however, somewhat undermining these simplistic assumptions. At the very least, the decisions that patients and carers have to make as part of these new treatments can be very difficult. For example, a newspaper article recounted the agony of the parents of little Alissa who signed up for clinical trials of a gene therapy for her rare condition. The pair who, as the article claimed, "decided to let hope conquer the fears", was "tormented by the idea that the therapy might only keep Alissa alive longer in the same miserable state. "Instead of having this cruel, short life of two, three, maybe four years, maybe she will just have a cruel, long life," [Alissa's father said]. That's what we were fearing" (*USA Today* 27/04/2021).

Gene therapies are still relatively untried, and their long-term (and even short-term) effects are not yet fully known. For instance, existing therapies do not always seem to provide long-lasting effects, or the effects are not quite on the level of a cure.[4] In the words of one American paediatrician, "I have some kids who are running around like little maniacs. [...] And I have some kids who still have some problems." This was because gene therapies, while dubbed life-changing, one-time treatments, did not necessarily mean they were "disease ending". "Some children who receive the therapies", the article argued, "still require significant support and never regain or develop certain capabilities" (*The Boston Globe*, 01/06/2021).

[4] An example here is the gene therapy for haemophilia, which "is effective, at least for several years, but the exact effect may be unpredictable and intensive monitoring for several months is needed" (Chou et al., 2023, p. 1101).

Another challenge is the price of these new therapies, and Zolgensma is not an outlier but much rather an indicator of a general trend towards increasingly expensive new compounds. Therapies are consistently coming in at between one and three million dollars per dose, and manufacturers are making a case for the economic viability of this price tag. Even at a treatment efficiency of approximately eight years, Hemgenix, the $3.5 million gene therapy for haemophilia, is still calculated to save "the US healthcare system $5–5.8 million per patient" (Salib, 2023, no pagination). Yet, these prices can also be a burden—for patients and for healthcare systems. In a personal account that appeared in *The New York Times,* American author and TV host Annabelle Gurwitch described her experience of undergoing an advanced molecular treatment for cancer:

> My health has stabilized and my oncologist tells me this portends well for long-term management. The medication I take every day retails at $500 a dose. I qualify for a subsidy, but only if I continue to purchase a premium insurance plan. Others in my cancer support group with lesser plans report paying between $1000 and $3000 a month out of pocket. Last week, I called Becca, a social worker and cancer counsellor. "I'm so grateful the therapy promises a longer life span," I told her, "but am I your only client who wonders if it wouldn't be better to check out earlier? I don't want to become a burden to my family." "No," she sighed. "I hear this all day." [...] I'm lucky for scientific advancements. But it will take more than luck for me to afford the lifetime of care this diagnosis requires. (*The New York Times*, 15/11/2020)

This experience puts quite another spin on the heroic success story of advanced molecular treatments and gene therapies. On the face of it, everything has worked out—the therapy is working, and funding is also forthcoming, at least as part of an insurance plan—but this has resulted in increased emotional pressure on the patient. A change of insurance plan (or of employment) could directly affect Gurwitch's access to treatment, and other patients are already paying substantial amounts of money on top of their existing insurance arrangements. This very striking example concerns a therapy that is not a "one-dose" treatment; rather, it will work only for a limited period, and does not address the underlying cause of the cancer: "Some people's cancer doesn't respond, some receive scant months of benefit, while others eke out five years before switching to protocols with potentially harsher side effects" (Gurwitch, 2022, B4). In the field of gene therapy and advanced molecular treatments such as gene-targeted therapy (precision medicine), the challenge of high costs looms large. Reality can thus be much more complex than heroic stories make it seem.

In the United Kingdom, the prospects—or the possibility—of funding an increasing number of expensive gene therapies through the NHS also prompted some (limited) critical coverage. Some went as far as to argue against the idea in general. Referring to a little boy offered a new gene therapy for his SMA condition, one commentator claimed that national health budgets could not afford it. In his words:

> Yes, it's wonderful this little boy's future has been transformed by a £1.8 m drug. But it raises troubling questions for the future of the NHS. [...] the NHS does not have unlimited resources to meet every new challenge. [...] It is not the fault of any particular government, but the reality of a publicly funded healthcare system. But the great danger with rationing, especially in an age of heightened emotion fuelled by social media and political opportunism, is that decisions can end up being influenced by sentimentality or outrage. [...] Images of a sick child chime more strongly with people than testimony on the suffering of an octogenarian in a residential home, or a schizophrenic abandoned in a community care flat. [...] I believe the NHS can survive and prosper if we are frank about its constraints. Wishful thinking and Utopianism are not the way forward. (*Daily Mail* 2/06/2021)

This account is diametrically opposed to heroic narratives of revolutionary change and possibility. Instead, it presents a (very jarring) argument against utopianism, and advocates against the availability of highly priced gene therapies on the NHS. It also acknowledges the difficulties that this line of reasoning encounters, as images of sick children hold strong emotive power that is not easily dispelled by relatively abstract references to the resources of the NHS. The emotions evoked by this piece show clearly how difficult it is to work against a heroic narrative: it is emotionally challenging, it must rely on less emotive imagery, and it also makes for a much weaker story, leaving the reader without a moment of resolution. Ultimately, an argument that ends with a simple denial of options may be problematising the underlying issues, but does not offer any perspective on the way forward.

The Blind Spots of Heroics

The most significant aspect of these examples of small-scale narrative arcs is what does *not* appear in heroic stories. As the last three examples show, the contrast between heroic accounts of gene therapies and more critical reports is relatively stark, both in register and in effect. One of the examples addresses this disparity as the difference between hope and fear, enlarging on the parents' decision to embrace hope and attempt the use of the gene therapy. The excerpt quoted above (about baby Alissa) does not disclose whether this treatment proved effective in the end, and the absence of this information remains in the mind afterwards, like an unanswered question. The longer version of the article discloses some preliminary results a number of weeks after treatment, and details the return of several abilities such as being able to laugh. This resolves the tension built around a high-stakes decision in the article, and creates hope. A later follow-up article then sadly reports that these gains were not to last (Weintraub, 2023). The title of this later article addresses the difficulty of reporting such a very sad ending: "Gene Therapy Offers Hope for Rare Diseases. But the Stories Don't Always Have Happy Endings". After reading about Alissa's improvements, it feels particularly poignant to know that this treatment did not provide a miracle, and to hear about the devastating situation faced by her family. Yet, even this article ultimately reverts to a note of hope, stating, for example, that "risk is a necessary part of progress", and quoting an interview with Dr Francis Collins on the longer view of the potential of gene therapies:

> Those pioneering patients and families who choose to go first in such trials demonstrate great courage, as they make sacrifices for the benefit of those who come after them. [...] I would want people to be optimistic that over

the coming years, we're going to get much better at this—and not to feel hopeless or despondent or pessimistic. (Weintraub, 2023)

The turn towards the notion of pioneers here introduces a different register from that of failure, and opens up another possible arc (and timescale) that may lead to success down the line. This approach seeks to avoid the pain of having no hope at all at the end of the article, which would encounter all of the issues raised earlier with regard to the argument against utopianism. Compared to the more enthusiastic earlier piece, this shows that a straightforward heroic story is much easier to tell, even if it does not represent real-life experience.

These narrative decisions, however, can also have deeper effects. Even the example of failed treatment does not problematise its likely price.[5] In all the examples cited above, it is noticeable that the more heroic tales tend not to problematise the actual payment for the medicine, preferring to subsume the extremely high prices per dose into hyperbolic reporting. In the case of "exponential growth" or similar reports of breakthrough, high prices become high returns, and the success of gene therapies is presented as an example of an economic fairy tale (or goldrush) that is about to take place. Only a very few critical pieces, cited towards the end of the discussion above, directly problematise different issues surrounding the payment of these prices, and these pieces are not as satisfactory as stories. In fact, as they leave the fundamental issues without resolution or even the possibility thereof, they feel jarring and ultimately a little less convincing than other grand visions of the future. Of course, the underlying issues of relative affordability and difficult new payment options do not change, never mind how they are presented. But the presentation itself does some work in directing discussion; even the term "affordability" keeps at arm's length the real-life implications of not having access to a potentially curative medicine. "Cost" is a more directly personal and emotive way of discussing this, as Annabelle Gurwitch's story shows.

But what makes it so difficult or inconvenient to cover the multifarious issues surrounding price and pricing in heroic stories such as those presented above? One indicator of this difficulty is the negative affect

[5] In this case, the trial was paid for by the pharmaceutical company.

generated by the excerpt containing the argument against utopianism. The register utilised in heroic patient stories is one of hope and potential miracles, of life and death; concerns surrounding money simply do not fit with these "higher" considerations. An example of this is the debate over designer babies—analysed by Petersen et al. (2005)—in which metaphors focused on the notion of "playing God", and the debate questioned whether humans should interfere where otherwise only God may do. This debate did not even consider money as a factor that needed to be discussed. Two decades later, the role of scientists is no longer questioned in the context of gene therapy. Instead, the notion of miracles and fairy tales introduces a different type of supernatural hope, which neither responds to, nor has anything to say regarding the very worldly concerns of healthcare cost. The health of a child is difficult to compute in terms of money, and it seems downright cruel if it is done in terms of metrics, such as Quality-Adjusted Life Years (QALYs), and found to be not cost-effective.

The heroic narrative also guides expectations in a more structural sense. The story of little Edward's journey towards receiving a dose of Zolgensma (see above) indicates that the high price of the treatment is under discussion here, but only as an obstacle to access. An online funding campaign brings together the "fantastic" sum of £120,000 in donations, but that only comprises a small part of the £1.7 million cost per dose. Edward's mother then is quoted as saying, "1.7 million is a lot of money but it is a one-off treatment [...]. We would love to see Edward walk and talk and this offers the best chance. Wouldn't all parents want the same?" In this case, the high price of the medicine functions like a hurdle that needs to be overcome in order to achieve the end of a quest: something that may seem unsurmountable, but later is shown to not have been as overwhelming as it appeared. Hope is a power that sets aside these obstacles, and when the treatment is found to be working, it makes all the effort worth it. This moment of resolution shows why health stories tend to lean towards the reporting of issues that can be solved, especially when contrasted with the story of the very sadly ineffective trial earlier on. The heroic, almost religious expectation of overcoming all obstacles guides attention away from mundane issues such as payment, which are not the main focus of the story; the heroic story's focus is on the radical reversal of an otherwise fatal condition, or, as Frank puts it, "restitution stories are about the triumph of medicine" (Frank, 2013, p. 115).

The decision on whether to discuss price is, of course, also part of the general framing of an article, as decided by the writer. There is an intentionality behind what to include, and how, but the argument made here is that there are also some underlying narrative structures that are hard to set aside, even with the best of intentions. It is difficult to discuss the affordability of a medicine (and the consequences of not being able to afford it) at the same time as describing it as a medical revolution that will give hope to those who never had it before. The moon shot, the Holy Grail, the fountain of youth—all of these mythical undertakings come at enormous expense, and can never be judged to be cost-effective, as there are no metrics that can accommodate their triumph. It often does not even seem relevant to discuss cost in the context of revolution, which is why many sensationalist accounts of the latest breakthroughs in gene therapy focus on their potential, and the description of a future in which these treatments will be commonplace. This is a move which determines public perceptions of ongoing research, and the context in which expectations of future realities are perceived—and cost is not the main factor in this perception. It is, however, one of the biggest issues that needs to be faced if and when gene therapies should become more commonplace. As Petersen et al. have argued, "New questions need to be asked, in our view, about the influence certain science fiction representations may have on public discussion and deliberation about scientific developments and their consequences for society" (Petersen et al., 2005, p. 338). The consistent underestimation of the complexities surrounding cost and pricing in the reporting of new gene therapies is undoubtedly one of the places where this discussion should start.

Conclusion

Narratives work in particular ways, and the hero's journey is undoubtedly one of the archetypes of storytelling. This hero undergoes many trials (Odysseus, Hercules, etc.) and often suffers greatly, yet the reader never truly doubts that the journey or quest will ultimately be crowned by success. This fundamental assumption is hard to set aside when focusing on a particular patient story. There are, of course, tragedies as well (Pollock, 2019, p. 134ff.), but they do not make for the most comfortable reading experience and are difficult to recount in a brief newspaper article. In fact, regular newspaper articles follow a relatively simple narrative structure, providing essential information in the first two paragraphs, followed by supporting information, then relevant questions and reflections for the future (Boyd et al., 2020, p. 7). Longer exposés go beyond this structure in their storytelling, and are more reflective of a "story" with a beginning, middle part, and end. This structure, however, necessitates some sort of conclusion at the end of the article, giving a sense of closure, in contrast to the simpler format's ending in unresolved questions. The need to close a narrative arc drives narrative decisions that seek to provide a resolution of the tension built during a story, and in the case of the hero's journey, this resolution most often takes the shape of an expectation of success that overcomes all challenges along the way and leaves the reader satisfied. Reality, however, does not always conform to this expectation, and many of the outcomes of gene therapy are still fundamentally too uncertain to assess the potential of this new medical technology with any accuracy.

Gene therapies have given rise to a huge variety of expectations, ranging from cures for rare diseases to a cure-all for common conditions. But as these technologies become more mundane, there are increasing challenges of making them a reality for a greater number of patients. Insurance payments and economic models for longer term sustainability are issues that are a lot more commonplace than the discovery of an entirely new way of seeing the body, as the White House press conference celebrated back in 2000. But the deciphering of the human genome has also conditioned (and directly promised) us to expect similar breakthroughs, and current media coverage supplies this recurrent celebration of a medical revolution. In a way, the original hype set the scene for this eventual moment of pay-off, of the long-anticipated miracle finally taking shape. This decades-long expectation of the future of medicine is difficult to pierce with mundane considerations of economies, but these mundane challenges are the long-term obstacle that gene therapies will have to overcome.

References

Aristotle (with Kenny, A.). (2020). *Poetics*. University Press.

Boyd, R. L., Blackburn, K. G., & Pennebaker, J. W. (2020). The narrative arc: Revealing core narrative structures through text analysis. *Science Advances, 6*(32), eaba2196. https://doi.org/10.1126/sciadv.aba2196

Chou, S.-C., Hsu, Y.-C., & Lin, S.-W. (2023). Gene therapy for hemophilia, a clinical viewpoint. *Journal of the Formosan Medical Association, 122*(11), 1101–1110. https://doi.org/10.1016/j.jfma.2023.05.008

Cooper, M. (2008). *Life as surplus: Biotechnology and capitalism in the neoliberal era*. University of Washington Press.

Dunbar, C. E., High, K. A., Joung, J. K., Kohn, D. B., Ozawa, K., & Sadelain, M. (2018). Gene therapy comes of age. *Science, 359*(6372), eaan4672. https://doi.org/10.1126/science.aan4672

Field, S. (2005). *Screenplay: The foundations of screenwriting* (Revised edition.). Delta Trade Paperbacks.

Fortun, M. (2008). *Promising genomics: Iceland and DeCODE genetics in a world of speculation*. University of California Press.

Frank, A. W. (2013). *The wounded storyteller: Body, illness, and ethics* (2nd ed.). The University of Chicago Press.

Gurwitch, A. (2022, March 13). The end of my life was killing me. *The Washington Post*. https://www.annabellegurwitch.com/_files/ugd/d499bc_852877190def47e48ca9260f863b8521.pdf

Hilberg, E., Stelmach, A., Kleinhout-Vliek, T., & Martin, P. (2026). Promises past and future - Gene therapy and the actualisation of future expectations. *Social Science & Medicine, 389*, 118794. https://doi.org/10.1016/j.socscimed.2025.118794

Hilgartner, S. (2017). *Reordering life: Knowledge and control in the genomics revolution*. The MIT Press.

Hogarth, S., & Martin, P. (2021). The ratio of vision to data: Promoting emergent science and technologies through promissory regulation, the case of the FDA and personalised medicine. *Regulation & Governance, 15*(3), 969–986. https://doi.org/10.1111/rego.12360

Jeffreys, S. (2019). *Playwriting: Structure, character, how and what to write*. Nick Hern Books.

Kay, L. E. (1993). *The molecular vision of life: Caltech, the Rockefeller Foundation, and the rise of the new biology*. Oxford University Press.

National Human Genome Research Institute. (2025, May 24). *Gene therapy.* NIH. https://www.genome.gov/genetics-glossary/Gene-Therapy#:~:text=Gene%20therapy%20is%20a%20technique,healthy%20version%20of%20that%20gene

Nerlich, B., Dingwall, R., & Clarke, D. D. (2002). The book of life: How the completion of the Human Genome Project was revealed to the public. *Health (London, England: 1997), 6*(4), 445–469. https://doi.org/10.1177/136345930200600403

Pavone, V., & Goven, J. (2017). *Bioeconomies.* Palgrave Macmillan.

Petersen, A. (2001). Biofantasies: Genetics and medicine in the print news media. *Social Science & Medicine (1982), 52*(8), 1255–1268. https://doi.org/10.1016/s0277-9536(00)00229-x

Petersen, A., Anderson, A., & Allan, S. (2005). Science fiction/science fact: Medical genetics in news stories. *New Genetics and Society, 24*(3), 337–353. https://doi.org/10.1080/14636770500350088

Pollock, A. (2019). *Synthesizing hope: Matter, knowledge, and place in South African drug discovery.* University of Chicago Press.

Priest, S. H. (1994). Structuring public debate on biotechnology: Media frames and public response. *Science Communication, 16*(2), 166–179. https://doi.org/10.1177/0164025994016002004

Riessman, C. K. (2005). Narrative analysis. In N. Kelly, C. Horrocks, K. Milnes, B. Roberts, & D. Robinson (Eds.), *Narrative, memory and everyday life* (pp. 1–7). University of Huddersfield.

Riessman, C. K. (2008). *Narrative methods for the human sciences.* Sage Publications.

Salib, V. (2023, May 24). The top 5 most expensive FDA-approved gene therapies | TechTarget. *Pharma Life Sciences.* https://www.techtarget.com/pharmalifesciences/feature/The-Top-5-Most-Expensive-FDA-Approved-Gene-Therapies

Stelmach, A., & Nerlich, B. (2015). Metaphors in search of a target: The curious case of epigenetics. *New Genetics and Society, 34*(2), 196–218. https://doi.org/10.1080/14636778.2015.1034849

The White House. (2000a, June 25). *President Clinton announces the completion of the first survey of the entire human genome.* Office of the Press Secretary. https://web.ornl.gov/sci/techresources/Human_Genome/project/clinton1.shtml

The White House. (2000b, June 26). *White House Press Briefing—2—Remarks by the President*. Office of the Press Secretary. https://doe-humangenomeproject.ornl.gov/white-house-press-briefing-2/

Turow, J. (with Michigan Publishing). (2010). *Playing doctor: Television, storytelling, and medical power* (New and expanded ed.). University of Michigan Press.

van Dijck, J. (1998). *Imagenation—Popular images of genetics*. Macmillan.

Weintraub, K. (2023, April 24). Gene therapy offers hope for rare diseases. But the stories don't always have happy endings. *USA TODAY*. https://www.usatoday.com/story/news/health/2023/04/24/gene-therapy-rare-disease-alissa-feldborg/11665320002/

Whitacre, R. (2024). 'Financial fallout' in the US biopharmaceutical industry: Maximizing shareholder value, regulatory capture, and the consequences for patients. *Social Science & Medicine, 344*, 116598. https://doi.org/10.1016/j.socscimed.2024.116598

16

Pathways from Genomic Innovations: Navigating Equity and Access in Global Healthcare to Avoid Ableism

Amarpreet Kaur

Introduction

This chapter explores attitudes towards navigating access to genomic innovations in an increasingly globalised society, the implications of inequitable access, and how access to such innovations translates into avoiding ableism. *Genomic innovations* is an umbrella term which I use to refer to technologies that utilise or depend on genomic data. In this chapter, I focus on those within the realm of human health and reproductive care in relation to genetic disease, and how the development and use of such innovations are becoming progressively mundane. Mundane practices which surround the interpretations and use of genomic data increasingly permeate people's expectations of genomic innovations to remedy or prepare for ailments, sometimes before the latter are known to exist or

A. Kaur (✉)
Health Services Management Centre, University of Birmingham, Birmingham, UK
e-mail: a.kaur.11@bham.ac.uk

V. Argudo-Portal et al. (eds.), *Mundane Genomics*,
https://doi.org/10.1007/978-981-95-4550-6_16

become symptomatic. Trying to remedy or prepare for ailments is a response to the ableism common in contemporary societies.

Wolbring (2008, pp. 252–253) defined ableism as "a set of beliefs, processes and practices that produce—based on abilities one exhibits or values—a particular understanding of oneself, one's body and one's relationship with others of humanity, other species and the environment, and includes how one is judged by others". He argued that "ableism reflects the sentiment of certain social groups and social structures that value and promote certain abilities" (p. 253). Thus, in the context of this chapter, ableism is taken to reference structures and values which favour perpetuating or increasing the prevalence of typical abilities within societies, that is, reducing the prevalence of (genetic) conditions which hinder hegemonic abilities. Ashley Eisenmenger (2019), a disability inclusion training specialist, wrote, "The world wasn't built with people with disabilities in mind, and because of that, the world we live in is inherently 'ableist'".

However, strides taken by disability activists over the last few decades have led to the greater acceptance, visibility, and valuing of diversity within societies. These strides can be seen to translate into genomics via the regulation that surrounds applications of genomic-based technologies, and the ethical debates that feature when genomic innovations are discussed: for example, in the United Kingdom (UK), preimplantation genetic testing (PGT) is only licensed for "serious diseases". While there is considerable ambiguity in what is classed as serious (Boardman & Clark, 2022; Wertz & Knoppers, 2002), there is a general consensus that such conditions are within the realm of those that cause premature death in conjunction with significant pain, suffering, and dependency: factors which, taken collectively, are considered to impede greatly a person's quality of life (Kaur, 2024).

Outside the realm of the "serious", there are many genetic conditions for which prevention or cure, via genomic intervention(s), is not legally possible. This restriction reflects the fact that societies are not seeking to eradicate every genetic disease in order to reach some utopian future in which only able-bodied people exist; in other words, they are trying to avoid ableism. In this context, in this chapter, I first provide an overview of the two main innovations that underpin pathways and access to other genomic technologies, namely genetic screening and genetic testing. I then explore remedial pathways, followed by prophylactic pathways, before discussing ableism in a genomic era. Here, I address the desirability of genomic innovations, objections to them, and how access to such innovations can avoid ableist narratives. I also explore how the mundane uptake of transnational care provides impetus for international regulation, and why such regulation could be pertinent to avoiding ableism. Finally, the financial implications of genomic innovations are considered, including the costs connected to enabling access to them that avoids ableism, and ethical matters relating to such considerations. The discussions also include findings from a three-phase research project I conducted between March 2018 and October 2019.

Methods

The research project received ethical approval from the University of Cambridge's Department of Sociology's Research Ethics Committee. The first phase of research comprised a mixed-methods online survey of a tailored sample of 521 British citizens aged 16–82 years. Parts of the survey focussed on potential uses of heritable human genome editing (HHGE) for reproductive purposes, and access to the technology. Findings from these parts of the survey are discussed in this chapter. The second phase of research consisted of semi-structured interviews with 13 professionals and experts who could cast light on the scientific, legislative, or ethical scope of HHGE in the UK. Data relating to the latter two, both of which implicitly include the themes of equity and access, are pertinent to discussions in this chapter. Finally, the third phase of research comprised interactive structured interviews with 21 people affected by monogenic conditions. The views of these participants are invaluable to discussions in this chapter as they highlight manifestations of ableism that are regularly experienced by people living with genetic conditions.

Qualitative data from the three phases of research were imported to NVivo 12, qualitative data analysis software, for reflexive thematic analysis (Braun & Clarke, 2021). Quantitative data from phases one and three were imported to SPSS, quantitative data analysis software, for descriptive statistical analysis. The data were then triangulated to interpret the findings further and to gain a more comprehensive understanding of them. The triangulated data pull together narratives, which are characteristic of ableism, and which highlight reflections on cost, equity, and access in relation to HHGE—presented later in this chapter—that could be translated to other genomic innovations and new reproductive technologies.

Given the sample sizes of the three respective phases of research, findings are not generalisable. Further, the views on HHGE and related experiences are confined to the UK healthcare system and, in many ways, cultures in the UK. This means that while the presented findings may be translated to other innovations, they may not be translatable to other cultures; this is important to note when reading the rest of this chapter. Other cultures and societies, outside of the UK, with different healthcare systems, politics, and ethical values are likely to think differently about genomic innovations and the issue of equitability (Ong, 2016). However, the richness of the findings in relation to attitudes and practices in the UK offer indicative insights that could inform further research, both in the UK and beyond.

Overview of Genomic Innovations

Genomic innovations, such as genetic screening (GS), genetic testing (GT), and genome editing (GE), continue to evoke attention across the globe (Cunha, 2017; Shendure et al., 2019), and are used to assist in directing treatment for personalised or precision medicine, and for exploring choices for current or future pregnancies (National Human Genome Research Institute, 2024). GS enables prophesies to be made about people's individual risks for specific genetic diseases and is commonly performed during pregnancies and on neonates (with parental consent). GT enables checking for and identifying changes to a gene that can cause disease, and is also commonly performed during pregnancy if genetic screening indicates a high risk of a genetic disease.

The distinction between screening and testing is that the former only indicates a risk while the latter is conclusive (unless a false positive has been reported). However, a conclusive diagnosis of a disease-causing variant of a gene does not indicate the extent to which an individual will experience symptoms of a given disease or the impact of the disease on an individual's quality of life (Kaur, 2024); this could be key to note when decisions relate to procreating or a pregnancy. These technologies have been in development and clinical use since the 1950s (Durmaz et al., 2015) and are readily available in most developed countries, but in some cases only via private healthcare or healthcare insurance; this could limit the demographics of those who can access them, particularly in the Global South and less developed countries. Nonetheless, the capacities of GS and GT across the globe continue to expand (Brittain et al., 2017; Vingill, 2023).

The National Human Genome Research Institute (2024) describes *personalised medicine* as "an emerging practice of medicine that uses an individual's genetic profile to guide decisions made in regard to the prevention, diagnosis, and treatment of disease". Given that genetic profiles are unique to individuals, such decisions can result in optimised efficiency or therapeutic benefit, hence the interchangeable term *precision medicine*. This branch of medicine is described as an emerging practice because it is still in its infancy and there is considerable genomic research to be done to develop it. The roles and functions of many genes, their variants, how they interact with each other, and how they interact with different environments, are still unknown. However, the means to make strides in such knowledge are now established, and scientists and health professionals utilise such knowledge to develop this field. New diagnoses, treatments, and preventative interventions continue to emerge.

Interventions include remedial and prophylactic innovations. Remedial pathways available to treat some genetic disease include gene therapy and somatic genome editing (SGE), while existing prophylactic pathways include GS and GT in relation to abortion, PGT, and carrier screening. Potential prophylactic technologies that are currently being developed include in utero gene therapy (IUGT) and HHGE. These innovations are detailed in the next sections.

Remedial Pathways

The first remedial pathway to emerge from genomic innovations was gene therapy (via viral transfer) in 1990. Scheller and Krebsbach (2009, p. 585) wrote that "gene therapy is defined as the treatment of disease by transfer of engineered genetic material into human cells". The technology was first used to treat a four-year-old girl born with severe combined immunodeficiency, and has since been used to treat haemophilia, cancers, heart diseases, and a certain form of blindness, to name a few genetic diseases (EuroGCT, 2020). However, the nature of this technology poses several risks. The person's body to which the engineered virus is introduced may consider the virus an intruder and seek to attack it; this can cause reactions which range from swelling to organ failure. The virus could also cause infection, or target the wrong cells. The latter is possible because viruses can affect more than one type of cell. Other forms of gene therapy have trialled the use of stem cells and liposomes (a fluid-filled sac), but these have had limited results (Mayo Clinic, 2024).

In the past decade, developments in GE technologies—techniques whereby specific parts of DNA can be added, removed, or replaced—have enabled the development of an alternative remedial pathway for some genetic diseases. GE became a viable option for clinical use from the mid-2010s with the advent of CRISPR-Cas (Eid et al., 2018). CRISPR is an acronym for clustered, regularly interspaced, short palindromic repeats; in short, this references how specific parts of DNA (i.e. genes) can be identified, and acts as a guiding system for numbered enzymes with the prefix Cas. Cas enzymes (a type of protein) can be programmed to identify specific parts of DNA so that it can create a break in its structure to enable edits to be made at a targeted point (Gaj et al., 2013). This technique could be used to treat or prevent the presentation of monogenic diseases for which the specific gene is known (Ormond et al., 2017), without some of the risks associated with gene therapies which involve viral transfer (Dunbar et al., 2018).

There are two forms of genome editing, somatic and germline. SGE is the editing of any adult cells that are not heritable, such as blood cells, and is legal for clinical applications in several countries. The nature of

SGE means that the treatment is limited to the person who consents to it and, thus, while the remedial treatment may relieve the individual of symptoms of a given disease, it does not prevent the transmission of the genetic disease to offspring. This point is salient to HHGE and is discussed later in the chapter.

Sickle cell disease (SCD) is one genetic condition for which remedial CRISPR-Cas-based treatment is available (Wong, 2023). It is a blood disorder that affects haemoglobin (Rees et al., 2010), a protein which carries blood through the body. SCD is characterised by crescent-shaped red blood cells; red blood cells are usually disc shaped. Crescent-shaped cells do not bend or move as efficiently as disc-shaped cells and can therefore block blood flow through the body. In December 2023, medical authorities in the UK and the United States approved the first CRISPR-Cas-based medicine, Casgevy, with the potential to cure SCD (U.S. FDA, 2023; Walsh, 2023).

The novelty of Casgevy means that, for now at least, access to it is limited for many. Treatments involving SGE are generally part of clinical trials which are not yet widespread across the globe, and the sites of innovation are often not located in areas with high populations of the targeted diseases. This can mean that treatments are limited to those who can access such trials geographically. Some people may be willing and have the means to travel nationally or even transnationally to access healthcare, but for others this may not be a viable option. Further, the approximate cost of Casgevy is currently £1.5 million per patient (Dixon, 2024). This price is expected to decline as the field expands and pharmaceutical companies compete. However, the cost does not include hospital admission to enable the preparation for, and the administration of, the therapy. In the UK, the national health service (NHS) is absorbing the costs of administering Casgevy; hospitals that are administering the treatment justify the one-off expense by contending that the cost of the medicine is outweighed by the cost of otherwise providing a lifetime of care to a person with SCD. This rationale is likely to apply to genomic-based drugs in similar scenarios should they become available for other genetic diseases. But healthcare systems in other countries, such as those in the Global South—many of which have a high prevalence of people with SCD—may not fund the cost of Casgevy itself nor that of associated hospital admission. Therefore, like access to GS and GT, this innovative pathway may only be available to people who can afford private healthcare or who have health insurance which will cover the costs, meaning that access to it is inequitable.

The lack of remedial treatments for genetic disease, the costs of novel treatments, and the cost of living with disease in general (Kaur, 2024) all help to explain why prophylactic pathways for genetic disease are popular innovations. However, again, these innovations are less likely to be readily available and/or accessible in less developed countries, such as many of those in the Global South.

Prophylactic Pathways

Clinical prophylactic pathways to avoid genetic disease originated in the 1960s with the combination of genetic screening and testing leading to abortions based on the presence of genetic disease. Abortion in the light of results from genetic testing has continued to expand, and in many countries, this is one of the few legal reasons accepted for abortion past 24 weeks of gestation (Centre for Reproductive Rights, 2022). Later, PGT was developed, a technology that can be used to circumvent some of the legal and moral stances that surround abortion. However, PGT involves the creation of embryos via in vitro fertilisation (IVF) specifically to test for a genetic disease. Embryos that test positive for the given genetic disease are deemed unviable for transfer to a uterus and discarded (Braude et al., 2002). Thus, some pro-life activists also have qualms about this technology on the basis that a potential life is being denied (Franklin & Roberts, 2006).

Another prophylactic pathway stemming from genomic innovations is the ability to diagnose carriers of genetic disorders, such as SCD. Carrier screening enables people to make reproductive decisions based on their own risk of transmitting a genetic disease, and the risk in combination with their partner's genomic data (Sparks, 2020). Usually, two carriers of the same genetic condition are deterred from procreating together if they want to avoid the high risk of their potential offspring being symptomatic with the condition. This pathway does not involve ending potential life, and may therefore be preferable to some, but it still aims to prevent the transmission of genetic disease to offspring via selective procreation. The desire to prevent a child being born with a genetic condition is therefore salient across all three existing prophylactic pathways. Reasons why such a desire exists and factors which influence it are explored later in this chapter. Prevention is a goal that drives the development of other prophylactic pathways; however, again, such technologies are less likely to be readily available and/or accessible in the Global South, where many genetic diseases are more likely to result in premature death, and effective treatment is harder to access than in developed and Western countries.

Ongoing developments in genomic innovations and advances in the capabilities of medical technologies have led to the emergence of a prophylactic pathway involving SGE. In utero gene therapy (IUGT)—that is, the delivery of gene therapy to a foetus while it is still being gestated in a uterus—is thought to have "therapeutic potential" (Peddi et al., 2022). Peddi et al. (2022) argue that IUGT could provide a minimally invasive approach to preventing genetic disorders, and that this could be more beneficial than remedial approaches. They argue this on the basis that the individual would never experience symptoms of the genetic condition being targeted by the therapy. Mattar et al. (2023) indicate that the concept of IUGT could be ready to move from "proof-of-principle disease models" to clinical translation; this would be a substantial step in progressing this potential pathway. The possibility of IUGT as a viable clinical pathway is, therefore, more tangible than other innovations also in development; yet, it is still likely to be restricted to specific sites of innovation.

Germline genome editing is the editing of sperm cells, egg cells, or very early embryo cells. Edits made to such germline cells can be inherited and are therefore far more controversial than SGE (Boardman, 2019; Gyngell et al., 2017); HHGE is not (yet) legally permitted in any country for any clinical applications, including trials. The distinct benefit of this potential innovation in comparison to existing pathways would be that individuals who undergo HHGE would not be able to transmit the targeted genetic disease to their offspring, who would thus not inherit that specific disease (Kaur & Border, 2020; Ormond et al., 2017). However, this technology would not prevent de novo presentations of genetic disease, or any other genetic disease that the edits do not target (Coller, 2019; Veltman & Brunner, 2012); this point is important to remember.

The availability of and access to these prophylactic pathways, much like remedial pathways involving genomic innovations, are crucial to navigating equity in global healthcare to avoid ableism. Differences in laws pertaining to abortion care across the globe have already led to some people taking desperate, risky, and dangerous measures to prevent a pregnancy being sustained, whether for medical or social reasons. Such practices are one of the leading causes of maternal mortality (Haddad & Nour, 2009). Further, transnational care for medical and reproductive purposes are both established arenas; however, they are often only accessible to people who have financial (and sometimes social) capital, and knowledge of their existence. The need to access transnational care to receive desired care is important to consider because this can also be a route to avoiding ableism.

Ableism in a Genomic Era

Restricting potential applications of HHGE to only serious diseases was a key factor in respondents' attitudes towards their support for the technology. Findings from my survey suggested that the severity of a condition was influential in terms of whether respondents would support the use of HHGE to circumvent a person being born with a genetic condition. For a scenario that described symptoms of many common ailments experienced in an average lifetime, 66.2% of respondents (345 of the 521) felt that HHGE should be permitted,[1] a scenario that acted as a baseline for others. In comparison, a scenario which mainly focussed on life expectancy, or premature death, elicited support for utilising HHGE from 86.8% of the respondents (452 of the 521). The latter scenario was the one which the highest proportion of respondents favoured allowing

[1] The common ailment scenario, which acted as a baseline for other scenarios, was detailed as follows: "Alecia and Bernardo have a genetic condition which is treatable but has no cure. If their child inherits the condition, it may experience some pain during its life as a result of the condition and will have to receive regular treatment throughout its life to manage the condition. The condition is not known to impair cognitive or physical abilities. On average, people with the condition have a slightly reduced life expectancy".

the use of HHGE to prevent genetic disease. Respondents' qualitative reflections on the scenario revealed that their support was also influenced by the interpretation that an individual's quality of life would be considerably reduced in such cases.[2] The following examples support this claim:

> This condition sounds like it will drastically affect quality of life and will be severely life limiting. Therefore I think the risks of GE would potentially be worthwhile.
> (Respondent 20, female aged 33, PhD researcher)

> The condition sounds very detrimental to the sufferer's quality of life, so if editing can improve that or prevent others from suffering then that would be better.
> (Respondent 158, male aged 28, administrator)

> The very intrusive medical treatments, and the duration of these, and the impact on quality of life seem like it would be very profound and interfering.
> (Respondent 269, female aged 27, unemployed)

These examples suggest that only in cases where an individual's quality of life will be "profoundly" adversely impacted by a genetic condition would mass support be generated for interventions to prevent the condition from being inherited. This suggests that people understand some level of ailment, including genetic ailments, to be somewhat normal and expected; this could be because of how mundane genomic conditions have now become. Thus, the increasing normalisation of diversity in abilities is drawing more attention to ableist structures and narratives throughout societies. The development of genomic innovations could have also instigated greater consideration of where societies want to draw a line between acceptable conditions and those undesirable enough to warrant prevention. Making such a distinction feeds into understandings and awareness of ableism.

[2] The life expectancy-focussed scenario was detailed as follows: "Javeria and Imran are carriers of a fatal genetic condition. People who inherit the condition have a reduced life expectancy and have to receive very intrusive medical treatments from birth until they die. Cognitive ability is not affected by this condition, but physical abilities may be impaired as physiological abilities deteriorate".

These findings, and those from other studies (van Mil et al., 2017; Wipperman & Campos, 2016), indicate that in the UK, mass support for applications of HHGE for reproductive purposes is subject to the innovation being restricted to use on severe diseases. Thus, a reasonable argument can be made that the development of this innovation is aimed at improving people's quality of life (and reducing suffering), rather than trying to eradicate all disease—a feat which will remain impossible due to the social construction of disease and de novo mutations. However, quality of life is also impacted by a person's mental health (Kaur, 2024), which is often affected by their social circumstances, social expectations of them, and the role they play in society (Franklin, 1997; Barili et al., 2023).

In many societies and cultures, having children is expected of maturing adults (Franklin, 1997) and can be a prominent desire for many people, although, of course, some people may want to remain childfree, regardless of whether they have a genetic condition. Nonetheless, during her interview, Professor Kathy Niakan, a developmental biologist, argued, "As medicine develops, people are living beyond their historical prognoses for a given disease, and people are now surviving to reproductive age. It will become a question for society what we can provide for them". Professor Niakan's comment is salient with regard to the quality of life for people living with genetic conditions because it could enable such people to fulfil the desire or social expectation to procreate. The future promise of HHGE to enable some people with genes that can result in severe genetic disease another reproductive choice is therefore a significant driver for this innovation. Both my own and other research studies suggest that this possibility provides hope for people living with genetic conditions who want to have children (International Commission on the Clinical Use of Human Germline Genome Editing et al., 2020; van Mil et al., 2017; Wipperman & Campos, 2016); this is because people with genetic conditions can struggle to secure their own child

There are several reasons for such struggles. First, the nature of some genetic diseases can mean that even if a pregnancy results in a live birth, the child may not survive through childhood. Yet, depending on the genetic composition of the contributing genetic parents, this outcome may not be avoidable without technological interventions (Kaur & Border, 2020)—a risk some potential parents are not willing to take.

Second, even if genetic relatedness is not important to aspiring parents, adoption can become even more difficult to navigate, and, in some cases, impossible if people have questionable health as adoption agencies have restrictions on who can adopt a child that often include considerations of age, income, and health status (Mutcherson, 2009, p. 324). This struggle was alluded to by a participant in my interactive structured interviews:

> It is quite hard to adopt in a family that's got a heritable condition because the adoptive agencies don't want to know because they've all got shortened life expectancies, so are you going to be there to see the child grow up? Does the child really want to witness families going through pain crises and hospital admissions and kidney failure and heart failure and strokes? And so they find it difficult to get in.
>
> (Miranda, aged 52, has a child with Pendred syndrome)

This quote from Miranda highlights several practical purposes for restrictions on who can adopt. Understandably, many restrictions serve to ensure the long-term, holistic well-being of adoptive children, but they are still inadvertently discriminatory towards people with various genetic conditions. Coupled with the fact that such people may not want to risk the guilt or judgement that could stem from passing on a severe genetic condition to a child (Franklin & Roberts, 2006; Kaur, 2022), these reasons provide impetus for the development and clinical utilisation of HHGE to create a pathway for them to secure a child. The creation of such a pathway could combat the ableist narrative of who is suitable to have and care for children, that is, that people with genetic conditions should not have genetically related offspring. However, there are different interpretations of the creation of such a pathway.

Objections to Genomic Innovations

A prominent ethical debate which applies to both remedial and prophylactic genomic innovations is the expressivist objection, based on the claim that the development of modes of intervention to remedy, ameliorate, or prevent genetic conditions indicates negative and damaging judgements about the value of people with disabilities (Buchanan, 1996, p. 28). The expressivist objection seems to align a person's value solely with disability and ignores other parts of a person's identity and the impact that living with a disability (as a result of disease) has on people's lives. In recognition of this debate, participants in my research were asked for their thoughts on people with genetic conditions being "edited out". Many of their views indicated that they did not locate people's sole value in their having a genetic disease or a disability, and that their greater focus was on the possibility of improving quality of life. The following quotes from participants are examples of this view:

> It's not editing people with disabilities out. It is giving people back quality of life. For me, my condition, and many people with my condition, our quality of life is very negatively affected. And it continues to be, it's progressive. So, it's a huge kind of mental health issue as well. Because we know our condition is progressing. And for me, it's really difficult when I hear people that don't have a disability, that don't have a disease, talk about the ethics around this sort of thing, because they don't really live day-to-day with what I'm having to deal with. So, actually, I do feel like there is a risk of editing these people out, in those words, but I'm one of those people, and I don't mind that.
>
> (George, aged 34, has Charcot-Marie-Tooth disease)

> I think that the germline editing therapy, if that could eradicate certain conditions, I think the person themselves is still going to be that person, they are still going to become that person, my daughter will still be my daughter, but without that worry of their health every day. [...] So, it is all those things that makes somebody's quality of life massively affected. Obviously, if we can change that for our future generations, and our future generations can be without the disease, and be without the health conditions, then if that works, I would welcome it.
>
> (Claire, aged 34, has a daughter with Hirschsprung's disease)

> I don't think we're being edited out. [...] I would still be me; I would just be me without having to deal with all the medication. Nobody is going to be like killed off.
>
> (Trevor, aged 30, has cystic fibrosis)

While the three quotes above indicate covert hopes that HHGE will be utilisable for the prevention of some genetic conditions, other participants highlighted that there are some genetic conditions, such as deafness, which are influential to a person's identity. They explained how editing out particular characteristics which are viewed as disabilities in mainstream societies could have negative impacts. The following quote illustrates this argument:

> I am pretty sure a deaf person would say that because the deaf community in particular see themselves as a linguistic minority and they have a cultural identity, they definitely don't see themselves as having a disease. A lot of deaf people can be vilified and so they would object. There was a lot about cochlear implants because of that reason; it didn't stop people from having implants but there was that view of the deaf culture being edited out so I think that would definitely be the case with this procedure.
> (Resina, aged 47, has beta thalassaemia and deafness)

To address the points raised by Resina, in cases where a specific characteristic is inherent to the cultural identity in which a person may be nurtured and reside, I assume that HHGE would not be sought. However, given the points raised earlier in this chapter in relation to severity of disease, I also question whether such conditions would even be considered severe enough to warrant prophylactic measures via HHGE. But, as argued in Professor Felicity Boardman's contribution to my research, "people with genetic conditions could become increasingly stigmatised and marginalised if the use of germline editing technologies were to become integrated into routine reproductive care". This point could be salient if applications of such technology were to become routine without public engagement and education on the limitations of genome editing (Kaur, 2021), or if societies were to try to force ableist values onto others. However, for people with conditions which they feel impede their quality of life, the pending possibility of HHGE can offer them hope.

Further, as the quotes at the beginning of this section indicate, living day to day with a genetic condition can be hard, and also impact people's mental health and their social lives. The struggles faced by people living with genetic conditions are far reaching, although many genetic diseases manifest in a variety of ways; some people can be less symptomatic than others with the same genetic disease, an argument often also cited by disability activists. There is much truth to the fact that the severity of disease cannot be determined by virtue of identifying the presence of a disease-causing variant of a gene. In prophylactic terms, while genomic data can provide information on the functioning of genes and predispositions, they do not reveal the extent to which a person may be affected by a genetic condition. This argument was stressed by another participant in my interactive structured interviews:

> Can you tell the severity of a disease from identifying the mutation? (No). So surely it comes down to individual decisions again. If you can't predict how badly a child will be affected by a disease, is it not better, I would think, then not to have the disease at all? […] A lot of people don't want to pass on to their future children the disorder or disease that they have got, so the idea of genome editing is appealing to them. Also, you just don't want your child to suffer, and it comes down to reducing the amount of suffering that is in the world, if you take into account all the other socio-economic factors.
>
> (Hannah, aged 31, has primary immunodeficiency disease)

This quote from Hannah highlights several points that can be influential in people's reproductive decisions: first, because how badly a child will be affected by a disease cannot be predicted, people making reproductive decisions have to consider whether they are willing to—or can—take the risk of having a child that could be severely affected by their condition; second, some people simply do not want to pass on a disorder; and third, people want to prevent suffering. This could be as a result of their own lived experiences or, as Hannah mentions, because of all the "other socioeconomic factors", such as those presented by living in an inherently ableist society. Similar sentiments in relation to these latter two points were expressed by Virginia, another participant in the same set of interviews:

> I am disabled myself, I have a genetic condition, and personally I would not want to have a child or bring another person into this world with the

> same genetic condition as myself. I think life with this disability is a struggle in this ableist world. Frankly, I would not choose to have this condition if I had the option, therefore I would not choose it for someone else either.
> (Virginia, aged 21, has muscular dystrophy)

Virginia was explicit about her feelings of not wanting to have the condition which affects her, nor wanting to pass the condition on to potential offspring, a view that was echoed by several of the participants. Such opinions surely counterbalance the expressivist objection. The sentiment that HHGE is potentially a humane innovation, insofar as it would utilise mundane genomic data to prevent severe diseases and the suffering associated with them, echoed across the three phases of my research. However, the potential good which could stem from HHGE was repeatedly overshadowed by concerns about who would be able to access the technology and how. Such concerns can also be translated to other genomic innovations.

Navigating Access to Genomic Innovations

Access to genomic innovations, both remedial and prophylactic, is one of the biggest barriers to whether and how they are or could be legally utilised in clinical settings. Across the globe, legislation surrounding applications of genomic innovations, particularly as they intersect with reproductive technologies, continues to differ vastly. These legal, regulatory differences in permitted clinical applications of genomic innovations are a key reason why some people seek transnational care. However, transnational care is not accessible to all; some people do not have the financial or social capital to navigate access to it, while others may not have the physical ability or health (as a result of their genetic condition or otherwise) to do so. These barriers essentially mean that, unless the ongoing development of genomic innovations encourages a global movement for the international regulation of reproductive and genomic technologies, their access would not be equitable. International regulation could permit or prohibit applications on a global basis to reduce the impetus for transnational care.

If clinical applications of genomic innovations were only to be (legally) available in select countries, such restrictions could influence people to travel to said countries to seek treatment or intervention, but this can be hard for people with genetic conditions—largely influenced by ableist structures. Often, people with genetic conditions and resulting disabilities require additional needs or measures to assist their capabilities, which can be more expensive than the basic needs of the average person. For example, when travelling abroad by aeroplane, people with additional needs may have to arrive earlier than other passengers to ensure that they can utilise passenger assistance at the airport (potentially adding to fatigue). While they may otherwise travel light, their needs may require them to check in luggage and pay any charges associated with heavier or non-cabin luggage. Other needs (not accommodated by ableist structures and regulations) may prevent them from travelling abroad in totality, regardless of their financial capital. People with genetic conditions may be limited in the accommodation options that are suitable for their needs and abilities, limitations that may also be applicable to dietary and medicinal requirements. All these "non-standard" measures can make travelling far harder, less palatable, if not impossible for some people with genetic conditions. Such practicalities can also restrict access to genomic innovations. On the other hand, international travel to access genomic innovations—transnational care, in other words—would not be necessary if every country were to have the same regulations, thereby mitigating the need for people to navigate ableist structures. Only if all who can benefit from genomic innovations are able to access them, can they be a route to avoiding ableism; otherwise, they may simply perpetuate ableism insofar as only the more able can access them.

However, the absence of legal access to clinical applications of such innovations is not the only barrier to their potential utilisation. Another key factor that influences access is the expense. Even if the innovations highlighted in this chapter were to be clinically viable (to a greater extent) across the globe, there is a high likelihood that such innovations would largely only be available via private care in many countries. Many people with genetic conditions may be reliant on national healthcare systems or insurance companies to afford their basic care. Genomic innovations may not be considered a necessity for anyone, including at their intersection with reproductive technologies; rather, they may be considered a lifestyle choice, regardless of social expectations. Thus, the impetus to fund access to genomic innovations publicly may be limited.

My survey included a question on whether HHGE should be available via the NHS (in the United Kingdom). Of the respondents, 39.9% (208 of 521) felt that it should, 16.3% (85 of 521) felt the converse, and the remaining 43.8% (228 of 521) were undecided. Among the corresponding qualitative responses to this question, respondents questioned whether the NHS could afford to fund HHGE. The following quote exemplifies this consideration:

> The NHS cannot currently afford to pay for the current cost of desired or required medical procedures—should we really be looking to escalate the costs further by adding even more expensive techniques to the repertoire?
> (Respondent 262, male aged 46, game designer and author).

Several respondents also considered the potential consequences of HHGE's not being available via the NHS. The following quote illustrates the associated reasoning:

> The NHS is already in a very precarious and unstable position and presumably genetic editing wouldn't come cheap. If anyone with a claim that genetic editing might improve their unborn children's lifestyle were granted genetic editing, then the cost of that would obviously be significant. On the other hand, privately funded genetic editing creates stronger divisions between the wealthy and the poor and sends the message that only those with money should be granted a pain free life, which is obviously extremely damaging.
> (Respondent 14, female aged 28, lecturer)

Many other respondents also considered the possible cost–benefit ratio to the NHS of funding HHGE versus funding a lifetime of care for different conditions, a perspective exemplified by the following view:

> There should be a cost benefit analysis done for the condition, the cost of genetic editing and the cost of caring for a person born with the condition in question. If the condition is severe or the cost difference for care is notable, that which is more beneficial to society, the person, and the parents should be considered.
> (Respondent 479, male aged 22, president of a union)

These corresponding factors in whether respondents thought HHGE should be available via the NHS indicate that people are conscious that national health services have limited funds. As Respondent 14 mentioned, genetic editing is not likely to be cheap. Thus, the ethical dilemma of funding a cheaper treatment or intervention for many over an expensive treatment or intervention for a select few is a factor that should also be considered. Further, since conducting my research, the world has also suffered from the unprecedented impact of the COVID pandemic, wars, and other disasters, all of which continue to force governments to implement yet more austerity measures. The UK is not exempt from such measures. Although the UK is renowned for its largely publicly funded NHS, its health services are becoming increasingly privatised.

Moreover, despite the range of support and hope which surrounds the development of HHGE as a legal reproductive choice, other findings from my survey suggest that few may actually be willing to utilise the technology. Only 35% of respondents to my survey thought that they would actually use HHGE if they could. This finding is illustrated in Fig. 16.1.

The corresponding qualitative responses to this question revealed that respondents were hesitant to using the technology (even hypothetically) because they did not want to be "guinea pigs" for it. An argument could be made that if initial applications of HHGE were successful, much like trends with the uptake of IVF and other innovations, more people may be open to using the technology.

However, considering the cost of developing such technology and the projected costs of making HHGE utilisable in clinical settings, would the cost of developing HHGE outweigh the potential benefits? This question can also be asked in relation to other genomic innovations and is often asked on the assumption that such innovations may only benefit niche populations. But, as genomic and scientific advances are made, several scholars have argued that potential applications of HHGE (as with other genomic innovations) could expand to be of benefit to a greater range of people if the technology were ever to be deemed safe (Savulescu et al., 2015; Cavaliere, 2018; Ormond et al., 2019). This could mean that such technologies have a wider catchment of people that they could help, with both remedial and prophylactic measures.

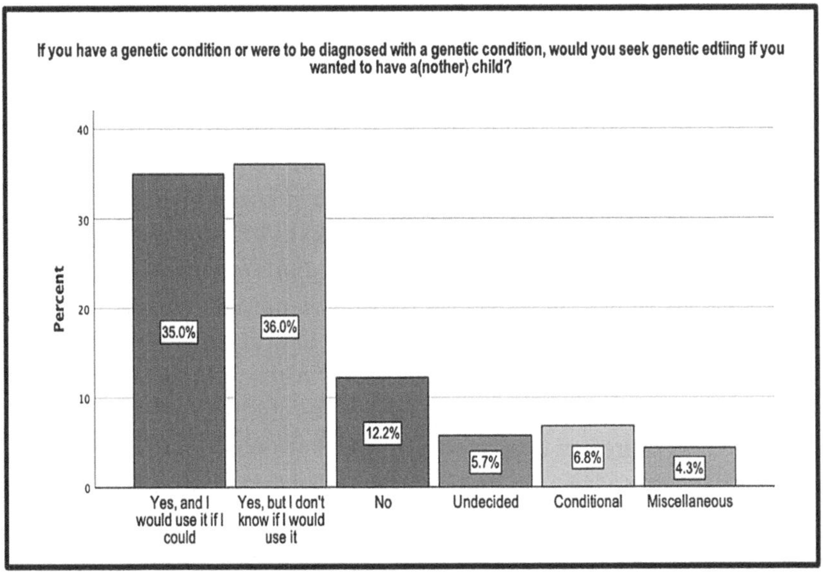

Fig. 16.1 Attitudes towards willingness to utilise HHGE

Yet, ableism is not exclusive to people with genetic conditions. There are also people within societies who have acquired disease, and disabilities, which are not a result of a genetic mutation, and they may also have a range of differences which require additional measures to assist their quality of life. Hence, an argument could be made that funding access to genomic innovations should not come at the cost to others in the wider society who require greater levels of social or other types of care—because ableist narratives extend beyond people with genetic conditions and are manifold. These arguments do not negate that the financial implications of funding genomic innovations can imply that only those with the abilities to access such innovations independently, whether physically or financially, are welcome to do so. Nor do they ignore that not being able to access treatments or interventions that could assist in enhancing quality of life could further social inequalities that exist in societies and perpetuate ableist narratives. However, they do sideline the issue that fighting healthcare systems or insurance companies, or trying to travel to places (or possibly a different country) where access to genomic innovations may be more affordable, can take energy that many people with genetic conditions may not (easily) be able to expend.

During the interim of awaiting the development of or access to genomic innovations, whether for remedial or prophylactic purposes, if they are desired, societies could increase social support for people living with disease, genetic or otherwise, making living with disease easier. Reshaping social structures and infrastructures to be more inclusive could improve people's quality of life to an extent that some people might not consider their disease necessary to remedy or prevent via technological interventions. This final point has a fairly political undertone; in many ways, whether the development of accessible pathways to genomic innovations is facilitated or the diversity in people's abilities is recognised and supported to a greater degree comes down to what governments and stakeholders are trying to achieve and why. As discussed in this chapter, genomic innovations could perpetuate or avoid ableism; governments and regulators can choose which pathways these innovations take.

Conclusion

In this chapter, I have argued that the severity of disease is pertinent to attitudes (in the UK) to whether genomic interventions are supported to prevent disease. Thus, that societies, in the UK at least, are not aiming to eradicate all disease—a feat which would be impossible due to de novo mutations and the social construction of disease. Moreover, societies are trying to improve the quality of life for people living with mundane genetic conditions; this includes giving them more choices to fulfil desires that may have previously been inconceivable or unattractive to them.

I explored how differences in legal, mundane, and clinical applications of genomic innovations across the globe could escalate the number of people seeking transnational care. I considered how transnational care supports ableist narratives, and suggested that preventing the need to seek it, via international regulation of the clinical application of genomic innovations, could avoid the perpetuation of ableism. I also argued that only if all who could benefit from genomic innovations, whether remedial or prophylactic, are able to access to them, can such innovations be a pathway to avoiding ableism. However, if access to genomic innovations remains unequitable, or is further privatised, this could exacerbate national and international social inequalities.

To this end, I conclude that if genomic innovations were to be equitably accessible and become a mundane intervention across the globe, the quality of life for people with genetic conditions could be enhanced. Such people could be provided pathways to improve their own holistic health and that of their children (should they want to have genetically related children). Further, the narrative that only those with hegemonic abilities and/or free from genetic disease (based on mundane genomic data) are suitable to have children could be avoided. However, I also recognise that such pathways could be expensive to facilitate and should not come at the expense of people living with acquired disease or disabilities.

Acknowledgements The research project which underpins this chapter was funded by the Economic and Social Research Council. Amarpreet remains eternally grateful to all who generously participated in her research. She would also like to thank the peer reviewers for this chapter; their feedback helped strengthen the final manuscript.

References

Barili, E., Grembi, V., & Rosso, A. C. (2023). Mental health between present issues and future expectations. *Health Policy, 128*, 42–48. https://doi.org/10.1016/j.healthpol.2022.11.010

Boardman, F. (2019). Human genome editing and the identity politics of genetic disability. *Journal of Community Genetics, 11*(2), 125–127. https://doi.org/10.1007/s12687-019-00437-4

Boardman, F. K., & Clark, C. C. (2022). What is a 'serious' genetic condition? The perceptions of people living with genetic conditions. *European Journal of Human Genetics, 30*(2), 2. https://doi.org/10.1038/s41431-021-00962-2

Braude, P., Pickering, S., Flinter, F., & Ogilvie, C. M. (2002). Preimplantation genetic diagnosis. *Nature Reviews Genetics, 3*(12), 12. https://doi.org/10.1038/nrg953

Braun, V., & Clarke, V. (2021). One size fits all? What counts as quality practice in (reflexive) thematic analysis? *Qualitative Research in Psychology, 18*(3), 328–352. https://doi.org/10.1080/14780887.2020.1769238

Brittain, H. K., Scott, R., & Thomas, E. (2017). The rise of the genome and personalised medicine. *Clinical Medicine, 17*(6), 545–551. https://doi.org/10.7861/clinmedicine.17-6-545

Buchanan, A. (1996). Choosing who will be disabled: Genetic intervention and the morality of inclusion. *Social Philosophy and Policy, 13*(2), 18–46. https://doi.org/10.1017/S0265052500003447

Cavaliere, G. (2018). Genome editing and assisted reproduction: Curing embryos, society or prospective parents? *Medicine, Health Care and Philosophy, 21*(2), 215–225. https://doi.org/10.1007/s11019-017-9793-y

Centre for Reproductive Rights. (2022, September 27). *The world's abortion laws*. Center for Reproductive Rights. https://reproductiverights.org/maps/worlds-abortion-laws/

Coller, B. S. (2019). Ethics of human genome editing. *Annual Review of Medicine, 70*(1), 289–305. https://doi.org/10.1146/annurev-med-112717-094629

Cunha, A. (2017). Genomic technologies—From tools to therapies. *Genome Medicine, 9*(1), 71. https://doi.org/10.1186/s13073-017-0462-9

Dixon, P. (2024, February 12). *Can high-cost drugs be good value? The case of Casgevy for sickle-cell disease and beta-thalassemia* [Blog Post]. https://www.cebm.ox.ac.uk/news/views/can-high-cost-drugs-be-good-value-the-case-of-casgevy-for-sickle-cell-disease-and-beta-thalassemia-1

Dunbar, C. E., High, K. A., Joung, J. K., Kohn, D. B., Ozawa, K., & Sadelain, M. (2018). Gene therapy comes of age. *Science, 359*(6372), eaan4672. https://doi.org/10.1126/science.aan4672

Durmaz, A. A., Karaca, E., Demkow, U., Toruner, G., Schoumans, J., & Cogulu, O. (2015). Evolution of genetic techniques: Past, present, and beyond. *BioMed Research International, 2015*, 461524. https://doi.org/10.1155/2015/461524

Eid, A., Alshareef, S., & Mahfouz, M. M. (2018). CRISPR Base editors: Genome editing without double-stranded breaks. *Biochemical Journal, 475*(11), 1955–1964. https://doi.org/10.1042/BCJ20170793

Eisenmenger, A. (2019, December 12). Ableism 101—What is ableism? What does it look like? *Access Living*. https://www.accessliving.org/newsroom/blog/ableism-101/

EuroGCT. (2020). *What conditions can currently be treated using gene and cell therapy?* https://www.eurogct.org/what-conditions-can-currently-be-treated-using-gene-and-cell-therapy

Franklin, S. (1997). *Embodied progress: A cultural account of assisted conception* (1st ed.). Routledge.

Franklin, S., & Roberts, C. (2006). *Born and made: An ethnography of preimplantation genetic diagnosis*. Princeton University Press.

Gaj, T., Gersbach, C. A., & Barbas, C. F. (2013). ZFN, TALEN, and CRISPR/Cas-based methods for genome engineering. *Trends in Biotechnology, 31*(7), 397–405. https://doi.org/10.1016/j.tibtech.2013.04.004

Gyngell, C., Douglas, T., & Savulescu, J. (2017). The ethics of germline gene editing. *Journal of Applied Philosophy, 34*(4), 498–513. https://doi.org/10.1111/japp.12249

Haddad, L. B., & Nour, N. M. (2009). Unsafe abortion: Unnecessary maternal mortality. *Reviews in Obstetrics and Gynecology, 2*(2), 122–126.

International Commission on the Clinical Use of Human Germline Genome Editing, The Royal Society, National Academy of Medicine, & National Academy of Sciences. (2020). *Heritable human genome editing*. National Academies Press (US). http://www.ncbi.nlm.nih.gov/books/NBK561519/

Kaur, A. (2021). Chapter 7: Human germline genome editing and its techsumptions. In R. Davis-Floyd (Ed.), *Birthing techno-sapiens: Human-technology co-evolution and the future of reproduction* (1st ed.). Routledge.

Kaur, A. (2022). *Human germline genome editing as a potential reproductive choice: An exploratory sociological study in the United Kingdom*. University of Cambridge. https://www.repository.cam.ac.uk/handle/1810/343067

Kaur, A. (2024). Severity in the genomic age: The significance of lived experience to understandings of severity. *European Journal of Human Genetics*, 1–6. https://doi.org/10.1038/s41431-024-01652-5

Kaur, A., & Border, P. (2020). *Human germline genome editing* (611; POSTnote). Parliamentary Office of Science and Technology.

Mattar, C. N., Chan, J. K., & Choolani, M. (2023). Gene modification therapies for hereditary diseases in the fetus. *Prenatal Diagnosis, 43*(5), 674–686. https://doi.org/10.1002/pd.6347

Mayo Clinic. (2024, April 23). *Gene therapy—Overview*. https://www.mayoclinic.org/tests-procedures/gene-therapy/about/pac-20384619

Mutcherson, K. (2009). Disabling dreams of parenthood: The fertility industry, anti-discrimination and parents with disabilities. *Law and Inequality, 27*(2), 311–364.

National Human Genome Research Institute. (2024, August 14). *Personalized Medicine*. https://www.genome.gov/genetics-glossary/Personalized-Medicine

Ong, A. (2016). *Fungible life: Experiment in the Asian city of life*. Duke University Press. https://doi.org/10.1215/9780822373643

Ormond, K. E., Mortlock, D. P., Scholes, D. T., Bombard, Y., Brody, L. C., Faucett, W. A., Garrison, N. A., Hercher, L., Isasi, R., Middleton, A., Musunuru, K., Shriner, D., Virani, A., & Young, C. E. (2017). Human germline genome editing. *The American Journal of Human Genetics, 101*(2), 167–176. https://doi.org/10.1016/j.ajhg.2017.06.012

Ormond, K. E., Bombard, Y., Bonham, V. L., Hoffman-Andrews, L., Howard, H., Isasi, R., Musunuru, K., Riggan, K. A., Michie, M., & Allyse, M. (2019). The clinical application of gene editing: Ethical and social issues. *Personalized Medicine, 16*(4), 337–350. https://doi.org/10.2217/pme-2018-0155

Peddi, N. C., Marasandra Ramesh, H., Gude, S. S., Gude, S. S., & Vuppalapati, S. (2022). Intrauterine fetal gene therapy: Is that the future and is that future now? *Cureus, 14*(2), e22521. https://doi.org/10.7759/cureus.22521

Rees, D. C., Williams, T. N., & Gladwin, M. T. (2010). Sickle-cell disease. *The Lancet, 376*(9757), 2018–2031. https://doi.org/10.1016/S0140-6736(10)61029-X

Savulescu, J., Pugh, J., Douglas, T., & Gyngell, C. (2015). The moral imperative to continue gene editing research on human embryos. *Protein & Cell, 6*(7), 476–479. https://doi.org/10.1007/s13238-015-0184-y

Scheller, E. L., & Krebsbach, P. H. (2009). Gene therapy. *Journal of Dental Research, 88*(7), 585–596. https://doi.org/10.1177/0022034509337480

Shendure, J., Findlay, G. M., & Snyder, M. W. (2019). Genomic medicine–progress, pitfalls, and promise. *Cell, 177*(1), 45–57. https://doi.org/10.1016/j.cell.2019.02.003

Sparks, T. N. (2020). Expanded carrier screening: Counseling and considerations. *Human Genetics, 139*(9), 1131–1139. https://doi.org/10.1007/s00439-019-02080-y

U.S. FDA. (2023, December 8). *FDA approves first gene therapies to treat patients with sickle cell disease*. FDA; FDA. https://www.fda.gov/news-events/press-announcements/fda-approves-first-gene-therapies-treat-patients-sickle-cell-disease

van Mil, A., Hopkins, H., & Kinsella, S. (2017). *Potential uses for genetic technologies: Dialogue and engagement research conducted on behalf of the Royal Society*. Findings Report 137.

Veltman, J. A., & Brunner, H. G. (2012). De novo mutations in human genetic disease. *Nature Reviews Genetics, 13*(8), 8. https://doi.org/10.1038/nrg3241

Vingill, S. (2023, November 15). *Four milestones powering the genome sequencing revolution.* https://www.ukri.org/blog/four-milestones-powering-the-genome-sequencing-revolution/

Walsh, F. (2023, November 16). Casgevy: UK approves gene-editing drug for sickle cell. *BBC News*. https://www.bbc.com/news/health-67435266

Wertz, D. C., & Knoppers, B. M. (2002). Serious genetic disorders: Can or should they be defined?*. *American Journal of Medical Genetics, 108*(1), 29–35. https://doi.org/10.1002/ajmg.10212

Wipperman, A., & Campos, M. (2016). *Genome editing technologies: The patient perspective* (p. 36). Genetic Alliance UK.

Wolbring, G. (2008). The politics of ableism. *Development, 51*(2), 252–258. https://doi.org/10.1057/dev.2008.17

Wong, C. (2023). UK first to approve CRISPR treatment for diseases: What you need to know. *Nature, 623*(7988), 676–677. https://doi.org/10.1038/d41586-023-03590-6

17

Storytelling, Facts, and Epigenetic Sensibilities: From Critique to Speculation

Stephanie Lloyd

In interviews after losing a loved one to suicide, the bereaved often recount the deceased's personal vulnerabilities, their difficult and often heartbreaking trajectories, and a death that was shocking, but, ultimately unsurprising. These accounts might be read as expressions, if ambivalent, of *neuroepigenetic sensibilities*, integrating elements of neuroepigenetic models of suicide risk gleaned from popular neuroscientific sources, as well as information picked up through participation in neuroepigenetic research. The words of the bereaved also inform neuroepigenetics research through the bereaved's participation in psychological autopsies that are used to classify the deceased's tissues for subsequent neuroepigenetics studies (e.g. suicide with or without abuse). Alongside the cultivation of neuroepigenetic sensibilities, this research process and the narratives that result from it convert people's lives into facts (e.g. early abuse, biomarkers) that reinforce neuroepigenetic models. But is this the

S. Lloyd (✉)
Department of Anthropology, Université Laval, Quebec City, QC, Canada
e-mail: stephanie.lloyd@ant.ulaval.ca

V. Argudo-Portal et al. (eds.), *Mundane Genomics*,
https://doi.org/10.1007/978-981-95-4550-6_17

only way accounts of the lives of people who die by suicide can be reconciled with neuroepigenetics research?

In this chapter, I draw on speculative accounts that depict limitless biomarker data points reflecting people's affective states and interviews with people recounting the stories of their parents' lives and deaths by suicide to examine current forms of knowledge and where research agendas are orienting future knowledge. I do so to consider the insights offered into people's lives and deaths by biomarkers and less certain ways of knowing, such as stories (Stevenson, 2014), as well as the limits of these insights. I end in a speculative mode, arguing for a possible future of neuroepigenetics in which narratives are populated as much by stories as facts.

Introduction

> Lieutenant Awn's face heated, her distress and anger plain to me. I couldn't read her mind, but I could read every twitch of her every muscle, so her emotions were as transparent to me as glass. (Leckie, 2013, p. 17)

In the epigraph above, science fiction novelist Ann Leckie describes the awareness of a starship-based artificial intelligence (AI), Justice of Toren, that operates human bodies (ancillaries) via implants, and accesses its officers' biological and sensory data. Through these means, it is aware of every action, breath, muscle twitch, hormone shift, and more, of its ancillaries and officers; without asking people a single question, it can detect stress, emotions, and a wide range of other states. The AI ship uses this information relatively indifferently, and can step in to support or intervene in the activities of people in ways they might—or might not—need before they think or dare to ask; a biomarker-based, phenotyping dream. A twist in the book, however, yields further insights. When the ship is destroyed and its AI thereafter trapped in a single ancillary body, it must live masquerading as a human. Under these circumstances, its engagement with people changes. It can intuit their biological states due to thousands of years of experience, and access them when technological interfaces are at its disposal, but this is not the same as its previous all-encompassing data access and accompanying relative detachment from individuals. Finding itself anchored in a body and in interpersonal relationships, it gains new insights and is struck by the lack of logic in people's actions—such as a past ship captain seemingly bent on self-destruction through drug use, but who fails to request medic-assisted suicide—and the complexity of emotions and how they drive behaviour such as self-deception. About these situations, Justice of Toren, now known by the name Breq, concludes, "There was a good deal I didn't understand, and nineteen years of pretending to be human hadn't taught me as much as I'd thought" (Leckie, 2013, p. 8).

What can we know about another person's emotional states? What can we infer from people's reactions and behaviours based on either an indifferent analysis of biomarkers and data points or stories of people's lives? A significant body of emerging bioscience research is currently investing in analyses of biomarkers and data points, with the goal of producing increasingly detailed and dynamic portraits of people's biosocial existence. For instance, certain researchers are collecting biosamples at multiple life points with the hope of correlating them with experiences and traits developed during this period. This collection includes biosamples and reports of adverse experiences during early life (at birth, then at one year old), and assessments of executive functioning at five years of age (Short et al., 2024). Other neuroscientific researchers aim to mobilize a staggering amount of data grounded in diverse disciplinary expertise (e.g. from behavioural specialists to molecular biologists) and methodological approaches and tools (e.g. studies of gene expression, a range of brain imaging techniques that are difficult to compare) by enrolling machine learning (e.g. BrainGPT) to detect patterns and predict fruitful research questions (Luo et al., 2025). At the core of these agendas is a desire to understand how experiences and their embodied causes and effects unfurl across a person's life.

Within this sphere of research, I am particularly interested in neuroepigenetics, a subfield of environmental epigenetics. Both focus on how life experiences affect gene expression; within this, neuroepigenetics seeks to correlate specific negative life experiences with changes in brain-based gene activity. While ostensibly situated as a science of plasticity, with the potential to capture the shifts in emotional states described by Leckie, environmental epigenetics, in practice, tends to focus on a few key measures of the environment and their integration. Researchers argue that this is necessary in order to have the greatest methodological control possible within their experimental paradigms to produce reliable results (Kular & Kular, 2018). Neuroepigenetics sits alongside other contemporary avenues of biosciences research that aim to understand the embodied consequences of exposure to pollution, stress, malnutrition, or adversity on various health profiles, including mental health, and the implications of these emerging models of health and disease (Lappé & Landecker, 2015; Meloni et al., 2016; Gibbon et al., 2024). Critiques of environmental epigenetics that claim it is reductive or that it is developing new models of determinism are widespread, from within and beyond the field (Müller et al., 2017). Over the years, these critiques have become repetitive, yet thus far, little has been done to overcome them. The result is a deadlock in which epigeneticists acknowledge the limitations and critiques of their fields, while largely maintaining the status quo. In the meantime, people's life stories are transformed into accounts grounded in a limited number of data points encapsulating events or experiences, and in the case of the research that interests me, the ability to understand suicide—as a visceral state and action—recedes.

As the future of neuroepigenetics research hurdles toward us, with aspirations that sufficient data points and pattern recognition will lead to the transparency of emotions described in the epigraph, it is worth considering the portrait of life sought in this research. Leckie's novels offer an instructive thought experiment (Le Guin, 2000) that allows us to perceive contemporary research anew: where it is and what it is doing. The speculative form of unparalleled access to individualized behavioural and biomarker data described by Leckie encourages us to consider what might remain unknowable if insights about people in the form of data points are prioritized and taken as essential representations of their lives, at the expense of other means of representation.

This chapter examines ambivalent *neuroepigenetic sensibilities*. I adapt this term from anthropologists Ayo Wahlberg and Laura Heinsen's concept of *genetic sensibility*, which they propose for analysing the "work of dosing, filtering and moderating genetic risk knowledge in order to guide children to be responsive towards what it means and takes to lead a surveillance life in the near or distant future" (Wahlberg & Heinsen, this volume) in the case of a cancer-predisposing genetic (germline) mutation. The term neuroepigenetic sensibilities reflects Wahlberg and Heinsen's interest in how risk knowledge is filtered and associated with surveillance, in reading life through the lens of molecular processes in the body, and providing information on how people conceive of and manage them. It departs, however, from genetic sensibilities in terms of the origin of risk, which in the case of neuroepigenetics entwines genetic and environmental factors. From a neuroepigenetics perspective, suicide is linked to psychopathology, which is in turn connected to distinct at-risk neurobiological profiles resulting from early adverse experiences. These acquired profiles are thought to represent the underlying risk factors for suicide and the impaired reasoning that turns normal negative life events into precipitating causes of suicide attempts (Lloyd & Larivée, 2020). Neuroepigenetic sensibilities are heightened for the loved ones of people who have died by suicide, who are invited to try to articulate the life trajectory of the deceased with neuroepigenetics research frameworks. The sensibilities affect how the lives of the deceased are envisioned retrospectively, as well as how people surveil themselves and their family members for signs of suicide risk. Both genetic and neuroepigenetic sensibilities speak to the core concept of this volume, *mundane genetics*, which "points to the ambivalent process through which genomic technologies and information have been normalized and made personally relevant to an increasing number of people, in and out of the medical realm" (Introduction in this volume). To trace people's ambivalent entanglement with neuroepigenetic sensibilities, as a form of mundane genetics, I examine people's appraisals of story-based and data point accounts of life and death—and the partial and uncertain portraits offered by each approach—to critically assess possible futures for neuroepigenetics research and propose the potential value of reconsidering knowledge production. In particular, loosening a tendency toward singular narrative control, whether grounded in biomarkers or stories, as it concerns explanations of suicide, might allow greater understandings of suicidality and respond to people's ambivalence associated with contemporary neuroepigenetic sensibilities.

Suicide and Its Contexts

Suicide is an individual act but one that most often occurs in clusters in specific subpopulations. Social scientists have documented the macro, population-level factors related to locally high levels of suicide, which are treated as indicators of individual pathologies and diagnostics of larger structural constellations (e.g. Marsh, 2013; Trovato, 1992). Such studies inscribe suicide in specific social, political, and economic contexts and within shared meanings (Widger, 2015; Stevenson, 2014; Krull & Trovato, 1994; Staples & Widger, 2012).

Suicide rates are also higher among marginalized and socially excluded groups (Silviken & Kvernmo, 2007), as well as among people suffering from psychiatric disorders (Renaud et al., 2008), with well-reported patterns of epidemiological distribution globally and by age, sex, gender, and ethnicity (Värnik, 2012; Clements-Nolle et al., 2006). For example, suicide is the second leading cause of mortality among youth (15–29) worldwide (WHO, 2014). The meanings and patterns of suicide (including methods, frequency of suicide, and suicide attempts) play out along gendered lines that are also often age-related (Saunders, 2016). Likewise, suicide (as opposed to suicidal ideation or attempts) is more common among boys and men than among girls and women (Mergl et al., 2015). Divergence from heteronormative ideals (Wool, 2015; Widger, 2015) and social exclusion, particularly among LGBTQIA2S+ youth (Wozolek et al., 2017) have equally been linked to increased suicide. Each dimension is accompanied by specific stories and experiences beyond associations with different rates of suicide. Documenting suicide is thus situated at the crossroads of complex and interacting contexts and narratives, extending deep into personal lives as well as collective pasts and futures.

By contrast, a different image of suicide and its contexts is emerging in neuroepigenetics, one in which individual experiences of adversity or stress are characterized as variables within models of risk. As neuroepigenetics researchers attempt to identify traits that might lead to long-term elevated suicide risk, they create models of specific kinds of suicide risk: for instance, impulsive or aggressive suicide risk, which may play out in different ways in different contexts, but is considered universal neurobiologically (Turecki, 2005). Genetics are not excluded from these models, but researchers' focus is on *gene expression* rather than *at-risk genes*. In these models, early life adversity (ELA) is understood as materialized in children's neurobiological traits and manifested later in life through a complex set of risk behaviours not only related to each other, but also aggravated by multiple mental health disorders and personality traits ranging from irritability to suicidal ideation and behaviours (Orri et al., 2018; Champagne, 2010). Critical links are thus envisioned between instances of ELA—identified through post-mortem interviews in psychological autopsies with people close to the deceased (see methods section)—and the development of negative mental health outcomes later in life, with neurobiological traits as the key correlating factors (Turecki & Brent, 2016). This "variable-oriented" (Séguin et al., 2011) approach produces specific understandings of suicide risk, while positing an ontologically flattened (Landecker & Panofsky, 2013) image of the impact of life experiences on a person's development, as multiple types of negative experiences (e.g. sexual or physical abuse carried out by people with different relationships to the child [e.g. parent vs. stranger]) are folded into the category of ELA.

The result is a new understanding of suicide, which continues to be negotiated after death as people's lives are recounted, interpreted, and reconstructed in their afterlives in neuroscientific research. Death by suicide later in life becomes, while not inevitable, perceived as the extreme endpoint of an unhealthy trajectory established during early life neuroplasticity, thus establishing the meaning of the suicidal act. In these explanations, there is little space for politics, suicides of resolve, or responses to social injustice (Lloyd & Larivée, 2020), which are supplanted by narratives of individual, negative, decontextualized experiences, pathological development, and risky decision making, seen as the result of acquired impulsive traits (e.g. Jollant et al., 2010). This research impacts a growing field of neuroscience that raises fundamental questions about humans and human volition.

In principle, this research—much like the epidemiological and social science research described earlier in this section—is attentive to how suicide is associated with specific profiles and contexts. Both share an interest in how a person's past affects their present and future, and offer the possibility of insights into how this affects people materially. However, much like genetics research before it, neuroepigenetics research converts life experiences into stable factors (e.g. ELA that is seen as measurable in neuroepigenetic biomarkers) and models of their relationships, an approach that risks naturalizing certain explanations and understandings of society–biology interactions. This naturalization within the biosciences, and in its impacts on the general public's understandings of suicide risk, reflects how neuroepigenetics models of the last 20 years are already becoming "mundane". Mundane neuroepigenetics and the models of suicide that result lose touch with the dynamic and relational aspects of people's lives that are irreducible to factors (Brives et al., 2016) and that might offer insights into what leads individual people to suicide.

Methods and Research Site

This chapter draws on research carried out since 2013 that tracks the work of neuroepigeneticists who are members of a multidisciplinary group renowned for translating findings about stress and adversity from model organisms (rats) to humans. This research is carried out in an institute/psychiatric hospital where researchers draw on multiple biosocial technologies (environmental epigenetics and specifically neuroepigenetics, epidemiological research, neuroanatomical models of plasticity) in their efforts to correlate specific life experiences with embodied neurobiological changes and subsequent psychiatric profiles to produce a coherent model of suicide risk. Our research sought to understand the production and reasoning within the group's neuroepigenetics research, which suggests that ELA, such as child abuse, leads to epigenetic changes affecting reactivity and other personal traits, increasing the risk of mental illness, addiction, and suicide (Lutz et al., 2021). They portray this as a subtype of suicide risk, one that does not explain all suicide, but explains a certain pathway of risk. The research in this group is but one instance of the increasing investment in and influence of neuroepigenetics research, with significant impacts on both scientific and personal understandings of neuropsychiatric risk.

Family members participate in two aspects of the group's neuroepigenetics research: first, by donating the family member's tissues to the group's affiliated brain bank, and second, by participating in psychological autopsies: retrospective accounts of the deceased's life, which integrate proxy-based clinical tools and interviews (Bifulco et al., 1994). While anyone close to the deceased can participate, it is most often kin who do so. Clinical tools assess the deceased's history of aggressive behaviours, major depressive episodes and personality traits, and experiences of abuse. These accounts supplement information gathered from medical charts, coroners' reports, details of prescribed medications, and reports from youth protection services. Drawing on this information, the group offers a posthumous *diagnostic impression* of the person—classifying their death, for instance, as suicide with depression or not—condensing a wide range of experiences of neglect, tension, physical, or sexual abuse into a categorization of suicide with or without abuse, based on reports of marked or moderate abuse.

Our research charts the tensions inherent in researchers' efforts to translate diverse data sets into a coherent model of neurobiological vulnerability to suicide risk. These multiple vantage points intersect and rely on one another to construct narratives of suicide risk, yet they are often founded on diverging or incommensurable epistemologies, including in terms of presumed hierarchies of causes (e.g. molecular, psychological, social) and different epistemological and ontological commitments.

As a team, we have carried out over 120 interviews (key informants were interviewed more than once) with researchers and trainees, clinicians, and people living with chronic suicidality, as well as over 300 hours of observation. Interviews with researchers—including principal investigators, advanced trainees, junior students, and lab technicians—focus on models of neurobiological risk and their production and translation, delving into specific research findings, as well as gaining insights into how research agendas are set and enacted in various laboratory spaces. Interviews with clinicians focus on their treatments and interventions, understanding of risk, and the ways that depression and suicide risk are managed in day-to-day clinical spaces (Larivée forthcoming). We have also interviewed the family members of people who die by suicide, asking them to reflect broadly on the deceased's life and how their understandings of suicide may have changed following their loved one's death and their participation in the psychological autopsy. The goal of these interviews is to understand the life and death of the deceased through narrative accounts and the complexity that underlies the relatively austere neuroepigenetics models.

Ways of Knowing, Stories, and Contexts

Neuroepigenetic Sensibilities

> "She told you then."
> "She didn't need to", replied Ship, in Seivarden's vision. "I do have some experience of the world, Lieutenant, and I see you very thoroughly." (Leckie, 2014, p. 299)

Above, Leckie portrays an exchange between Lieutenant Seivarden and the AI-run starship Mercy of Kalr, who has detected subtle gestures and physical signs suggesting stress. The lieutenant is worried about dangers to the fleet captain, Breq, but strives to maintain a calm and authoritative presence on the ship in Breq's absence. In this moment, Seivarden's desire for the emotion-dulling drug Kef—to which she was previously addicted and which almost led to her death—appears to return. When Mercy of Kalr alludes to this desire, Seivarden concludes that Breq must have informed the ship of her past problems. The ship counters that its thorough understanding of her—made possible by reading her bodily signs and its knowledge base—allowed it to ascertain her drug problem independently. Seivarden is well known as a former captain who lost her ship and crew—thousands of people—in a military operation, and subsequently her family. Based on Mercy of Kalr's analysis of Seivarden's biological and behavioural signs, it suggests she should see the ship's medic about her drug problem. Seivarden insists that it is her present concerns that are the cause of her anxiety, and her stress is indeed mitigated with the resolution of the crisis she feared, related to Breq. At this point, Mercy of Kalr's pressure on Seivarden lessens. By drawing on data points about Seivarden's past history, the starship drew a conclusion that seemed coherent with the information available to it, but which diverged from the meaning Seivarden herself associated with her story and the reason for her stress in the present moment—although, of course, had the situation ended otherwise, the relative weight or ultimate source of Seivarden's problem may have been read differently.

Much like Mercy of Kalr's AI, through our increasing familiarization with neuroscientific explanations that pervade research and media discourses, we have become accustomed to associating biochemical traits, behaviours, and at-risk trajectories with addiction and mental health challenges. Stories of past tragic events leading to bad personal outcomes are equally commonplace. People who have lived through a loved one's mental illness or suicide have substantial reasons to contemplate these narratives. Such was the case in interviews with people who had lost a loved one to suicide, when family members recounted stories of personal vulnerabilities, difficult and often heartbreaking trajectories, and deaths that were, in the words of a woman who had lost her daughter to suicide, "shocking, but, in the end, not surprising" (Lloyd & Larivée, 2021). While common in our interviews, this perspective did not reflect the assessment of everyone with whom we spoke who had lost a loved one to suicide. Nonetheless, the commonness of the sentiment may in part be the result of the situations that led people to donate the neural tissues of their family member; often, these people's kin had lived with long-term suicidality and mental illness, which they wanted to understand better. Moreover, the people we spoke to had already participated in psychological autopsies and, therefore, had already engaged in retrospective, narrative-making processes whose end point—suicide—was already known. Therefore, while the mother's statement above is reflective of the views of people with whom we spoke, I do not want to suggest that this represents responses to suicide more generally.

In our previous work, my team and I focused on the bereaved's accounts as they relate to the relational nature of the trauma studied in neuroepigenetics and the thick web of sentiments that family members share beyond trauma (Lloyd & Larivée, 2021). Here, I am interested in how family members engage with neuroepigenetic sensibilities—a form of thinking that resonates with emerging neuroepigenetics models of suicide risk: how, in retrospect, these people report signs of suicide risk among their kin as plainly as Mercy of Kalr read Seivarden's risk of addiction and mental illness, and how participation in neuroepigenetic research further solidifies these perspectives.

I focus on the stories of the lives and deaths of two people who were described in one interview, but reflected divergent pathways to suicide. Each aligns with neuroepigenetic sensibilities in different ways—one more classically, the other tentatively—with the hope that epigenetics research will reveal a cause of suicide not yet fully reflected in existing neuroepigenetics model of risk. The interview was with two people, a couple, discussing their parents' suicides. I refer to them as F and H. They responded independently to my recruitment flyer, but asked to be interviewed together. F and H described what led to their participation in research on suicide and their reflections on their parents' suicides. F had only recently completed the clinical interviews that accompanied the donation of her mother's brain to the brain bank and H had done so with regard to his father the previous month:

> F: My sister, me… we knew it [their mother's suicide] would happen, we just didn't know when... So that's why it was a shock, because we didn't want it to happen… It's the third suicide in my family, something that I've experienced since I was young. It's something that we hear about all the time in my family, and that we talk about all the time. It's not the same thing as H… for me, I have the impression that it's something that can be much more anchored at the genetic level… For H, it's isolated.
>
> H: Me, on my father's side... maybe there are genetic links [they will find] in the research, I don't know… Let's take my father's case… I have the feeling that he didn't feel well, he felt overwhelmed, and the pain he had was very strong, very deep. Maybe if he had been in his right mind, he would have said something... it's not that he had personal problems... Well, in any case, not that we know of... But [without depression] he would have overcome them, he would have gotten past them…
>
> F: For me, it's more like… she [her mother] was born with this *mal de vivre* [pain of living]... It wasn't her first attempt… she's made attempts in the last fifteen years… I think she was crying out for help… but at the same time she didn't have the psychological strength to get through it or to seek help to try to heal it… My grandfather—my mother's father—my 14-year-old cousin [died of suicide]. Then my great-grandmother also

made several suicide attempts, but in the end, she didn't die of suicide... yes, that's it, all on my mother's side.

F and H recounted scenarios consistent with at risk trajectories for suicide: a high number of familial suicides pointing to hereditary or genetic factors, mental illness, loss of hope, and cumulative fatigue, making it difficult to get through a rough period. F and H agreed that there must be something in F's family; genetics were specifically evoked. Yet, F hesitated to attribute her mother's suicide solely to genetics, describing experiences that shaped her mother's life and situations she felt cast doubt on a gene-based explanation. F stated that her aunt, who was 11 months older than her mother, had a successful career and her life overall was completely different from that of her mother. The aunt did not follow the family's pattern. Moreover, although F considered it possible that her mother experienced lifelong angst, she pointed to a traumatic experience in her mother's adolescence as deeply affecting her life trajectory. As a young teen, F's aunt had asked her own mother how her father died, and was informed of her father's suicide or at least received enough information to deduce what happened. By contrast, F's mother learned of her father's suicide from a peer at school. F described her mother's experience of learning about her father's suicide in this way as "a trauma" that amounted to losing her father and no longer believing she knew him. "It was an abandonment, the suicide, it was hard for her. She went off on a high, became a total rebel. I think there were scars, then more scars".

F's description advances multiple lines of reasoning and uncertainty that are common in narratives, reflecting the possibility that there is no simple or consistent story to uncover in people's accounts (Mattingly & Garro, 2000, p. 10). Nonetheless, multiple elements of F's account resonate with neuroepigenetics models, which consider not only potential predispositions to mental health problems (perhaps genetic) but also their amplification or activation through exposure to traumatic events (even if, in this case, these did not occur on the timeline of neuroplasticity commonly studied in epigenetic models). These models also suggest that some people may be resistant to the effects of traumatic events.

In contrast to F, who could point to substantial suicidality and acknowledged risk factors in her family, H was left wondering if the research would yield information on genetic factors implicated in isolated instances of suicide, like his father's. Yet, the two agreed that shifts in thinking or reasoning accompanied experiences of mental illness preceding suicidality. F reflected on this based on her own experiences and those of H's father:

> Having gone through a depression, I can understand how it is really, really easy to let go and no longer see what's going on around you… As H said, his father, if he had had a small window that he had opened for two seconds... he would have thought of us, he would never have done that, never. That's my perception of suicide, that people, they're aware of their actions, but not 100% of the consequences… They think that they can't get through it, they think that it's too late, life is over.

F believed that her mother's reflections were similarly affected by mental illness: "My mother, if she had been normal... she's not a cowardly person, she's not a selfish person, she's a super generous person who would never have wanted us to suffer and to hurt us [by her suicide]". F and H's deliberations on the lives of their family members reflect neuroepigenetic sensibilities in that they provide a lens through which a constellation of biological and behavioural signs can be read as constituting the risk that someone might act in ways other people—or themselves, in the absence of psychopathology—might not, thus responding to an ordinary negative circumstance unusually. In this case, family members saw the deceased, prior to death, as somewhat alienated from themselves and their actions.

During our conversation, F and H spontaneously referenced sources of information on neuroscientific research encountered in their day-to-day and professional lives (F worked in the health sector). Like all people who donate family members' tissues to the brain bank, they are informed of the research that uses brain bank materials. The brain bank's website emphasizes the crucial role of brain tissue to understand psychiatric disorders, including suicide. It details the cells, proteins, and genes that researchers aim to identify to uncover the causes of mental and neurological illnesses and develop new treatments. Brain donations are described as vital for this research, permitting local and international studies. The information on neuroscientific research made an impression on F and H and elicited restrained hope on their parts:

> F: When they dissect the brain, it's sure that under the microscope, they'll see things...
>
> H: What they do is compare the brain with... a normal one ... Given that these are psychological problems, at the brain level, it's sure that [they'll see something]... I find it really interesting that they have the organ itself to try to find [evidence of suicidality]... I don't know how many years they've been doing it, and at the beginning, you find nothing, but at a given point, by digging deeper, you find something and identify patterns.

They elaborated on the information provided by the brain bank:

> F: I saw that they had already discovered ... It's a cell that they discovered that there was more of in people who were not suicidal...
>
> H: In the video, yes.
>
> F: Well I found it interesting that... they were saying they had found perhaps the marker of [suicide risk]... Then, that they would be able to know, according to each person's marker, who was missing it or who had it, to know the right medication with the right dose for each person. Because each person is different. That, I was happy to know.

Donors also learn about the research group's way of thinking about suicide through the questions asked in psychological autopsies, such as those about early adversity and mental illness. These are the life experiences whose material traces researchers attempt to identify in donated brains, and questions about them led to answers that surprised both F and H:

> F: I was surprised with the profile they did of my mother, the personality tests; I thought borderline would have come out first, but it was major depression... there were personality tests, they asked you 150,000 questions.
>
> H: Yeah... at the end, with everything she [psychological autopsy interviewer] was saying... my father, he came out in everything. In everything... [He] had all nine out of nine. When it came out, we couldn't believe it...

She was just doing the interview. Then at the end, she said, "There are nine major criteria compared for that, and your father had all nine." There, we looked at each other, it's like: what the fuck. Oh shit. OK.

F: His father told us, during the holidays, that he was having a little depression.

H: My father, that was always it, with him, everything is… a little car, a little… way of saying things.

F: A little depression.

H: Yeah, a little depression. But it was when he told me, during the holidays or a little bit after… we were talking on the phone and he told me, "I love you." He had never told me I love you… It's like: wow… that's what scared me. In my head, it's like: wow, he's depressed. But I wouldn't have thought he would attempt suicide… I just thought he really wasn't feeling great.

The psychological autopsies led to unexpected results in terms of the diagnostic categories proposed and the severity of their parents' problems. The biographies that F and H had previously attributed to their parents, about borderline personality disorder and a "little depression", were both rewritten to become major depression. A sense of not having known or understood their loved one settled in them and remained.

For F and H, taking part in this research involved untangling their loved ones' life stories in a way that could articulate with neuroepigenetics research methods, oriented by data and results sought by the team. They described the eventual results they envisaged from the research:

H: Knowing that there are chemical things that happen in the brain, in my opinion, it would be possible, eventually, to have a medication, something for this dark side, to not do it [attempt suicide]… depending on the cases you have, in my opinion, in suicides, there must be several types of suicide, at the hormonal level, there are some that are chemical... it's hard, some people aren't thinking straight, then something happens, but there are others… who are sane and will do it out of pride...

F: In fact, they may not discover what's going on in the brain; they'll discover something more at the pharmaceutical level.

H: Maybe.

F: I don't think that analysing, basically, how the brain works or anything, that, I think that it will take a very, very long time, but I think that at the pharmaceutical level, they will develop a drug that will patch the problem.

H: ... Will it be people who have depression, and in the drugs for depression, there will be a little bit of medication for that [against suicidality]? Because it would be unusual that someone just wants to take medication to not attempt suicide all their life, it's like... OK, it's weird. But that's how we think about it, no idea, we'll see in 200 years.

F and H oscillated between faith in the research and tepid expectations of eventual understandings of the neural correlates of suicidal behaviour. Their experiences convinced them that something is going on in the brain, but they doubted that it would be shared by all suicidal people. Despite being informed of and impressed by discoveries already made by the research group, and their anticipation that more patterns will be found at a molecular scale, they expressed limited hope that these findings and patterns would lead to significant understandings of people and help beyond pharmaceutical "patches".

F and H's ambivalent neuroepigenetic sensibilities at least partially align with neuroepigenetics researchers' hopes that, with enough data, they will be able to understand what leads some people to suicide and not others, and develop interventions to help people deemed at risk. These aspirations drive a research process in which people's lives are converted into facts (e.g. mental illness, early abuse), data points to be collected, compiled, and monitored. The resulting portrait of life and risk—the outcome of investments by researchers and the families that contribute biological materials and participate in psychological autopsies—reinforces existing developmental models as people's lives are mined for factors present in the long-standing psychological theories that inform epigenetics research. In a circular manner, aspects of these models inform how family members describe their lost loved ones, as neuroepigenetic sensibilities seep into day-to-day knowledge.

However, are neuroepigenetic sensibilities enough to understand a person's suicide? After all, while research correlates the experience of early adversity and mental illness with elevated suicide risk, most people who experience abuse or depression will not die of suicide. Are neuroepigenetic sensibilities enough to account for the descriptions of suicide offered by social scientists or in epidemiological research? Is it possible that a focus on a limited—or even eventually large or almost infinite and dynamic—number of facts about people who have died by suicide prevents us from understanding what led to their deaths?

Life Stories

> Memory is an event horizon; What's caught in it is gone but it's always there. (Leckie, 2014, p. 273)

After a blunt exchange about annexations, injustices, and the impossible position of people living in these circumstances, Breq accompanies a young woman accused of violence—violence which she admits to perpetrating—against the oppressive owners of the plantation where she and her family work, to a hearing with a local magistrate. The hearing is more than the woman, Queter, would otherwise receive and allows her to enunciate the wrongdoings of the plantation owners formally. While they walk to the hearing, Breq sings the lyrics of the song above. Memories saturate the moment, as they do much of Leckie's Imperial Radch trilogy. Loss, both past and anticipated, haunts the characters: Queter's loss of family members during acute moments of oppression, anticipation of more loss in the likelihood she will be taken from her remaining family and undergo psychological re-education. It is also a moment of reflection for Breq. Breq herself is the remains of a ship's AI implanted into the body of a person stolen and enslaved during an annexation, a person who became the raw material for an AI's ancillary. She is aware of this past and its horrors but has no memories of the person she was before her body was taken and her mind implanted. What she does remember is her 2000-year AI existence as a ship, omnipotent in its knowledge of every breath, movement, action of its officers; of being one of a set of ancillaries and seeing and feeling through their bodies; of having the rest of itself around it, never alone. Breq's past is thus littered with trauma, loss, and regret, but it is her enforced solitude in one body that haunts her continually: loss of data input (e.g. multiple eyes through which to see a situation, biomarker input) and a life of companionship and *being* multiple people (e.g. always being with the rest of herself). Yet, over the course of its life as both starship and ancillary, Justice of Toren/Breq realizes that neither data input nor stories are sufficient to understand a situation or experience. The former informs us of states that might be fleeting (such as affective waves) or lasting (such as durable biomarkers), and the latter offers insights into how situations or experiences are felt and remain in our memories. Both are event horizons as described by Leckie—gone but always there—yet differently.

I remain with the words of F and H as they describe their parents' lives and deaths and their experiences creating their parents' profiles for the brain bank. While F and H at times recounted the stories of their parents in ways that *could* be read as neuroepigenetic sensibilities, in this section, I consider how their ambivalence suggests that, as people's lives are condensed into a series of facts, it exhausts the possibility of knowing (Stevenson, 2014, p. 34) what led to their deaths by suicide. In doing so, to the previous discussion of neuroepigenetic sensibilities, I add a consideration of the challenges of reconciling life stories and reasons for suicide with neuroepigenetic research narratives. I additionally reflect on the challenges of capturing or even understanding life stories and their frequent inconsistencies or incoherence and the ultimate unknowability of other people's lives. I propose the potential value of renegotiating knowledge production and loosening narrative control as it concerns suicide and suicidal behaviour.

In our exchange, F and H readily pointed to the potential of neuroscientific research, even as they indicated the preliminary nature of current findings. They were aware of how their parents' donated tissues would be used in researchers' attempts identify biomarkers associated with suicide risk. F succinctly described the research process as follows:

> Well, my mother, I know that she has become a number. Her brain is a number… So the researchers do not have my mother's name. They have a profile... Basically, the interview is to put a profile on the brain… for the research, to catalogue and then to [be able to take out]… let's say, all the

people of a certain age, men, women, and who are all borderline, really to do the research in relation to [those categories].

H echoed his previous comments that the researchers will look for patterns, look for similarities and differences. This is the conclusion H and F came to through the process of donation, but it was not their initial understanding of the research to which they foresaw contributing:

> H: Initially, the interview [i.e. psychological autopsy], the first time we heard about it, I thought it was like they were doing research on my father's brain... to try to give us answers... but in the end, it wasn't that. And it's logical... by studying this [patterns], it's not about a specific case that you want to understand exactly, it's research...
>
> F: But I still... I was hoping all the same to have a certain answer that... not necessarily with my mother's brain, but to speak with a researcher or whoever was able to tell me whether all the suicides that I have in my family or whatever the case, that someone would tell me: yes, it's genetic, it's hereditary, and should I worry about myself, my sister, my brother or our children?... I don't see it from the same point of view as the others [researchers], because I'm more worried about those who are alive than those who are dead...
>
> H: Can it be confirmed if it can be genetic or not? Because I never thought that... maybe in all families there are some who have tried and we don't even know it, because they haven't talked about it.

The patterns identified in neuroscientific research were less compelling to F and H than the potential of a biologically grounded explanation of their loved ones' profiles and its implications for their families. They did not disvalue the search for patterns and generalizations, but rather questioned its ability to explain what, in their family members' lives, led to their deaths. They knew that the limited explanatory value of neuroscience research was partly due to its preliminary nature, but they suspected other limitations that current lines of research would not overcome. They wondered if researchers would miss significant aspects of suicidal behaviour and multiple paths to suicide, given the profiles of people whose tissues are donated and whose stories are recounted. They also wondered whether the research captured their parents' portraits sufficiently to inform existing research questions. F and H discussed F's mother's posthumous diagnosis of depression and F said, "I don't know how… Yeah, that's it, I was kind of surprised by that." She returned to the centrality of borderline personality disorder in her mother's life trajectory several times in the interview, as though she could not quite convince herself of the posthumous assessment.

> [The psychological autopsy interviewer] told me… my mother's case is a very heavy one. When you look at the whole profile that we tried to do for her. I know it's a very serious case, but... It's just that, I don't want them to tell me the reasons why my mother died of suicide, I think that I'm capable of finding the answers to that myself, but it's really the genetic marker that I really want to have an answer about.

From F's perspective, the research might identify a biological factor linked to suicide, but it would only in part account for what leads to suicide. F's assertion resonates with anthropologist Lisa Stevenson's critique of *anonymous care*, a term she uses to describe bureaucracies and public health organizations that are caring, "but indifferently" (Stevenson, 2012, p. 593). This indifference allows comparisons of categories of life experiences and correlated biomarkers (Lloyd & Larivée, 2021). By contrast, F linked her mother's suicide to specific mental health problems, and considered the latter irreducible to a biomarker, and especially one of depression. She associated her mother's psychological difficulties with traumas, a feeling of being abandoned by her father and suspicions that she never knew him—the specific weight of which is not accounted for in neuroepigenetic research. These experiences were exacerbated by F's mother's estrangement from her family, as F and her siblings grew tired of dealing with their mother's mental health and addiction problems. F was left oscillating between a neuroepigenetic sensibility and maintaining space to repudiate scientific accounts of her mother's life and death. She was eager to learn from the neuroepigenetics risk data, but also shared Stevenson's critique that experiences are not reducible to facts (i.e. neuro-epigenetics explanations), nor can they be grasped through a lens of indifference, which ultimately exhausts ways of knowing what leads to suicide.

F herself had a history of suicidal thoughts. She was acutely aware of how living with mental health problems made suicidality possible, which shaped her impressions of suicidal behaviour.

> The good thing about all this is that I had my depression when I was young, it puts you back on solid ground, you rebuild strong foundations and after four years of psychotherapy, you come back with a solid footing which means that I have more tools to equip myself in life, even if I am a person who tends to be anxious by nature, at least I have tools to help me get through it and I don't think I'm at risk of that [suicide].

F regarded psychotherapy as playing a key role in her life, pulling her off the suicidal trajectory that her mother routinely anticipated for her and her siblings, a trajectory F feared being on once diagnosed with depression. H, too, evoked the potential benefit of psychotherapy. But in recounting his psychological autopsy participation, H described reflecting, while taking the elevator in the mental health institute associated with the brain bank, "My father would never have gone there, he would have turned on his heel immediately, at the elevator, he would have changed course and asked, 'What am I doing here?'" Although he wished his father had had support, he also noted the reluctance of some people, or even generations, to avail themselves of it, pointing to the complexity of assessing the contribution of different environmental factors; each must be considered within a person's life to be understood.

Beyond the benefits of psychotherapy, F considered her awareness of the trauma her mother experienced as a result of her father's suicide to be instrumental in resisting suicidal thoughts: "I mean, if you were a young child, then your father [died by suicide]... I was raising my daughter alone, so I knew that if something happened to me, my daughter would find herself alone, so I clung to the idea that my daughter depended on me, if I [had suicidal thoughts]." In these moments, F echoed the words of other participants. C, a woman we spoke to following her daughter's suicide and whose father also died of suicide, recounted memories of how difficult it was to live through her father's death and what it did to the family:

> When I was depressed, it's true that I didn't want to stop living, but I wanted to stop suffering… It goes through our heads [suicidal ideation], flashes… But I've never tried anything... Because it's flashes… I don't live far from the water's edge, and it's all well and good to go to the water's edge [thinking about suicide], but to do it, that's another matter… I thought about my children, my husband. And having experienced it, that it was difficult for those who remain… "No", I said to myself, "I'm capable of getting through it."

Neither F nor C attempted to label the way the past, present, and future merge in moments of suicidal thought and how they drew on this to pull themselves through, though other people did. T, a woman to whom I spoke after her mother's suicide, recounted a long family history of mental illness, addiction, and suicide, yet pointedly rejected a clinical, or clinic-adjacent, explanation of her own avoidance of serious mental health problems and suicide. A researcher to whom she had spoken about her mother's death and her family told her that she was very resilient. Reflecting on this during our conversation, T said, "I have something else, I don't have resiliency… I have a survival instinct".

The act of pulling through might be explained by their genes or other biomarkers, but it is also explained by their stories. These stories, if reduced to facts, would suggest that F, C, and T have elevated suicide risk because of familial suicides. Yet, when these stories are left in narrative form, they are forceful and revealing, speaking to emotions felt, what people hold on to or let go, and why people act or not based on experiences, feelings, and contexts. They also offer a framework that might permit scientists to understand the meaning or significance of the biomarkers they currently seek sometime in the future.

In a continuation of her critique of anonymous care, Stevenson argues that facts offer limited insights into the lives and deaths of people. Instead, she argues for less certain ways of knowing, which can be better conveyed through modes such as stories or storytelling (Stevenson, 2014). In interviews, F and H noted increasing knowledge and societal conversations about suicide, but that "those who have not experienced suicide will never be able to put themselves in our shoes". Knowing about suicide, in their view, was different from living it, and only people who had intimate experience of suicide could recount its effects. It was by drawing on these intimate experiences that F and H related the life experiences, environmental factors, and personality traits that they believe led to their parents' deaths, and in the case of F, how the particularities of her own life pulled her away from suicidal thoughts.

Yet, stories are not infallible. F and H readily admitted that they did not have a global view of their parents' lives. In the psychological autopsy, F was accompanied by her stepfather, sister, and brother-in-law. She affirmed that without all their contributions, significant parts of her mother's experiences would remain wholly absent from the story of her life collected by researchers. H agreed and they pointed to several challenges to producing a life narrative:

> H: When they asked my dad's girlfriend, "How did he do that [die by suicide]?" She came out with stuff. Ah yeah, shit, OK, it's not at all the same idea [I had of him].
>
> F: Yeah, it's surprising to know how we cannot know people… It's like dissecting our parent, about everything and nothing… After, I was talking with my sister, and it was like... we were talking about our own perception, but we could be completely off base, too.
>
> H: Sometimes [during the psychological autopsy], we all had different opinions, because… It's a perception, how he was… Because my sister doesn't have the same vision as me, because we don't have the same orientation to my father, my father was different with me than with my sister. So sometimes, we ask ourselves questions…

F and H judged the narrative they produced about their parents' lives to be valuable but incomplete, and they remained unnerved by the multiple perspectives presented of their parents' lives. Each was accurate in its own ways, but they did not know which, if any, of the accounts might have most thoroughly reflected their parents' lives or guided their actions. Furthermore, they readily noted the self-deception that can reign in our day-to-day responses to or readings of situations and the decisive effects of retrospection on interpretations of a person's life:

> H: The last few weeks that we saw him… he worried me, but I would never have thought that he would do that, take his own life, never ever ever. But

I already had a flash: I hope not… But I would never have thought that. But you could see [now] that he was more… much more sensitive, there were signs.

F: We don't see them, the signs. We see them…

H: We see them afterwards.

F: We always see them afterwards.

In their exchanges overall, F and H oscillated between neuroepigenetic narratives about suicide, a prioritization of life stories over biomarkers, and doubt about what we can know about a person or what led to their suicide because of the limits imposed by our partial perspective and self-deception. One could conclude that drawing exclusively on either biomarkers or stories might fail to provide coherent information. Data points slip over the event horizon, bereft of meaning although reflecting a physiological state at the moment of their measure—the physical effects of a wave of emotion—or a physiological trend. They offer transparency, like Justice of Toren's reading of Lieutenant Awn's emotional state, but only the ability to infer, rather than know, the experience associated with her emotions and their subject/object. Or like Mercy of Kalr's thorough reading of Seivarden, which led it to the wrong conclusion and a misunderstanding of the situation, given its outcome, despite its reliance on accurate biological data. The source of memories might escape the event horizon even if the memories remain part of our life's story, but that story is always partial, subject to change based on who is telling or when it is told.

Following Stevenson, I would suggest that facts offer limited insights into suicidal behaviour, but so do stories, each in their own ways. Yet, we are drawn to facts. Stevenson describes the calmness of the idea of facts, lying uncontested and waiting to be discovered (2014, p. 32). She further elaborates:

> Joan Didion, in *The Year of Magical Thinking*, describes wanting desperately to piece together the medical facts of her husband's death, as if then she could somehow do something to prevent it from having happened. If facts are calming, they are also magical, functioning as a kind of epistemological anxiolytic. (Stevenson, 2014, p. 32)

The draw of facts may aid in understanding people's ambivalent engagement with neuroepigenetic sensibilities and may limit doubts about neuroepigenetic explanations and their incompleteness, in part overshadowing other ways of understanding suicide. Epigenetics explanations, after all, have relatively easily slipped into mundane ways of understanding risk in the twenty-first century, following decades of reasoning about genetics. These models are increasingly producing powerful narratives about life experiences and how we are coming to understand suicide. Yet, facts do not have a thoroughgoing calming effect. Suicidology "misfit" Jennifer White and others have critiqued efforts to produce coherent visions of suicide and suggest that:

> "the very attempt to prevent and control differences, anomalies and death, [potentially means] that life itself, in all of its creative and unpredictable character, is smothered" (Kouri & White, 2014, p. 182). Katrina Jaworski makes much the same point and suggests that the implicit demand to read suicide through a singular lens, is "an interpretive act that is itself deathlike, continually denying the autonomy and agency of divergent voices and experiences that deserve to be heard and recognized in our understandings of suicide". How might we reflexively engage with these paradoxes and contradictions to recognize how we are potentially contributing to the very problems that suicidology has been set up to solve? (White, 2015, p. 2)

White's argument aligns with F's refusal to accept the singular narrative authority of the very research to which she was contributing, casting doubt on its ability to fully account for her mother's suicide. It was her mother's life story that, in F's opinion, also explained her death. Yet, as F and H both highlighted, stories of people's lives have limits too, fraught with oversights, unknowns, and self-deception. Both facts—and the patterns they might establish—and stories populate understandings of people's experiences and actions, in different ways. What remains to be seen is how they might meaningfully shape a person's life when read together.

What If? Speculating on a Future of Mundane Epigenetics

> Seven Issa [officer on another ship] frowned, and made a doubtful gesture with her left hand, awkwardly... "Ships have feelings."
> "Yes, of course." [one of Justice of Toren's ancillaries answered for it] Without feelings insignificant decisions become excruciating attempts to compare endless arrays of inconsequential things. It's just easier to handle those with emotions. (Leckie, 2013, p. 88)

At the outset of this chapter, I noted how commonplace critiques of environmental epigenetics (including neuroepigenetics) have become. Indeed, critique is essential to take stock of where we are and build more liveable techno-scientific worlds (Kenney, 2024); however, critiques of environmental epigenetics have led to little change in the practice of science. This is not particularly surprising; it is difficult to change long-entrenched research priorities and practices, and epigenetic research inherited the knowledge and tools of the much-less-plastic genetic research that preceded it. Yet, rather than focusing on critique in this text, I have drawn instead on Leckie's work to engage with a possible future of biomarker-informed knowledge about suicide, working in a speculative mode. To draw out the particularities of a biomarker-based approach, I have also considered the potential of stories to provide insights into why, beyond risk factors, people believe their loved ones died by suicide. If family members invested in neuroepigenetic sensibilities, they also expressed ambivalence about the direction in which increasingly mundane neuroscientific understandings of life might ultimately lead in terms of understanding why their family member died.

The speculative mode offers the possibility of looking at situations from a new perspective. This mode is already present, if unenunciated, in many existing forms of research. For instance, neuroethics has been critiqued as "pointless speculation about unlikely scenarios [that] result[s] in unnecessary ethical solutions" (Gilbert & Russo, 2024, p. 115). Arguments that fall into this category are described as following an "if-then-therefore" style of reasoning that starts with hypothetical phrases, speculative results, and assertions that certain actions need to be taken

(Gilbert & Russo, 2024). In contrast to such approaches, which often imply that speculative futures are imminent in a quasi-deflection of the speculative mode, I suggest we own and repurpose speculation to consider the future that we want, actively and imaginatively. As anthropologist Elizabeth F. S. Roberts suggests, we make speculative leaps all the time: "speculation is extremely important for asking questions, and we all make assumptions as we speculate" (Roberts, 2021, p. 364). And the assumptions in speculations—or ways of thinking and orienting ourselves towards a situation—are powerful. Speculations and fictions expand our "visions of what is *possible*" (Benjamin, 2016, p. 2, italics in original) or possible futures (Wolf-Meyer, 2019, p. 3).

So, what if? What if neuroepigeneticists' research was populated as much by stories as facts? For instance, stories of the paths by which people arrive at different points in their lives, of the intersections of relationships that might be both terrible and joyful, of positive, negative, or indifferent encounters with institutions meant to support them, along with a consideration of economic and political factors that lead to the marginalization of some people more than others. What if, alongside critique, we imagined what neuroepigenetics could be? This is the generative role of speculative fiction. As feminist science studies scholar Martha Kenney has argued:

> The speculative mode is always a bit improper. It is not about what is but what might be. Because it is not in accordance with accepted reality, speculation therefore also feels dubious, embarrassing, and unseemly. However, risking speculative impropriety, I argue, might be important in getting ourselves unstuck from the usual patterns of biological storytelling and activate different possibilities for attending and responding within our more-than-human world. (Kenney, 2022, p. 17)

I would like to imagine a neuroepigenetics that could propose biomarkers that reflect where people's loved ones fit into the molecular patterns of certain types of suicide risk, but that would leave space for what people identify as the reasons for their kin's deaths by suicide or, in other cases, why they did not die by suicide. A neuroepigenetics that could do more than identify biomarkers for the same old categories—of, for instance, certain psychopathologies—that have long been studied in genetics and psychiatric research, which would focus instead on how environments, broadly speaking, affect people's lives. An inescapable implication of this type of research would be that, while medications or efforts to reverse epigenetic states will continue to play a role in biomedical research and practice, scientists would also demonstrate how environments can be sustained, promoted, or changed to support people's flourishing. Such research would necessarily include but move beyond the lab, reassess dichotomous categories of suicide with or without depression or abuse, and consider multiple perspectives in order to understand adversity and its effects. It would also take seriously what people who have experienced suicidal ideation contend kept them from moving towards suicidal behaviour. In a context where the loved ones of people who died by suicide increasingly seek meaning in neuroepigenetic reasoning to make sense of their loss, it would seek to document or study many of the intense life states that people note as integral to suicide. It would be attentive to emotions and self-deception, without relegating them to unreason and thoughtlessness resulting from mental illness. It would naturally integrate a consideration of relations between people, communities, environments, and structures, among other elements. These are not, then, just more data points. This is about making "better numbers" (Roberts, 2021) in scientific research, about what shapes a person's life and emplaces them in the world.

This kind of neuroepigenetics could represent what Sandra Harding called a *successor science* (Harding, 1986; Kenney, 2022). The research and approaches of this neuroepigenetics could be wide ranging. It might veer close to environmental epigenetics as it is currently practised (e.g. in methods or sought-after outcomes), to identify targeted sources of influence and their effects. It could also engage in methodological approaches that integrate naturalistic explorations of environments and their effects, but that, in current research paradigms, risk being rejected by researchers as too uncontrolled. However, it would, fundamentally, recognize that while reductivism is integral to experimentation, it is not a good basis for ontological claims (i.e. how results might relate to people's overall profiles or life trajectories) (Lloyd & Kenney, 2023). It would value data points and stories for the different ways they inform us of people's lives, acknowledge their limits, and seek to understand what is consequential in the diverse paths that might lead someone to death. Instead of stories being rendered into facts, epigenetics—the science of plasticity—could become a science at the service of storytelling. It could depict what leaves a lasting impression at the event horizon.

References

Benjamin, R. (2016). Racial fictions, biological facts: Expanding the sociological imagination through speculative methods. *Catalyst: Feminism, Theory, Technoscience, 2*(2), 2. https://doi.org/10.28968/cftt.v2i2.28798

Bifulco, A., Brown, G. W., & Harris, T. O. (1994). Childhood Experience of Care and Abuse (CECA): A retrospective interview measure. *Journal of Child Psychology and Psychiatry, 35*(8), 1419–1435. https://doi.org/10.1111/j.1469-7610.1994.tb01284.x

Brives, C., Marcis, F. L., & Sanabria, E. (2016). What's in a context? Tenses and tensions in evidence-based medicine. *Medical Anthropology, 35*(5), 369–376. https://doi.org/10.1080/01459740.2016.1160089

Champagne, F. A. (2010). Epigenetic influence of social experiences across the lifespan. *Developmental Psychobiology, 52*(4), 299–311. https://doi.org/10.1002/dev.20436

Clements-Nolle, K., Marx, R., & Katz, M. (2006). Attempted suicide among transgender persons: The influence of gender-based discrimination and victimization. *Journal of Homosexuality, 51*(3), 53–69. https://doi.org/10.1300/J082v51n03_04

Gibbon, S., Elizabeth, F., Hardy, R., Dominique, P., Martha, M., Goncalves-Soares, A., Mathers, R., Penkler, M., Fraga, S., Kim, A., Pentecost, M., Tabor, E., Jeffries Hein, R., Lappé, M., Borra, C., Rossmann, S., Lloyd, S., Filipe, A., & Silva, S. (2024). New frontiers for Biosocial Birth Cohort Research: Interdisciplinary approaches to exposure, harmonisation and collaboration. *Wellcome Open Research, 9*, 355. https://doi.org/10.12688/wellcomeopenres.21734.1

Gilbert, F., & Russo, I. (2024). Neurorights: The Land of speculative ethics and alarming claims? *AJOB Neuroscience, 15*(2), 113–115.

Harding, S. (1986). The instability of the analytical categories of feminist theory. *Signs, 11*(4), 645–664.

Jollant, F., Lawrence, N. S., Olie, E., O'Daly, O., Malafosse, A., Courtet, P., & Phillips, M. L. (2010). Decreased activation of lateral orbitofrontal cortex during risky choices under uncertainty is associated with disadvantageous decision-making and suicidal behavior. *NeuroImage, 51*(3), 1275–1281. https://doi.org/10.1016/j.neuroimage.2010.03.027

Kenney, M. (2022). Daphnia and Apollo: An epigenetic fable. *Catalyst: Feminism, Theory, Technoscience, 8*(2), 2. https://doi.org/10.28968/cftt.v8i2.36728

Kenney, M. (2024). Care in technoscience. In U. Felt & A. Irwin (Eds.), *Elgar encyclopedia of science and technology studies* (pp. 426–434). Edward Elgar Publishing. https://www.elgaronline.com/display/book/9781800377998/ch44.xml

Kouri, S., & White, J. (2014). Thinking the other side of suicide: Engagements with life. *International Journal of Child, Youth and Family Studies, 1*, 180–203.

Krull, C., & Trovato, F. (1994). The quiet revolution and the sex differential in Quebec's suicide rates: 1931–1986. *Social Forces, 72*(4), 1121–1147. https://doi.org/10.1093/sf/72.4.1121

Kular, L., & Kular, S. (2018). Epigenetics applied to psychiatry: Clinical opportunities and future challenges. *Psychiatry and Clinical Neurosciences, 72*(4), 195–211. https://doi.org/10.1111/pcn.12634

Landecker, H., & Panofsky, A. (2013). From social structure to gene regulation, and back: A critical introduction to environmental epigenetics for sociology. *Annual Review of Sociology, 39*, 333–357.

Lappé, M., & Landecker, H. (2015). How the genome got a life span. *New Genetics and Society, 34*(2), 152–176. https://doi.org/10.1080/14636778.2015.1034851

Le Guin, U. K. (2000). Introduction. In *The left hand of darkness*. Penguin.

Leckie, A. (2013). *Ancillary Justice*. Orbit.

Leckie, A. (2014). *Ancillary Sword*. Orbit.

Lloyd, S., & Kenney, M. (2023). *Storytelling, facts, and epigenetic sensibilities: From critique to speculation*. Society for the Social Studies of Science.

Lloyd, S., & Larivée, A. (2020). Time, trauma, and the brain: How suicide came to have no significant precipitating event. *Science in Context, 33*(3), 299–327.

Lloyd, S., & Larivée, A. (2021). Shared relations: Trauma and kinship in the afterlife of death. *Medical Anthropology Quarterly, 35*(4), 476–492.

Luo, X., Rechardt, A., Sun, G., Nejad, K. K., Yáñez, F., Yilmaz, B., Lee, K., Cohen, A. O., Borghesani, V., Pashkov, A., Marinazzo, D., Nicholas, J., Salatiello, A., Sucholutsky, I., Minervini, P., Razavi, S., Rocca, R., Yusifov, E.,

Okalova, T., et al. (2025). Large language models surpass human experts in predicting neuroscience results. *Nature Human Behaviour, 9*(2), 305–315. https://doi.org/10.1038/s41562-024-02046-9

Lutz, P.-E., Chay, M.-A., Pacis, A., Chen, G. G., Aouabed, Z., Maffioletti, E., Théroux, J.-F., Grenier, J.-C., Yang, J., Aguirre, M., Ernst, C., Redensek, A., van Kempen, L. C., Yalcin, I., Kwan, T., Mechawar, N., Pastinen, T., & Turecki, G. (2021). Non-CG methylation and multiple histone profiles associate child abuse with immune and small GTPase dysregulation. *Nature Communications, 12*(1), 1132. https://doi.org/10.1038/s41467-021-21365-3

Marsh, I. (2013). The uses of history in the unmaking of modern suicide. *Journal of Social History, 46*(3), 744–756. https://doi.org/10.1093/jsh/shs130

Mattingly, C., & Garro, L. C. (2000). *Narrative and the cultural construction of illness and healing*. University of California Press.

Meloni, M., Williams, S., & Martin, P. (2016). The biosocial: Sociological themes and issues. *Sociological Review Monograph, 64*(1), 7–25. https://doi.org/10.1002/2059-7932.12010

Mergl, R., Koburger, N., Heinrichs, K., Székely, A., Tóth, M. D., Coyne, J., Quintão, S., Arensman, E., Coffey, C., Maxwell, M., Värnik, A., van Audenhove, C., McDaid, D., Sarchiapone, M., Schmidtke, A., Genz, A., Gusmão, R., & Hegerl, U. (2015). What are reasons for the large gender differences in the lethality of suicidal acts? An epidemiological analysis in four European countries. *PLoS One, 10*(7), e0129062. https://doi.org/10.1371/journal.pone.0129062

Müller, R., Hanson, C., Hanson, M., Penkler, M., Samaras, G., Chiapperino, L., Dupré, J., Kenney, M., Kuzawa, C., Latimer, J., Lloyd, S., Lunkes, A., Macdonald, M., Meloni, M., Nerlich, B., Panese, F., Pickersgill, M., Richardson, S., Rüegg, J., et al. (2017). The biosocial genome? Interdisciplinary perspectives on environmental epigenetics, health and society. *EMBO Reports, 18*(10), 1677–1682.

Orri, M., Perret, L. C., Turecki, G., & Geoffroy, M.-C. (2018). Association between irritability and suicide-related outcomes across the life-course. Systematic review of both community and clinical studies. *Journal of Affective Disorders, 239*, 220–233. https://doi.org/10.1016/j.jad.2018.07.010

Renaud, J., Berlim, M. T., McGirr, A., Tousignant, M., & Turecki, G. (2008). Current psychiatric morbidity, aggression/impulsivity, and personality dimensions in child and adolescent suicide: A case-control study. *Journal of Affective Disorders, 105*(1), 221–228.

Roberts, E. F. S. (2021). Making better numbers through bioethnographic collaboration. *American Anthropologist, 123*(2), 355–369. https://doi.org/10.1111/aman.13560

Saunders, K. E. A. (2016). Risk factors for suicide in children and young people: Common yet complex. *The Lancet. Psychiatry, 3*(8), 699–700. https://doi.org/10.1016/S2215-0366(16)30102-X

Séguin, M., Renaud, J., Lesage, A., Robert, M., & Turecki, G. (2011). Youth and young adult suicide: A study of life trajectory. *Journal of Psychiatric Research, 45*(7), 863–870.

Short, A. K., Weber, R., Kamei, N., Wilcox Thai, C., Arora, H., Mortazavi, A., Stern, H. S., Glynn, L., & Baram, T. Z. (2024). Individual longitudinal changes in DNA-methylome identify signatures of early-life adversity and correlate with later outcome. *Neurobiology of Stress, 31*, 100652. https://doi.org/10.1016/j.ynstr.2024.100652

Silviken, A., & Kvernmo, S. (2007). Suicide attempts among indigenous Sami adolescents and majority peers in Arctic Norway: Prevalence and associated risk factors. *Journal of Adolescence, 30*(4), 613–626. https://doi.org/10.1016/j.adolescence.2006.06.004

Staples, J., & Widger, T. (2012). Situating suicide as an anthropological problem: Ethnographic approaches to understanding self-harm and self-inflicted death. *Culture, Medicine, and Psychiatry, 36*(2), 183–203.

Stevenson, L. (2012). The psychic life of biopolitics: Survival, cooperation, and Inuit community. *American Ethnologist, 39*(3), 592–613. https://doi.org/10.1111/j.1548-1425.2012.01383.x

Stevenson, L. (2014). *Life beside itself: Imagining care in the Canadian Arctic*. University of California Press.

Trovato, F. (1992). A Durkheimian analysis of youth suicide: Canada, 1971 and 1981. *Suicide & Life-Threatening Behavior, 22*(4), 413–427.

Turecki, G. (2005). Dissecting the suicide phenotype: The role of impulsive-aggressive behaviours. *Journal of Psychiatry & Neuroscience, 30*(6), 398–408.

Turecki, G., & Brent, D. A. (2016). Suicide and suicidal behaviour. *The Lancet, 387*(10024), 1227–1239. https://doi.org/10.1016/S0140-6736(15)00234-2

Värnik, P. (2012). Suicide in the world. *International Journal of Environmental Research and Public Health, 9*(3), 760–771. https://doi.org/10.3390/ijerph9030760

White, J. (2015). Shaking Up suicidology, Jennifer White. *Social Epistemology Review and Reply Collective, 4*(6), 4.

Widger, T. (2015). Suicide and the "poison complex": Toxic relationalities, child development, and the Sri Lankan self-harm epidemic. *Medical Anthropology*. https://doi.org/10.1080/01459740.2015.1012616

Wolf-Meyer, M. J. (2019). *Theory for the world to come: Speculative fiction and apocalyptic anthropology*. University of Minnesota Press.

Wool, Z. H. (2015). Critical military studies, queer theory, and the possibilities of critique: The case of suicide and family caregiving in the US military. *Critical Military Studies, 1*(1), 23–37. https://doi.org/10.1080/23337486.2014.964600

World Health Organization. (2014). *Preventing suicide: A global imperative*. https://www.who.int/publications/i/item/9789241564779

Wozolek, B., Wootton, L., & Demlow, A. (2017). The school-to-coffin pipeline. *Cultural Studies—Critical Methodologies, 17*(5). https://scholarsphere.psu.edu/resources/34bb2e16-9056-4ef2-b4a6-0667d62a19e9

18

The Mundane and the New Normal: Translating Microbiome Research to "Tell Hong Kong's Story Well"

Mankei Tam and Roberta Raffaetá

Introduction

"I'm waiting for another donor to give me a stool sample," Sarah said when I[1] encountered her after interviewing her fellow microbiome scientists. Microbiome research is a burgeoning field that reconfigures the human body as a symbiotic system containing a multitude of microbiotic communities, considered epigenetic mediations of health and illness. As a research assistant at WAND[2]—a university research institution focusing on gut microbiome science—Sarah collects samples and stores them

[1] We use "I" to denote the first author in this chapter.

[2] To protect the confidentiality of the interlocutors, we use "WAND" as a pseudonym of the lab. All proper names of the PIs, Co-PIs, and other lab members are also pseudonymized in this chapter.

M. Tam (✉) • R. Raffaetá
Department of Philosophy and Cultural Heritage, Ca' Foscari, University of Venice, Venezia, Italy
e-mail: tam.mankei@unive.it; roberta.raffaeta@unive.it

V. Argudo-Portal et al. (eds.), *Mundane Genomics*,
https://doi.org/10.1007/978-981-95-4550-6_18

in lab refrigerators before they become sequenced data. Sequencing stools is a technology used in scientific experimentation on the gut, and linked to biomedical innovations from personalized medicine to apps that offer health monitoring and dietary advice to users, and other commodified health practices; it is also embedded in the "gut's metabolic political economies" (Raffaetà & Ferrari, 2025).

Funded by the Hong Kong (HK) Government since 2019, WAND's innovations include a series of probiotics formulas targeting COVID-19, autism, and early-life health matters. This chapter examines the mundane aspect of genomics demonstrated by our microbiome study, illustrating the multiple ways in which its normalization interacts with techno-national culture and regulations, local biologies and healthcare systems, and techno-global imaginaries, reasserting the promises of genomics within the changing geopolitical landscape in Asia. Drawing on our ethnographic research at WAND, we explore not only how the mundane is enacted by microbiome research but also how it intersects with a new socio-political normal enacted by the HK Government and Chinese State.

In the era dating from the 2019 mass protests and the COVID pandemic, the mundane has become a set of practices that are not only scientific but also biological and political; scientists are encouraged to "tell good stories about Hong Kong" (說好香港故事) despite its accelerating integration with mainland China. Since 2019, HK's political situation has affected scientific practices, changing the meaning of transnational research as scientists are enlisted as allies of nationalist biopolitics. "From chaos to order, from order to prosperity" (由亂入治 由治及興)—Ka-chiu Lee (Chief Executive of the HK Government) and pro-government politicians chanted this slogan to highlight the returning normality of citizens' lives. This normality works to define the ramifications of two historic events—the 2019 mass protests and the COVID pandemic—while presenting itself as a platform for subsequent projects that rework social norms and political arrangements in HK, which was transferred from British to Chinese sovereignty in 1997. King, WAND's PI, is enthusiastic about microbiome science in HK but uncertain about the city's future. "For me, it's sad to see. I treated HK as my second home…When I was a PhD student, everywhere I went, I could be proud of being a resident in HK, but it is hard now." Originally from Southeast

Asia, King gained his PhD at an HK university, then migrated to Canada before joining WAND in 2022. Many of his team members from mainland China have mixed feelings and opinions about normalizing HK society.

I lived through the mass protests and the pandemic lockdown. Post-pandemic, HK entered a prolonged economic downturn, while political fear drove a population outflow. Contrasting with the government's optimistic tone, public sentiment lamented the setbacks in HK. Some called Exchange Square, the building housing the stock exchange, "a relic of history" (South China Morning Post, 2024), while an internationally renowned HK film director said, "Both spaces for breathing and imagination have been narrowed" (Mingpao, 2024). Nevertheless, the Chief Executive keeps urging citizens to "tell good stories" about HK. In this chapter, we ask: how do microbiome scientists echo the call "to tell Hong Kong's story well" through their translational practices, while capitalizing on new hopes that HK will be repositioned as Asia's microbiome hub? How can microbes, the forms of life contemplated by microbiome scientists, address the question of HK's specificity? What new kinds of knowledge and forms of value are produced by microbiome scientists to relate HK citizens' health to China within the new normal?

We proceed with the theoretical framework and methodological settings that help us understand how gut microbiome knowledge production occurs at the intersection of academic capitalism and the changing local/national biopolitics in the post-pandemic, post-2019 protest era. The rest of this chapter is divided into three sections. The first describes the establishment of WAND and presents an overview of the mundane knowledge practices of microbiome scientists. The second traces how WAND scientists mark off gut microbiota as an "organ" interacting with an external environment enacted by a sovereign biopower during the COVID-19 pandemic. Microbiome science blurs the boundary of humans as individual organisms, reconceiving certain microbes as valuable and profitable while shifting the line of defense from the border that separated mainland China and HK to the gut microbiota. The third section examines WAND's research on the early-life microbiome, illustrating how scientists become "hybrid figures" that attribute Chinese-ness to data and samples to enable global knowledge and biovalue circulation. In the mundane landscape of genomics, we suggest that HK's hybrid space has been reformulated as an innovation site by HK–Chinese collaborations at the HK–China border.

Theoretical Framework

The Microbiome as Epigenetic Factor

The gut microbiome is the ecological community of genomes living in the gut. The scientific and public popularity of the gut (Raffaetà & Ferrari, 2025) is part of the wider "microbial turn" in life (Gilbert & Knight, 2017; Gilbert et al., 2018) and the social sciences (Benezra et al., 2012) alike. Microbes are the primary drivers of terrestrial biodiversity and functions and also play a fundamental role in shaping and maintaining human health. Between half and nearly all the cells in the human body are microbial, as they are essential for digestion, immune responses, and hormone regulation; they even influence our emotions and behavior. Our microbiomes are shaped by factors such as diet, birth method, antibiotic use, and social contact, which in turn affect our overall wellbeing. From this perspective, health is not just about defending the body against disease—it depends on our interaction with microbes which bring "into" the body what is "outside" it (Kelty & Landecker, 2019). Bodies, in this light, can be seen as ecosystems, and health as a product of ever-changing biological relationships, rather than a fixed state within individual organisms. Microbiome science challenges the idea of a boundary defining humans as individual organisms (Dupré, 2010; Dupré & O'Malley, 2007; Ironstone, 2019), sparking the potential for novel approaches to examining the entanglements among microbiomes across diverse hosts and habitats (Berg et al., 2020; Hou et al., 2022; O'Toole, 2024). In this respect, microbes have inspired social studies approaches to microbes as desired models of non-competitive social dynamics (Brives et al., 2021; Haraway, 2016; Hird, 2009; Sagan, 2011).

This microbial turn is a relatively recent development, one of the major results of the Human Genome Project (HGP), which identified "a second self" made of microbial "fellow travellers" (The Human Microbiome Project Consortium, 2012), positioning microbes as epigenetic factors (Capua, 2019; Krautkramer et al., 2016; Pepken et al., 2024; Woo & Alenghat, 2022) in health and illness. HGP provided not only the motivation, but also the technoscientific infrastructure to study the microbiome, offering increasingly affordable sequencing technology. Until a few decades ago, Western microbiology primarily studied microbes by growing microbes on plates. However, a significant limitation was that the majority—approximately 99 percent—of microbes on Earth could not be cultivated in a lab, a phenomenon known as the great plate anomaly (Robinson et al., 2010). The sequencing technology offered by the HGP has eliminated the necessity of cultivating microbes for study; metagenomics is the analysis of the genome of microbial communities (microbiomes) that have been directly sampled in natural environments.[3]

Hybridity and Mobility in Microbiome Research in Hong Kong

WAND research is situated within these trends. Manipulating the 99 percent of microbial life within the human body involves a new path of innovation. WAND incorporates translational research, a model of practicing science defined by the US National Institutes of Health in the mid-2000s (Au, 2021; Rubio et al., 2010), whereby practitioners move research findings from the researcher's bench to the patient's bedside by developing new diagnostics and therapeutics through the pharmaceutical industry. WAND's first innovation is a series of probiotic formulas targeting long COVID-19 syndromes and infants' gut health. This chapter builds on Western STS analysis of translational research, as WAND represents another exemplification of academic capitalism and scientific entrepreneurism. The two concepts denote a process in which scientists become semi-entrepreneurs, creating and extracting value by economizing life and its management within a neoliberal market (Helmreich, 2008; Sunder Rajan, 2006, 2012). In the biocapital production cycle,

[3] The "meta" in "metagenomics" signifies that DNA analysis occurs at the ecosystem or microbial community level rather than the individual microbial species level.

academic reputation and credibility constitute epistemic capital, a strategic asset for creating economic capital (Fochler, 2016). In this process, the temporal dimension and the promissory future become significant, as economic capital not only involves short-term monetary accumulation through commodification but also future values that could be materialized in financial markets through "assetization" (Birch & Muniesa, 2020); the latter is contingent on the valuation apparatus and the "associated epistemic, narrative transformations that turn health into a future-oriented 'fact'" (Roy, 2020, p. 98). Translating microbiome science into a business is not linear because knowledge and value co-produce each other (Pinel, 2021; van Wichelen, 2023).

In the field of healthcare and wellbeing, microbiome research fuels a flourishing market for innovative start-ups and personalized medicine solutions aimed at an elite public through public academia–private industry collaborations (Ferrari et al., 2024; Kotliar & Grosglik, 2023; van Wichelen, 2023; Widmer, 2021). However, these studies focus on the biocapitalization of microbiome science in the Western neoliberal health governance framework. In Asian countries, as Ong (2010) has observed, life sciences and genetic biotechnology development have been driven by the state as a means of decolonization. In Asian STS studies, Anderson (2012) advocates seeking new pathways to understand "local technoscientific nationalism, to area studies, to postcolonial critique, and to the interrogation of globalization" (p. 445). Like Anderson, we reject the idea of an "Asia" as a geographical region of shared cultural values; instead, we recognize it as a changing geopolitical landscape of a "remarkably cosmopolitan, heterogeneous, and hybrid" character (ibid. p. 448; see also Chen, 2010). In a similar vein, recent studies from scholars such as Ibata-Arens (2019) have proposed a framework combining the default ideas of an open and networked technoglobal scenario and a technonationalist alternative to analyze models of scientific practice and knowledge transfer. In the Chinese context, a case in point is the Beijing Genomics Institute (BGI) in Shenzhen, a metropolitan city near the border of HK. Against the backdrop of China's being referred to as the world's factory, Stevens (2018) examines the BGI as a hybrid institution that challenges conventional distinctions between laboratory, factory, and startup. Meanwhile, Ong (2016) has analyzed how the BGI's hybrid model intertwines national identity, ethnic diversity, and future health strategies as it evolves to a "global DNA assembly factory."

Taking inspiration from this body of work, we regard HK as a particularly interesting case within the Asian geopolitical landscape. We suggest that the border separating HK from mainland China is biologically and politically significant in the place-making of HK, expanding our understanding of human–microbe interactions in China within the new mundane genomics.

Politically, after the handover of HK from British colonial rule to China in 1997, HK citizens were assured of a 50-year transitional period during which HK's autonomy would be manifested through its own judicial, political, and economic systems. This constitutional infrastructure has distinguished HK from other Asian countries or cities in mainland China; flows of capital and talent characterize HK via its proximity to China (Siu & Ku, 2008), while it is also a hybrid space of cultural production and freedom of expression (Abbas, 1997). This "specificity of Hong Kong" (Choy, 2011), which is underscored by comparisons—British colonial governance and Chinese rule, East and West, Asia and Europe/North America, urban development and the conservation of nature—has played a vital role in the place-making of HK since it came into being in the nineteenth century.

The anti-extradition bill movement, also known as the 2019 mass protests, was the most extensive series of protests in HK under Chinese rule, beginning in March 2019 and continuing until mid-2020. To crack down on it, the Chinese government implemented the National Security Law (NSL) in June 2020, which led to widespread arrests of political activists and legislators engaging in the city's democratic movement.[4] Sociologist Ho-fung Hung (2022) posits that HK's political specificities were changed by the NSL, which has brought "an end to 'one country two systems' twenty-seven years before its time" (p. 201). However, Hung holds that HK's specificities will evolve and continue as it is "a hybrid and contentious space" with a "geohistorical character as a vibrant economic hub at the edge of empires" (p. 198). He argues that the place-making of

[4] As a manifestation of this tightening grip, HK police brutality during the protests was seen by the majority of HK citizens as an extension of Chinese State violence (Cheng, 2020).

HK is embedded in the tension between the city's continuous global connectedness and China's expanding economic and political presence.

HK society has been put back together again as a constitutive normal in which two Chinas are operating: the sovereign and the global. While sovereign China is bounded by a homogeneous "national" entity, global China is characterized by China's "economic expansion and globalizing strategy in other domains" (Lee, 2017), such as migration, education, science, and technologies (see Deng, 2024; Hubbert, 2019; Zhang, 2022). In geopolitical terms, the integration of HK into Chinese State governance is shown by the Greater Bay Area (GBA) initiative, which combines HK and the ten cities in or near the Pearl River Delta to form the most populated megapolis in the southern coastal area of China. Meanwhile, HK's financial role in China's Belt and Road Initiative (BRI), a strategy to globalize China's influence by connecting Asia with Africa and Europe via land and sea networks, is increasing due to growing US–China tension over trade and human rights (HKSAR, 2023). As a global financial center, HK is positioned to serve China's state capitalism via the tightening political control of the Chinese Government and the increasing economic influence of state enterprises.

HK's new socio-political hybridity is relevant to understanding the new mundane in genomics in China and its surroundings. WAND's geopolitical position facilitated the emergence of what Hauskeller and Beltrame (2016) have called *hybrid space*, in which elements of both public and private interests coexist and interact while actors, practices, and resources from both realms are entangled, fostering complementary relationships with low competitiveness. Beyond that, by hybridizing the border, WAND's scientific practices also blur ethical, institutional, infrastructural, and biological boundaries. Likewise, WAND scientists become hybrid figures—individuals or entities that embody the complex interplay between biological and technological, local and global, human and nonhuman elements—in the moment of HK's political transition to a new normal that mixes geopolitical tensions with post-pandemic aspirations and anxieties.

On the other hand, long before this new phase of transition, the health practices of HK citizens had been transformed by environmental pollution and disease outbreaks originating in China. Anthropologist Ayo

Wahlberg (2018) introduces the concept of *exposed biology* to describe how modernization in China has led to environmental degradation and widespread concerns about its impact on human health, highlighting how bodily vulnerabilities have arisen as a result of exposure to pollutants, intersecting with health technologies and state policies. We borrow this concept to study WAND's scientific practices during and after COVID-19, taking HK citizens' gut as an exposed biology to the virus and other environmental pollutants whereby the limits and conditions of the scientists' knowledge-making practices are defined, and their concern for the future of HK is represented. Inspired by Sunder Rajan and Leonelli (2013), who characterized "mobility [as] central to the idea (and ideal) of knowledge that animates translational research" (p. 465), we trace how WAND scientists move knowledge from the lab to the market and also across the HK–China border to facilitate new scientific collaborations while negotiating with a biopower aiming to regulate the population within this normalizing society.

Methodology

We investigate the roles of the gut microbiome, enacting the mundane as a set of knowledge practices that are both scientific and biopolitical in post-2019, post-COVID Hong Kong. Located in the Hong Kong Science Park, WAND has a wet lab, a mouse experimental facility, and a dry lab. It is a leading actor in the field in HK, evidenced by its organization of annual microbiome summits since 2023. We focus our analysis on King's lab, where ethnographic fieldwork took place between December 2023 and April 2024. The team comprises lead scientists, Co-PIs, postdocs, bioinformaticians, and PhD students whose expertise includes microbiology, virology, epidemiology, veterinary science, and bioinformatics. Li, Zheng, and Lu are Co-PIs, assisting King in leading a group of 16 members. During the period of our fieldwork, the lab focused on the early-life microbiome, the gut–brain axis, and mental health; work on COVID-19 and anti-microbial resistance was being phased out. I attended dry lab meetings in person and participated in regular lab meetings through Zoom. I also accompanied lab members to conferences and seminars and observed their interview sessions with research participants. Twenty semi-structured interviews were conducted with PIs, Co-PIs, and junior researchers, ranging in duration from 45 to 120 minutes. Follow-up interviews were conducted in person and online with Co-PIs after the fieldwork period until March 2025. The interview questions examined the participants' career trajectories, their roles in WAND, the nature of their work, and their perspectives on the development of microbiome science within the changing geopolitical landscape in Asia. In addition, we analyzed the scholarly and non-scholarly materials produced by and relating to WAND, which included academic publications, press releases, social media activity, website content, conference speeches, media reports, and public interviews. These materials complemented our interviews and offered valuable insights into WAND's research strategies and concerns during COVID-19, and collaborative efforts on the early-life microbiome between HK and mainland China. Data analysis was performed through thematic analysis to identify key patterns and themes.

Due to time constraints, Lam and Wong, WAND's directors, were unavailable for interview. Instead, we relied on featured interviews in public media outlets and the commentaries written by Lam, a prolific writer regularly published in newspapers, and his personal social media, for our understanding of his vision of WAND and the aspirations of HK microbiome scientists. Regarding government policies on biomedical innovations, we compiled news articles, policy and institutional documents, and content on official websites during and after fieldwork to make sense of the policy shift from a market-oriented to a state-oriented approach over the past decade (see also Luk, 2016 for another study of an HK-manufactured anti-cancer drug for comparison). The first author conducted interviews in English or Mandarin, which the interviewees preferred; only one interlocutor was interviewed in Cantonese, reflecting that most members of the team come from mainland China. Each lab is unique in its way: WAND's workplace environment is characterized by its quietness; there were not many verbal conversations, and communications mostly took place via WeChat and emails. Also, as WAND has been expanding, I was not assigned a workspace and had to rely on regular meetings to observe the team's daily interactions. The research followed GDPR guidelines and standard procedures for managing personal data.

All participants provided informed consent; complete confidentiality and anonymity were always assured.

WAND: Hong Kong's "Asia's First of Its Kind" Microbiome Hub

Since 2017, the HK government has made the life sciences, artificial intelligence, and robotics strategic fields for political investment, and a new initiative, the InnoHK, has been designed to "transform the city into a global innovation powerhouse," placing these fields within the clusters Health@InnoHK and AIR@InnoHK. Another goal is for the innovation and technology sector to "contribute to the realization of the country's long-range objective of becoming one of the top innovative nations in the world and the great rejuvenation of the Chinese nation, as an acknowledgment of the national support and a move to live up to the great expectation of our country on the I&T development in Hong Kong" (HKSAR, 2022, p. 4). Investment in the life sciences is embedded in state-led objectives and China's global ambition, in sharp contrast to the British colonial laissez-faire policy and free-market ideology embraced by most HK citizens (Luk, 2016).

As Ong (2010) has observed, development in the life sciences and biotechnology in Asian countries has been driven by the state; indeed, the Singaporean Biopolis' biotech initiatives were partially powered by a postcolonial nationalist drive "to overcome past humiliations and to restore national identity and political ambition" (p. 3). HK scientists have also been enlisted as allies of nationalist efforts and biopolitics, but guided by the government's agenda of integration with mainland China. WAND scientists have responded to this call by invoking Chineseness in their knowledge production and value propositions, thereby building legitimacy and credibility for their ongoing projects and future funding proposals. Essentially, WAND must build biological assets that embody political, social, and ethical values. Let me further explain through the vision of WAND's Director Lam and its scientific model of translational knowledge-making.

Lam states that WAND is "Asia's first of its kind": a site of transnational knowledge flow that combines university, government, and entrepreneurship to form "the only microbiome R&D center in this part of the world [Asia]." Lam uses "Asia" as a global positioning device (Ong, 2016) to highlight the interdependence between WAND, Western, and Asian scientific institutions, while WAND's translational practices produce innovations that substantiate Lam's claim of WAND's being "the first of its kind." Scientists from Australia, Canada, the United Kingdom, and the United States are invited to WAND's annual summits to provide international perspectives, while experts from HK, China, and other Asian countries present case studies. On WAND's website, emblems of renowned universities occupy a prominent space, highlighting WAND's strong international network and collaboration with Western academic institutions. In framing WAND as "Asia's first," Lam slid into the familiar narrative that engenders the city's postcoloniality as a hybrid space serving cultural and financial exchanges between East and West (Choy, 2011). In this narrative, the skyscrapers alongside Victoria Harbor iconize HK as "Asia's World City" (HKSAR, n.d.).

Lam is sensitive to the geopolitical tensions between the United States and China that affect HK; he has tried to "hold on to facilitating [academic] exchanges among overseas experts to maintain its international biomedical standards." Meanwhile, he emphasizes HK scientists' local responsibilities, as they "cannot rely solely on adopting Western scientific journals and guidelines to improve medical quality; instead, developing local scientific and research capacity is the key." With HK positioned as an Asian innovation site, it is also engaged in a competitive relationship with the West. As Ong (2016) has explained, Asia signifies "an emerging geo-biosocial space" that includes "billions of Asian bodies overlooked in Euro-American research" (pp. 52–54).

Meanwhile, WAND's other task is to turn scientific findings into publications to attract more funding from the biotech and pharmaceutical industries (Fochler, 2016). Such findings must be pertinent to the HK population and translated into effective diagnostics and therapeutics. As King explained, "We are doing innovation. We are making microbiome science more translatable and more accessible to society… You need to take a near-future perspective and think very smartly, what are the needs in society or clinical settings and how can we fill this gap right away. This is a different way of thinking about microbiome research." WAND's biomedical innovation must be evaluated for its ethical potential to contribute to the health of HK and Asian citizens, while WAND scientists have the "dual goals of evolving the scientific enterprise to extend into other domains of 'social trust'" (Ong, 2016, p. 52).

The "different way" King refers to is translational research. From "bench to bedside," the translational process involves transferring knowledge from basic to clinical research. Findings from clinical trials are used to develop new diagnostics and therapeutics, which are deployed in public hospitals or further enhanced to create personalized or precision medicine. Here, clinician-scientists play a crucial role in understanding patients' needs and dissolving the institutional boundaries between universities and hospitals to facilitate collaborations that allow data collection, sampling, lab analysis, and clinical trials (Fischer, 2020). While in other countries precision medicine may raise social justice issues (Dupras & Ravitsky, 2016; Green et al., 2023), this does not apply in HK as the public health system has a robust infrastructure complemented by private services (Schoeb, 2016), with the government providing the citizenry with universal access to healthcare services at affordable costs. In WAND, both directors are clinician-scientists, and the institution is linked to the university's teaching hospital. In these settings, scientists like King have "a sense of responsibility to consider fast-tracking useful results in the form of translational applications" (Benezra, 2023, p. 169), which is also expressed in the notion of innovation. However, how does WAND's knowledge production effectively and efficiently foster innovation? In the following two sections, I further explain their knowledge-making process and how they mobilized gut microbes during the COVID pandemic to make microbiome science translatable and valuable.

"Innovate with Guts": WAND's Translational Knowledge-Making Practices

"Innovate with Guts" became WAND's slogan in 2022, signaling the focus of its scientific and entrepreneurial efforts. These ventures include studying the holistic functioning and associations between the gut microbiome and diseases, such as obesity and inflammatory bowel disorders, from which Chinese and Asian populations are increasingly suffering. Through new concepts of gut–brain, gut–liver, and gut–lung axes, WAND scientists postulate the gut as a central organ, linking infants' gut microbiota with mothers' behavior and environmental exposure, and the wellbeing of human bodies with a healthy gut microbiome. WAND popularizes its science on social media by presenting the gut as a second brain (see also Hadhazy, n.d.) that influences people's mood and connects to depression, autism, and other mental illnesses. With human genomes constituting only 1% of our body, it claims, "We are more microbes than humans!" This highlights WAND scientists' efforts in experimenting with gut microbial genomes to modulate human health, manage individual risk profiles, and identify biomarkers to help enable new microbiome medicine for HK and Asian citizens. To further this initiative, they adopt an ecosystemic framework to understand interactions and relations between humans, microbes, and the environment. Co-PI Li, a virologist, commented, "It's like studying the whole forest rather than looking for a single pathway from point A to B in the forest as I was trained [to do] in virology."

WAND scientific practices start by collecting and analyzing mundane human stools. As a material proxy for the gut, stools are so crucial that WAND gives souvenir stool figures to guests and Lam jokes that they are a team of "shit busters," a joke common in other microbiome labs (Raffaetà, 2022). As van Wichelen (2023) notes, "Shit is ubiquitous and free" (p. 384), which helps to lower the ethical–legal hurdles for its collection, processing, and transference. Most donors consider stools waste, unlike tissue or a swab extracted from organs, making the consent process relatively straightforward. Samples are often collected in a clinical environment such as public hospitals, where donors put them in vials; they then become fecal samples, flexibly manipulated in the wet lab for research purposes, and stored in refrigerators until their value is exhausted and they become waste again.[5] In the wet lab, fecal samples are processed and the genetic contents inside the gut microbiome are metagenomically analyzed. While metagenomics is not a universally accessible technology, given its cost, WAND has governmental support, meaning it can afford expensive animal facilities, a metabolizer, a sequencer, and supercomputing power, allowing its scientists to achieve results in 2–3 weeks from sampling to sequencing. This enables them to move knowledge efficiently from the lab to the market to create new chains of biovalue.

Large-scale datasets are grouped into cohorts of which WAND has an extensive network, with participants enrolled through the university, the teaching hospital, and local non-governmental organizations. In longitudinal studies, building cohorts enables scientists to trace individual health and population effects within broader social, political, economic, and environmental exposures. Metagenomics and bioinformatics provide a new platform to test hypotheses and make statistically significant associations between exposures and outcomes at the molecular level, although such associations do not necessarily demonstrate a cause–effect link but rather a directional and relational one (Richardson, 2020; see also Pentecost & Meloni, 2020). WAND's cohort studies link various groups' gut microbiome profiles to disease risks, infant resiliency in later life, and conditions that shape mental and neurodevelopmental disorders in local and national populations. Their goal is to develop *microbiome medicine* that can modify risk profiles, improve treatment outcomes, and even cure diseases by manipulating the gut microbiome. The biggest cohort under construction is "Mama," which comprises mother–infant pairs recruited from HK and mainland China. This cohort rewrites the material and geopolitical boundaries of WAND's science (see below). One of the institution's first cohort studies focused on COVID-19 and a probiotic formula that redefined the line of defense in the human gut.

[5] Thus, fecal samples are boundary objects (Star & Griesemer, 1989) that move from donors to scientists and from clinical or domestic environments to the wet lab.

Gut Microbiome as an Exposed Biology Across a Porous Boundary During COVID-19

"One of WAND's achievements during the pandemic was that we produced the symbiotic formula that shook the market… We are not only doing science but also doing innovation," King said. To tell a good HK story, WAND scientists mobilized their research on probiotics, marking the gut microbiome as a site for complementary treatment following draconian governmental measures, including vaccination and quarantine.

During the pandemic, HK citizens needed to be protected from the virus encroaching from the mainland. HK and mainland China are connected by the Shenzhen border—a political device of "One Country, Two Systems" that has historically performed biosociality and biosecurity—but this did not suffice as a virus barrier. Previous disease outbreaks and zoonotic threats have evidenced the significance of border control (Centre for Food Safety, n.d.), which forms part of HK's *local microbiologies* in HK. Coined by Erin Koch (2011), the term extends Margaret Lock's notion of *local biologies* (1993, 2001) to cover microbes and human–microbe relations; these are "part of unique social matrices, and their biochemical attributes are only made meaningful in particular contexts" (Koch, 2011, p. 83).

China has been at the epicenter of avian flu, SARS, COVID-19, several influenza outbreaks, and other viruses and diseases that have spread across the planet and threaten global public health, partly owing to the close human–animal interactions in intensive livestock farming and practices involving the handling of live animals in wet markets. Thus, HK has been a World Health Organization (WHO) outpost for monitoring and investigating disease outbreaks and zoonotic threats since the 1970s (Fearnley, 2020b). After the change of sovereignty in 1997, HK gained a "new identity as a sentinel post under Chinese sovereignty," issuing warning signals at regional and global levels (Keck, 2020, p. 81; see also Ong, 2016). However, during the SARS outbreak in 2003, the delayed report from Chinese authorities caught HK health experts and officials unprepared, resulting in a death toll of 299 citizens including several medical workers (Leung et al., 2009). Many HK citizens joined a mass protest in

July 2003, blaming the HK government for their poor response and lack of intervention in the Chinese Government's cover-ups. As an outpost or biosentinel, HK is on the Chinese border; thus, the government's capacity to gather adequate, timely reports of emerging outbreaks is constantly tested and scrutinized.

However, when the first COVID-19 case appeared in HK in December 2019, that border did not hold. Medical workers in HK had a radically different response from those in China and the United States (Au et al., 2022a, 2022b). Invoking a distrust narrative, they joined lawmakers in pressuring the HK government to implement more stringent measures against the outbreak. This distrust related not only to the border's being too porous to halt COVID-19 but also to HK's ambivalence toward increasing Chinese control which was "rooted in their mishandling of SARS in 2003, the perceived deference of the H.K. government to Mainland authorities, and the 2019 political upheaval in H.K. over the Extradition Law Amendment Bill" (Au et al., 2022b, p. 734).

The combined history of outbreaks and recent political turmoil led HK society to act to counter the imminent threat from the Chinese mainland; this added to the moral responsibility of medical experts to protect citizens' bodies and the role of the HK healthcare system as the *biosentinel infrastructure* (Ong, 2016). COVID-19 led to unprecedented lockdowns, draconian quarantine measures, and controversies over vaccination. In February 2021, the Secretary for Food and Health launched a population-wide vaccination program, authorizing the use of the inactivated vaccine Sinovac and mRNA vaccine BNT162b2, developed by China-based Sinovac and German-based BioNTech, respectively (HKSAR, 2021a, 2021b, 2024). However, distrust of the government led to media controversy over the approval process of Sinovac and its immune efficacy (HKSAR, 2021c).

For the first time in HK's history, the labor union representing public hospital medical staff held a 5-day "anti-epidemic strike" in February 2020, calling on policymakers to close the borders with mainland China and supply more masks to citizens. The strike was considered a sequel to the 2019 mass protests, as the leadership of the labor union comprised young clinicians and medical workers who had been outspoken during the movement, mixing biological and political concerns. One leader became a candidate for the legislative council and was later arrested and charged with subverting state power through the National Security Law.

WAND scientists played the strategic role of providing supplementary treatment for both vaccines. They created a cohort of local citizens vaccinated from April to August, one-third of whom received Sinovac (HKSAR, 2024). In a study conducted in March 2020, WAND's co-founder Wong sequenced fecal samples collected from a patient cohort during the first wave of the pandemic to study changes in the fecal microbiome among COVID-19-infected patients. In a global first, Wong associated an altered gut microbiome profile with immune dysfunction and more severe diseases, showing that the microbiome profile determines whether a patient will become very sick from COVID-19. In other words, WAND scientists marked gut microbiota as an exposed biology consequential to later health, allowing citizens to dissociate from compulsory vaccination. WAND scientists posit that the gut microbiome—the new organ embedded in an "ecosystem of connections" (Raffaetà, 2024)—interacted with the external environment during lockdown, conditioned by the spread of COVID-19, the choice of vaccines, and adaptation and change in behavior and diet.

Several bacterial species, including *Bifidobacterium adolescentis* and *Roseburia hominis*, were identified as microbiome biomarkers that could predict immune response to vaccination and the persistence of antibodies. WAND's follow-up studies demonstrated that these biomarkers also enacted post-acute COVID-19 syndrome (PACS) within the microbiome. More commonly known as 'long COVID', PACS refers to the long-term brain fog, fatigue, impairment in concentration, and degraded mobility suffered by patients recovering from COVID-19, which was ignored as a serious illness early on in the pandemic (Au et al., 2022a, 2022b, 2023). WAND scientists claimed that patients' state of health and risk of long COVID were predictable and could be improved and lowered through gut microbiome modulation.

The COVID-19 pandemic opened up a peculiar time and space for WAND scientists, one in which biological materialities and the social-political environment came together to make the gut microbiome and *Bifidobacterium* meaningful in public health. However, probiotic *Bifidobacterium* is not curative when the mechanistic links underlying the therapeutic effects of these microbiota-based interventions are not provided. As Liew, a research assistant from mainland China who joined WAND during the pandemic, reflected, "Whenever my friends feel not very well, physically or mentally, I'll advise them to take some [WAND] probiotics. It does no harm. I also have discount coupons to buy them." Probiotic *Bifidobacterium* does not cure long COVID as a chronic disease, although it contributes to users' general wellness and improves long COVID symptoms. However, as Helmreich (2008) writes, "Value in the market sense and value in the ethical sense co-constitute one another in biocapital" (p. 465). While they are not officially recognized drugs, WAND has fast-tracked scientific findings on the microbiome to probiotics that are readily available in the consumer market and relatively affordable for HK citizens.

Politically, by handling COVID-19 as a microbial issue, WAND scientists reconfigured the first line of defense as the gut microbiome, enlisting symbiotic microbes to enhance the immune system while overcoming biological differentiations from the choice of vaccines.[6] As Helmreich (2009) suggested, labs are "devices for creating significance, for separating figure from ground, for adjusting what counts as text and context" (p. 57). WAND scientists created a biologically and statistically *normalized biomedical body* (Lock, 2018), defined by the pre-COVID microbiome profile in their cohorts. Through this, they claimed the gut microbiome could be restored to its pre-COVID normal. However, this molecularized body has to gain new attributes to work, which we explore through WAND's research on the early-life microbiome and its normalization in Chinese biopolitical and biomedical contexts.

[6] Vaccination in HK was not compulsory, but the government imposed restrictions on unvaccinated citizens.

The Mama Cohort: Normalizing the Chinese–HK Early-Life Microbiome

> In the 4th–5th century, Chinese medicine doctors had used human feces to treat diarrhea. That prescription was called the Yellow Dragon Soup (黃龍湯). China was the first to use gut microbes as medicine!—a WAND scientist during a conference.

The Yellow Dragon Soup, a story I heard many times during my fieldwork, reverses the US mainstream narrative about the direction of innovation (from the US to Asia to vice versa), showing an intimate legacy linking traditional Chinese medicine and Western microbiome therapeutics. WAND is not the first scientist group invoking an "affective notion of unified Chineseness" (Ong, 2016, p. 68). While Ong observed an oscillation in public discourse between "Asian" and "Chinese" among genetic scientists in Singaporean Biopolis, WAND demonstrated an unwavering commitment to "Chineseness" to claim their scientific practices as "Asia's first of its kind," as demonstrated by one of WAND's central themes—the early-life microbiome.

As WAND's brochure reads, "The infant's microbiome establishes gut health in later life. A faulty microbiota in infancy can lead to diseases such as obesity, autism, allergy, diabetes, metabolic diseases, cancer, depression, and inflammatory bowel disease." It is also King's area of expertise. "When I shifted from animal to human microbiome, my first study was related to the early-life microbiome. It has always been my interest because I'm a believer in DoHAD theory." DoHAD stands for Developmental Origins of Health and Disease theory, which hypothesizes that environmental exposures during early life trigger epigenetic adaptations that can increase the risk of chronic disease and permanently influence health in adulthood (Pentecost, 2024). Lam and Zheng, the Co-PIs, popularized the DoHAD theory by mobilizing the Chinese proverb, "Three years old determines eighty (三歲定八十)," when discussing the significance of WAND's works on the early-life microbiome. By its very nature, the proverb marks childhood as the formative years of individuals. It represents traditional cultural beliefs about parenting and personhood, which microbiome scientists have appropriated to associate potential biomarkers in an infant's microbiome with the health of Chinese adults. Likewise, "three years old determines eighty" relates Chinese parenting to global pathways of therapeutics and diagnostics created by microbiome science.

According to HK politicians, a good story about Hong Kong is tethered to the Chinese Government's call to tell China's story well. To form what is called the "Mama cohort," WAND aims to recruit 10,000 mother–infant pairs between 2019 and 2027 for a longitudinal study from pregnancy to the seventh birthday. This data is mainly from mainland China; HK mother–infants constitute only a tiny proportion. The database contains clinical and demographic information about both parents; meanwhile, samples of the mother's breast milk and the infant's feces are collected to store in a biobank, and will be sequenced for analysis in the future.

Here, the compatibility of HK microbiome science with China is not only cultural but also material and informatics-related. A recent survey by Li et al. (2024) concluded that mother–infant cohorts in mainland China are disparate and varied in number of participants, sample collection protocol, and demographic distribution, rendering them incomparable.[7] The scientific uniqueness of the Mama cohort lies primarily in building bridges between HK and Chinese research institutions to enable sharing samples and data. Thomas, the wet lab manager, states, "Mama cohort looks into every detail of the fathers, mothers, and babies. Fingernails, breast milk, hair, saliva, and stools are collected. All participants in the mainland and HK are required to give these samples [with their consent]."

With standardized protocols, WAND scientists affirmed their leading role among Chinese collaborators in the Mama cohort. Each collaborator represents a dot on the map of China, utilizing HK as a hub where Chinese, HK, and international scientists work together. Within and between groups in the cohort, data and samples generate new forms of difference that are defined, sampled, and analyzed. The Mama cohort

---

[7] For example, the China National Birth Cohort, a nation-based cohort, does not collect fecal samples from mothers or infants, making microbiome research challenging if not impossible.

becomes fungible through "the coding and alignment of variations" (Ong, 2016, p. 13), categories, and standards that can travel across China, including HK. Chineseness, in the project, has the properties of what Bruno Latour (1986) calls the immutable mobile—it remains stable across the borders of HK and China, rendering the different datasets and samples in the Mama cohort "presentable, readable and combinable with one another" (pp. 7–9). Yet, this may go beyond integrating Chinese and HK biologies, as Chineseness is acquiring a new political meaning that also includes HK.

Hybridizing the Border: The New Mundane of Creating (Bio)value

Under "One Country, Two Systems," HK citizens can cross the border to mainland China almost effortlessly. However, Chinese citizens require visas to enter HK. WAND scientists can routinely travel to Shenzhen facilities to access biological samples collected in mainland China. However, due to national security concerns, the export of biological materials and human genetic resources from China to HK is subject to strict "bio-sovereignty" regulations (Fearnley, 2020a). From this perspective, HK is outside the jurisdiction of the Chinese regime.

In 2019, at around the same time as the founding of WAND, the Greater Bay Area (GBA) initiative was launched by the Chinese Government (HKSAR, 2019) with the aim of linking the Southern Chinese cities—HK, Macau, Guangzhou, Shenzhen, Zhuhai, Foshan, Zhongshan, Dongguan, Huizhou, Jiangmen, and Zhaoqing—into an integrated economic area. HK and the ten cities around the Pearl River Delta have a total population of 86 million; geographically, the GBA is the most populated megapolis in the southern coastal area of China and has blurred the border between HK and mainland China. To facilitate the integration of HK into the GBA, one of the new infrastructures at the border is the HK-Shenzhen Innovation and Technology Park (HKSIT) located in the Lok Ma Chau Loop; WAND was among the first research institutions and enterprises to operate lab facilities there. The northbound movement of scientists has transformative effects on knowledge production and policy decisions (Fearnley, 2020a).

First, WAND's disposition facilitated the emergence of a hybrid space (Hauskeller & Beltrame, 2016) inhabited by different collaborators, fostering complementary relationships with low competitiveness. As King explained, "In China, when researchers have the budget for the cohort, it is only for one or two years, which is not enough to conduct microbiome science...We can follow the same protocol and provide funding to maintain the cohort." He also noted that Chinese researchers cannot do multisite research, partly because of competition and hierarchy among national and provincial universities, adding, "However, there are many universities in China, and as scientists from HK, we can treat everybody at the same level; we are not competing with mainland universities because we are internationalized. We do not want to be the king in China." Moreover, with a higher ranking in the West than most universities in China, WAND's translational practices render data categorized as "Chinese" a more "promissory asset" (Martin, 2015) with a more profitable future. In sequencing the Chinese early-life microbiome, HK's hybridity is reshaped by the accelerating integration with Chinese institutions.

Second, as Fearnley (2020a) has noted, Chinese scientists are not allowed to share or exchange viral samples with international scientists even during emergencies, such as influenza outbreaks or the COVID pandemic. However, by positioning WAND under the GBA initiative, HK scientists are hopeful of accessing such samples by hybridizing the border between HK and Shenzhen. King believed that the HKSIT would benefit HK scientists because it includes many laboratories where HK scientists are working on Chinese samples. "HK is very small, and if we study only for the people of HK, who cares? But if we study samples from mainland China, we're making a big impact." Through genomics, HK's new hybridity, characterized by its Chineseness, is becoming mundane as the center of knowledge and biovalue creation moves north toward the border between HK and China.

Mobilizing the GBA initiative, Lam highlighted the potential biovalue generated by studies that mix biology and politics:

> The microbiome market is expected to grow at a compound annual growth rate [CAGR] of 30 to 40 percent per year. We urge the government to streamline the regulatory approval of our innovations in the Greater Bay Area… I believe such a statutory body would effectively attract pharmaceutical companies and talent and build HK into a new pharmaceutical innovation hub… We advocate establishing a modern version of 'Noah's Ark' to process and store gut microbiome biosamples in the Lok Ma Chau Loop Area.

"Noah's Ark," the name of the biobank, suggests not only the future economic but also the ethical value of life in public health (Dumit, 2012). While perhaps not Lam's original intention, the name also reminds us that microbiome science originated in the West. More importantly, WAND scientists stand at the border as hybrid figures (Bauer et al., 2013) and enablers of global knowledge and biovalue circulation, remaking HK into a new microbiome innovation hub comprising not only new biological and political forms of the mundane but—as transformations are always relative to one's positioning—a new social form of the mundane as well.

Alex, a HK-born PhD student in London, observes, "Well, WAND may have big data on Chinese people to make medicines that cater to their needs. But in terms of research agenda, HK is lagging behind us [Western scientists]." After Alex left WAND, only two HK members remained, a pronounced minority; the majority comprised young researchers and scientists from mainland China. Thus, the iconic HK story of the East–West connection can be used to re-interpret WAND's hierarchy vertically. PIs trained in the West lead the organization, while the lower level postdocs and RAs are almost all Chinese. Zheng joined WAND after graduating in HK, remaining in HK rather than working abroad in order to care for his parents. Aware of the Mandarin-speaking majority in WAND, he insisted on conducting meetings in English to practice his language skills. Lu, meanwhile, had worked for BGI in Shenzhen. Leaving her parents and husband there, she brought her son to HK for his education. Zhao and Liew, also BGI ex-employees, found speaking Mandarin advantageous for their career development in HK. Zhao was recently married; he lived in Shenzhen and crossed the border daily for work. Moderated by hybrid figures like King, this migration pattern—Chinese students and scientists coming to HK universities in search of a new life—has been normalized during the integration of HK into the GBA. Meanwhile, institutions like WAND act as powerhouses in which Chinese scientists play a majority role in remaking Hong Kong as "Asia's first of its kind" innovation hub.

Conclusion

In this chapter, we explore mundane aspects of genomics in an HK university research unit, illustrating how its normalization oscillates between academic research and biovalue creation, techno-global practices and techno-nationalist initiatives, and local and exposed biologies, meanwhile reasserting the promises of genomics within the changing geopolitical landscape in Asia. In particular, we narrated a specific instance of mundane genomics: microbiome research in WAND, a lab funded by the HK Government in 2019, a crucial year that marked a new transition of HK under Chinese sovereignty and the beginning of the COVID pandemic. We showed how, by becoming mundane, microbiome research contributes to creating a new biopolitical normal within the hybrid and mobile space of HK–mainland China interactions.

We compared two WAND projects during the pandemic, and an early-life microbiome study of mother–infant pairs in HK and China. Among WAND's many microbiome research projects, these two are selected because, first, they allowed us to raise issues relating to sovereignty, biocapital, and healthcare; and, second, we can examine how the translational practices and innovations of microbiome scientists become mundane by reconfiguring a molecularized body embedded in the biopolitical governance and biosocial space carved out by the Chinese State in post-2019 HK.

Current analyses of STS literature on microbiome scientific knowledge-making and innovation predominantly focus on the process of biocapitalization and the enterprising of academic entities using the Western neoliberal framework of health governance. In the Asian context, previous cases (Singaporean Biopolis and Chinese BGI) have given us insights into how the life sciences and genomics have embraced the state agenda and national identity to navigate their research and assetization strategies. Our analysis of WAND contributes to, first, how (post-)genomics becomes mundane, entering the public domain through an emergency (COVID-19) and everyday life concerns (early-life healthcare). This chapter was written at the end of WAND's initial funding phase (2019–2024). I witnessed the scientists writing proposals to the university grant council, the health bureau of the government, and independent foundations to sustain operations. WAND has produced a remarkable number of publications on COVID-19 and the early-life microbiome, which have helped in gaining further grants and attracting the interest of the pharmaceutical industry. However, we have demonstrated that WAND is not a superficial exemplification of academic capitalism, as many STS studies have discussed in the neoliberal Western science context. Its efforts are political, transcending the scientific or economic, which regulates the political interactions with mainland China in the post-2019 transition. As a government-funded institution, its practices are embedded in Chinese State-led objectives and global ambition, pushing its scientific entrepreneurial traits and presenting knowledge pertinent to HK and Asian society.

Second, we analyzed how scientists make their science possible through mundane practices at the intersection of state power, economic policy, and scientific innovation. This new mundane is characterized by scientific practices that enable HK scientists to move across national borders and create new sites and value chains; likewise, mainland scientists are moving south. New patterns of simultaneous relatedness and incompatibility, as well as genomic indebtedness to mainland China, are emerging as WAND strives to be "Asia's first of its kind." In this process, the iconic HK story of the East–West connection becomes uncertain.

Many scholars consider the new normal of post-COVID HK to be characterized by the increasing sovereign power of the Chinese regime. As Fearnley (2020a) has noted, the pandemic also triggered a shift from biosovereignty based on a knowledge-control regime created for the exchange and regulation of viral samples to a new model based on sharing open-access data and a hybrid culture "between the moral economies of credit, authorship, and property in the experimental and collecting sciences" (p. 484). Both have been confirmed by HK's integration into China through the GBA initiative. Although the East–West connection that characterized HK's postcolonial hybridity has become uncertain since the 2019 mass protests and COVID pandemic era, a new form of hybridity is emerging. Examining WAND's research on the early-life microbiome, we showed how scientists, as "hybrid figures," translate bio-sovereignty into transnational scientific collaboration by enhancing the compatibility and fungibility of data and samples. HK's new specificity has materialized by hybridizing the boundaries among Chinese research institutions and the border between HK and Shenzhen.

Third, as we see during the pandemic, WAND's science tackled the question of Chinese sovereignty and HK's autonomy by shifting the first line of defense from the border to citizens' bodies, with scientists enlisting symbiotic microbes to stabilize immune responses to vaccines and establish a new normalized body through gut microbiome biomarkers. Microbiome scientists blur the boundary of humans as individual organisms, reconfiguring health as the modulation of microbes cohabiting in the human body. In both COVID-19 and the early-life microbiome research, WAND scientists have addressed the asymmetry of biopolitical power through translational practices to create new biovalue in commodities and data. WAND's microbiome science has invested in being both "HK" and "China," bringing HK's local and exposed biologies together for the good of the citizenry. This relates microbiome science to the formulation of healthy bodies while redefining the specificity of HK within the accelerating integration of HK into China and the changing geopolitics in Asia. The challenge here is the coming together of biology and politics framed within a nationalist vision—the "great rejuvenation of China." Essentially, this also challenges microbiome scientists: can they still discuss mundane, nation-based research when the biosocial is reconfigured by billions of microbic species and their connections on both local and planetary scales?

Acknowledgement This chapter is part of a project that has received funding from the Italian Ministry of University and Research (MUR) - FARE Programme (ID R20A8R2WH4 SHK_HealthXCross). I acknowledge the project ERC-

HealthXCross (GA n. 949742) funded by the European Union's Horizon 2020 Research and Innovation Programme for the fruitful collaboration.

References

Abbas, A. (1997). Hong Kong: Other histories, other politics. *Public Culture, 9*(3), 293–313.

Anderson, W. (2012). Asia as method in science and technology studies. *East Asian Science, Technology and Society, 6*(4), 445–451.

Au, L. (2021). Recent scientific/intellectual movements in biomedicine. *Social Science & Medicine (1982), 278*, 113950.

Au, L., Capotescu, C., Eyal, G., & Finestone, G. (2022a). Long covid and medical gaslighting: Dismissal, delayed diagnosis, and deferred treatment. *SSM. Qualitative Research in Health, 2*, 100167–100167.

Au, L., Fu, Z., & Liu, C. (2022b). 'It's (not) like the flu': Expert narratives and the COVID-19 pandemic in mainland China, Hong Kong, and the United States. *Sociological Forum (Randolph, N.J.), 37*(3), 722–743.

Au, L., Capotescu, C., Curi, A., Gonçalves Leonel da Silva, R., & Eyal, G. (2023). Long Covid requires a global response centred on equity and dialogue. *Global Health Action, 16*(1), 2244757–2244757.

Bauer, S., Blacker, S., Nils, G., & Martina, S. (2013, December 23). *The racehorse on the runway*. Nautilus. https://nautil.us/the-racehorse-on-the-runway-234715/

Benezra, A. (2023). *Gut Anthro: An experiment in thinking with microbes*. University of Minnesota Press.

Benezra, A., DeStefano, J., & Gordon, J. I. (2012). Anthropology of microbes. *Proceedings of the National Academy of Sciences, 109*(17), 6378–6381.

Berg, G., Rybakova, D., Fischer, D., Cernava, T., Vergès, M.-C. C., Charles, T., Chen, X., Cocolin, L., Eversole, K., Corral, G. H., Kazou, M., Kinkel, L., Lange, L., Lima, N., Loy, A., Macklin, J. A., Maguin, E., Mauchline, T., McClure, R., et al. (2020). Microbiome definition re-visited: Old concepts and new challenges. *Microbiome, 8*(1), 103–122.

Birch, K., & Muniesa, F. (2020). *Assetization: Turning things into assets in technoscientific capitalism*. The MIT Press.

Brives, C., Rest, M., & Sariola, S. (Eds.). (2021). *With microbes*. Mattering Press.

Capua, I. (2019). *Salute Circolare. Una Rivoluzione Necessaria*. Egea.

Centre for Food Safety. (n.d.). *Import inspection of live food animals*. Retrieved October 31, 2024, from https://www.cfs.gov.hk/english/import/import_ifa.html

Chen, K.-H. (2010). *Asia as method: Toward deimperialization*. Duke University Press.

Cheng, S. (2020). Pikachu's tears: Children's perspectives on violence in Hong Kong. *Feminist Studies, 46*(1), 216–225.

Choy, T. K. (2011). *Ecologies of comparison: An ethnography of endangerment in Hong Kong*. Duke University Press.

Deng, G. T. (2024). *Chinese Espresso: Contested race and convivial space in contemporary Italy*. Princeton University Press.

Dumit, J. (2012). *Drugs for life: How pharmaceutical companies define our health*. Duke University Press.

Dupras, C., & Ravitsky, V. (2016). Epigenetics in the neoliberal 'regime of truth': Hastings center report. *Hastings Center Report, 46*(1), 26–35.

Dupré, J. (2010). The polygenomic organism. *The Sociological Review (Keele), 58*(s1), 19–31.

Dupré, J., & O'Malley, M. A. (2007). Metagenomics and biological ontology. *Studies in History and Philosophy of Science Part C: Studies in History and Philosophy of Biological and Biomedical Sciences, 38*(4), 834–846.

Fearnley, L. (2020a). Viral sovereignty or sequence etiquette? Asian science, open data, and knowledge control in global virus surveillance. *East Asian Science, Technology and Society, 14*(3), 479–505.

Fearnley, L. (2020b). *Virulent zones: Animal disease and global health at China's Pandemic epicenter*. Duke University Press.

Ferrari, L., Raffaetà, R., & Beltrame, L. (2024). Capitalization and the production of value at the nexus of academia and industry: The case of a microbiome startup. *New Genetics and Society, 43*(1).

Fischer, M. M. J. (2020). Lively Biotech and Translational Research: In memoriam: Judah Folkman, teacher, healer, scientist. In *Lively capital* (pp. 385–436). Duke University Press.

Fochler, M. (2016). Variants of epistemic capitalism: Knowledge production and the accumulation of worth in commercial biotechnology and the academic life sciences. *Science, Technology, & Human Values, 41*(5), 922–948.

Gilbert, J., & Knight, R. (2017). *Dirt is good: The advantage of germs for your child's developing immune system*. St. Martin's Press.

Gilbert, J. A., Blaser, M. J., Caporaso, J. G., Jansson, J. K., Lynch, S. V., & Knight, R. (2018). Current understanding of the human microbiome. *Nature Medicine, 24*(4), 392–400.

Green, S., Prainsack, B., & Sabatello, M. (2023). Precision medicine and the problem of structural injustice. *Medicine, Health Care, and Philosophy, 26*(3), 433–450.

Hadhazy, A. (n.d.). Think twice: How the gut's 'second brain' influences mood and well-being. *Scientific American*. Retrieved September 30, 2024, from https://www.scientificamerican.com/article/gut-second-brain/

Haraway, D. (2016). *Staying with the trouble. Making kin in the Chthulucene*. Duke University Press.

Hauskeller, C., & Beltrame, L. (2016). The hybrid bioeconomy of umbilical cord blood banking: Re-examining the narrative of opposition between public and private services. *BioSocieties, 11*(4), 415–434.

Helmreich, S. (2008). Species of biocapital. *Science as Culture, 17*(4), 463–478.

Helmreich, S. (2009). *Alien Ocean: Anthropological voyages in microbial seas*. University of California Press.

Hird, M. (2009). *The origins of sociable life: Evolution after science studies*. Palgrave Macmillan.

HKSAR. (2019). *CE attends meeting of leading group for development of Guangdong-Hong Kong-Macao Greater Bay area*. Retrieved July 11, 2024, from https://www.info.gov.hk/gia/general/201911/06/P2019110600764.htm

HKSAR. (2021a). *SFH authorises COVID-19 vaccine by Fosun Pharma/BioNTech for emergency use in Hong Kong*. Retrieved October 7, 2024, from https://www.info.gov.hk/gia/general/202101/25/P2021012500829.htm

HKSAR. (2021b). *SFH authorises COVID-19 vaccine by Sinovac for emergency use in Hong Kong*. Retrieved October 7, 2024, from https://www.info.gov.hk/gia/general/202102/18/P2021021800495.htm?fontSize=1

HKSAR. (2021c). *Government responds to approval of vaccine by Sinovac for emergency use*. Retrieved October 7, 2024, from https://www.info.gov.hk/gia/general/202102/19/P2021021900029.htm?fontSize=1

HKSAR. (2022). *Hong Kong innovation and technology development blueprint*. Retrieved July 9, 2024, from https://www.itib.gov.hk/en/publications/I&T%20Blueprint%20Book_EN_single_Digital.pdf

HKSAR. (2023). *John Lee to tell HK's good stories*. Retrieved July 9, 2024, from https://www.news.gov.hk/eng/2023/10/20231013/20231013_143440_008.html

HKSAR. (2024). *DH special bulletin on the fight against COVID-19*. Retrieved July 9, 2024, from https://www.chp.gov.hk/files/pdf/chp53_tc.pdf

HKSAR. (n.d.). *Asia's world city*. Retrieved October 17, 2024, from https://www.info.gov.hk/info/sar5/easia.htm

Hou, K., Wu, Z.-X., Chen, X.-Y., Wang, J.-Q., Zhang, D., Xiao, C., Zhu, D., Koya, J. B., Wei, L., Li, J., & Chen, Z.-S. (2022). Microbiota in health and diseases. *Signal Transduction and Targeted Therapy, 7*(1), 135–128.

Hubbert, J. (2019). *China in the world: An anthropology of Confucius Institutes, soft power, and globalization*. University of Hawaii Press.

Hung, H. (2022). *City on the edge: Hong Kong under Chinese rule*. Cambridge University Press.

Ibata-Arens, K. C. (2019). *Beyond technonationalism: Biomedical innovation and entrepreneurship in Asia*. Stanford University Press.

Ironstone, P. (2019). Me, my self, and the multitude: Microbiopolitics of the human microbiome. *European Journal of Social Theory, 22*(3), 325–341.

Keck, F. (2020). *Avian reservoirs: Virus hunters and birdwatchers in Chinese sentinel posts*. Duke University Press.

Kelty, C., & Landecker, H. (2019). Outside in: Microbiomes, epigenomes, visceral sensing, and metabolic ethics. In J. Niewohner (Ed.), *After practice: Thinking through matter(s) and meaning relationally* (pp. 53–65). Panama Verlag.

Koch, E. (2011). Local microbiologies of tuberculosis: Insights from the republic of Georgia. *Medical Anthropology, 30*(1), 81–101.

Kotliar, D. M., & Grosglik, R. (2023). On the contesting conceptualisation of the human body: Between 'Homo-Microbis' and 'Homo-Algorithmicus'. *Body & Society, 29*(3), 81–108.

Krautkramer, K. A., et al. (2016). Diet-microbiota interactions mediate global epigenetic programming in multiple host tissues. *Molecular Cell, 64*(5), 982–992.

Latour, B. (1986). Visualisation and cognition: Drawing things together. *Knowledge and Society: Studies in the Sociology of Culture and Present, 6*, 1–40.

Lee, C. K. (2017). *The specter of global China: Politics, labor, and foreign investment in Africa*. The University of Chicago Press.

Leung, G. M., Ho, L. M., Lam, T. H., & Hedley, A. J. (2009). Epidemiology of SARS in the, 2003 Hong Kong epidemic. *Hong Kong Medical Journal, 15*(Suppl 9(6)), 12–16.

Li, L., Li, K., Zhou, X., & Knowles, R. L. (2024). Maximising the potential of Chinese birth cohort studies: A systematic review of mother–baby cohorts in mainland China. *Public Health, 227*, 119–130.

Lock, M. (1993). *Encounters with aging: Mythologies of menopause in Japan and North America*. University of California Press.

Lock, M. (2001). The tempering of medical anthropology: Troubling natural categories. *Medical Anthropology Quarterly, 15*(4), 478–492.

Lock, M. (2018). Mutable environments and permeable human bodies. *The Journal of the Royal Anthropological Institute, 24*(3), 449–474.

Luk, C. Y. L. (2016). Biotech in Hong Kong: How biologist-entrepreneurs pursued 'Hong Kong's bioscience dream'. *East Asian Science, Technology and Society, 10*(3), 291–313.

Martin, P. (2015). Commercialising neurofutures: Promissory economies, value creation and the making of a new industry. *BioSocieties, 10*(4), 422–443.

Mingpao. (2024). *An Interview with Johnnie To: What is Film-making? Part 1*. Retrieved October 31, 2024, from https://news.mingpao.com/pns/%e5%89%af%e5%88%8a/article/20241027/s00005/1729959115561/%e4%bb%80%e9%ba%bc%e4%ba%ba%e8%a8%aa%e5%95%8f%e4%bb%80%e9%ba%bc%e4%ba%ba-%e5%b0%88%e8%a8%aa%e6%9d%9c%e7%90%aa%e5%b3%af-%e9%9b%bb%e5%bd%b1%e5%89%b5%e4%bd%9c%e6%98%af%e4%bb%80%e9%ba%bc-%ef%bc%88%e5%89%8d%e7%af%87%ef%bc%89

O'Toole, G. A. (2024). We have a community problem. *Journal of Bacteriology, 206*(4), e0007324.

Ong, A. (2010). Introduction: An analytics of biotechnology and ethics at multiple scales. In *Asian Biotech: Ethics and communities of fate* (pp. 1–52). Duke University Press.

Ong, A. (2016). *Fungible life: Experiment in the Asian city of life*. Duke University Press.

Pentecost, M. (2024). *The handbook of DOHaD and society: Past, present, and future directions of biosocial collaboration* (1st ed.). Cambridge University Press.

Pentecost, M., & Meloni, M. (2020). 'It's never too early': Preconception care and postgenomic models of life. *Frontiers in Sociology, 5*, 21–21.

Pepken, M. L., Hansen, S. B., & Limborg, M. T. (2024). Unraveling host regulation of gut microbiota through the epigenome–microbiome axis. *Trends in Microbiology, 32*(12), 1229–1240.

Pinel, C. (2021). Renting valuable assets: Knowledge and value production in academic science. *Science, Technology, & Human Values, 46*(2), 275–297.

Raffaetà, R. (2022). *Metagenomic futures: How microbiome research is reconfiguring health and what it means to be human*. Routledge.

Raffaetà, R. (2024). Cognition and intelligence after the post-human turn insights from the brain-gut axis: Insights from the brain-gut axis. *Jolma, 2*.

Raffaetà, R., & Ferrari, L. (2025). From gut feelings to data assets: Ethnographic explorations of the gut's metabolic political economies. *Humanities & Social Sciences Communications, 12*(1), 429–428.

Richardson, S. S. (2020). Maternal bodies in the postgenomic order: Gender and the explanatory landscape of epigenetics. In *Postgenomics* (pp. 210–231). Duke University Press.

Robinson, C. J., Bohannan, B. J. M., & Young, V. B. (2010). From structure to function: The ecology of host-associated microbial communities. *Microbiology and Molecular Biology Reviews: MMBR, 74*(3), 453–476.

Roy, V. (2020). A crisis for cures? Tracing assetization and value in biomedical innovation. In K. Birch (Ed.), *Assetization: Turning things into assets in technoscientific capitalism* (pp. 97–124). MIT Press.

Rubio, D. M., Schoenbaum, E. E., Lee, L. S., Schteingart, D. E., Marantz, P. R., Anderson, K. E., Platt, L. D., Baez, A., & Esposito, K. (2010). Defining translational research: Implications for training. *Academic Medicine, 85*(3), 470–475.

Sagan, D. (2011, November 18). *The human is more than human: Interspecies communities and the new 'facts of life'*. Society for Cultural Anthropology. https://culanth.org/fieldsights/the-human-is-more-than-human-interspecies-communities-and-the-new-facts-of-life

Schoeb, V. (2016). Healthcare service in Hong Kong and its challenges. The role of health professionals within a social model of health. *China Perspectives, 2016*(2016/4), 51–58.

Siu, H. F., & Ku, A. S. M. (2008). *Hong Kong mobile: Making a global population*. Hong Kong University Press.

South China Morning Post. (2024, January 5). *South China Morning Post*. https://www.scmp.com/comment/opinion/article/3247033/its-not-beijing-thats-turning-hong-kong-just-another-chinese-city

Star, S. L., & Griesemer, J. R. (1989). Institutional ecology, 'translations' and boundary objects: Amateurs and professionals in Berkeley's Museum of vertebrate zoology, 1907–39. *Social Studies of Science, 19*(3), 387–420.

Stevens, H. (2018). Starting up biology in China: Performances of life at BGI. *Osiris (Bruges), 33*(1), 85–106.

Sunder Rajan, K. (2006). *Biocapital: The constitution of postgenomic life*. Duke University Press.

Sunder Rajan, K. (2012). *Lively capital: Biotechnologies, ethics, and governance in global markets*. Duke University Press.

Sunder Rajan, K., & Leonelli, S. (2013). Introduction: Biomedical transactions, postgenomics, and knowledge/value. *Public Culture, 25*(3 (71)), 463–475.

The Human Microbiome Project Consortium. (2012). Structure, function and diversity of the healthy human microbiome. *Nature, 486*(7402), 207–214.

van Wichelen, S. (2023). Shit, in silico: On the postcolonial materiality of bio-information. *Public Culture*, 379–391.

Wahlberg, A. (2018). *Good quality: The routinization of sperm banking in China*. University of California Press.

Widmer, A. (2021). Positioning human microbiome DTC tests on the search for health, data and alternatives amid the financialisation of life. *Medicine Anthropology Theory, 8*(2), 1–12.

Woo, V., & Alenghat, T. (2022). Epigenetic regulation by gut microbiota. *Gut Microbes, 14*(1).

Zhang, J. Y. (with Datta Burton, S.). (2022). *The elephant and the dragon in contemporary life sciences: A call for decolonising global governance*. Manchester University Press.

19

Conclusion. Coproducing Genomics or Why We Need to Take the Mundane Seriously

Vincenzo Pavone, Ayo Wahlberg, Violeta Argudo-Portal, and Mauro Turrini

In this volume, we have collectively shown that we need to take the *mundane* seriously in analytical terms. After the initial euphoria—driven by grand visions of precision medicine, predictive health, and revolutionary diagnostics—what followed was not a linear progression of innovation but rather the messy integration of genomics into everyday life. This post-hype phase, marked by fragmentation and the absence of centralized

V. Pavone (✉) • M. Turrini
Instituto De Políticas y Bienes Públicos (IPP), Consejo Superior de Investigaciones Científicas (CSIC), Madrid, Spain
e-mail: vincenzo.pavone@csic.es; mauro.turrini@csic.es

A. Wahlberg
Department of Anthropology, University of Copenhagen, Copenhagen, Denmark
e-mail: ayo.wahlberg@anthro.ku.dk

V. Argudo-Portal
Department of Social Anthropology, University of Barcelona, Barcelona, Spain
e-mail: vargudoportal@ub.edu

V. Argudo-Portal et al. (eds.), *Mundane Genomics*,
https://doi.org/10.1007/978-981-95-4550-6_19

control, invites a different analytic orientation: one that pays attention not to the spectacular but to the ordinary, not to the breakthrough, but to the drift.

The mundane matters not only for capturing this slow unfolding of genomics in practice but also for aligning our work with a rich stream of scholarship in Science and Technology Studies (STS), innovation studies, and healthcare research. This literature reminds us that technologies do not succeed or fail in binary terms, but are continuously negotiated, adapted, and transformed in their encounters with everyday practices, infrastructures, and institutions.

There are at least two key ways in which the mundane becomes analytically productive. First, when technologies become mundane—when they slip into ordinary routines, become unnoticed, unremarkable, and domesticated—they also become leaky; that is, they begin to exceed their original boundaries and intended purposes. Like mobile phones morphing into smartphones, their integration into daily life spawns uses and meanings far removed from those that designers or regulators could have imagined. Genomic technologies, once heralded as the key to a new era of biomedicine, now show up in settings and practices that are diffuse, informal, and frequently unspectacular: ancestry testing kits in living rooms, pharmacogenetic alerts buried in electronic health records, consent forms rewritten to accommodate biobank withdrawals, or algorithms triaging cancer risk scores.

This leads to the second point: when genomics becomes mundane, it reveals its contingent and emergent character. The effects of genomic technologies are not predetermined; they are shaped by the ways people interpret, ignore, repurpose, or contest them. This everyday engagement produces *affects*—hopes, anxieties, fatigue—as well as *effects*—organizational shifts, new forms of labour, emerging inequalities, social identities, and infrastructures. These affective and material dimensions do not merely accompany genomics; they constitute it.

In this sense, the different "mundane as" categories—mundane as ordinary, routine, institutional, modest, and leaky—help us to decentre genomics. They displace the focus from centralized projects, high-level policy debates, and the promises of biotech marketing to trace instead how genomics takes shape across disparate sites and actors. They allow us to see genomics not as a coherent field with a singular trajectory, but as a set of practices and artefacts that generate multiple, sometimes contradictory, trajectories. It is in the mundane that genomics is both domesticated and disrupted, both stabilized and unravelled. Attending the mundane is not about lowering the analytical stakes—it is about understanding where the stakes actually are.

The Consequences of Genomics Becoming Mundane

As genomic knowledge and technologies increasingly integrate into the fabric of everyday life, their normalization—or banalization—brings significant and far-reaching consequences. What was once the domain of cutting-edge science or extraordinary medical intervention now pervades parenting, reproduction, criminal justice, identity formation, and global governance. The shift towards *mundane genomics* does not dilute its power; rather, it diffuses and deepens it, embedding genomics into everyday decisions, responsibilities, and imaginaries in ways that can both empower and constrain.

In parenting, as Wahlberg, Heinsen, and Vendel Petersen show, genomics is negotiated not through formal protocols but through everyday rhythms of care, protection, and trust. Risk, in this context, is not simply a statistic but a relational and emotional experience, one bound up with when a child is deemed "ready" to know about a genetic condition, or when a parent feels prepared to disclose it. The mundane does not remove the moral weight; it relocates it into the intimate, subjective sphere of family life.

This deep embedding also reconfigures how we interpret human behaviour. As Lloyd warns, the rise of neuro-epigenetic explanations for phenomena like suicide, risks reducing complex social acts to predetermined biological outcomes. In this framing, experiences such as trauma or socio-political injustice are overshadowed by molecular narratives. As genomics becomes a default lens, it can marginalize alternative explanations and silence context, personal agency, and systemic critique.

The incorporation of genomics into ancestry narratives—as explored by Romijn—further reveals how genetic knowledge reshapes the everyday. Genomic ancestry becomes a tool for crafting identity and belonging in the present, often through emotionally charged stories that compress history into a "shallow time". While these narratives may appear personal, they are entangled in collective memory and moral reckonings, particularly around race, heritage, and colonialism.

Genomics' reach into reproduction is especially fraught. Pavone, Bosch, and Turrini illustrate how patients navigating reproductive technologies like PGT-A find themselves on a high-stakes, emotionally charged, and economically draining journey. As genomic screening becomes routinized or even "compulsory", it generates a logic of escalation—each cycle and test promising yet one more chance, in a tantalizing path increasingly harder to exit. Hope itself is engineered through technical means, driving repeated interventions.

Writing on related matters from a different perspective, Kaur emphasizes that mundane genomics reflects and reinforces social norms, particularly regarding disability and human worth. As genomics becomes routine, it risks naturalizing structural ableism unless robust governance ensures equity and ethical oversight. It is not just a technical process—it is a socio-cultural force reshaping how we value health and life.

This routinization also has regulatory consequences. Arias points out that in Spain, the state's reluctance to regulate certain genomic technologies stems not from oversight, but from a strategic normalization that privileges consumer choice and private enterprise. Once a technology is normalized, it becomes harder to question or legislate against. Lawmakers lose both the political will and imaginative capacity to propose alternatives.

In this context, Hoffmann and Tenenberg's illustration of how tools like polygenic risk scores circulate despite questionable clinical utility is especially daunting. These scores foster speculative behaviours—such as lifestyle adjustments, health tracking, or future-oriented anxieties—that individualize responsibility. Rather than offering clear medical guidance, they embed genomics into health habits and reinforce a neoliberal ethos of self-care and self-surveillance.

At a deeper level, this normalization disperses ethical reasoning. As mundane genomics becomes infrastructural, it no longer provokes public debate; it pre-empts it. Genomic governance now operates through workflows, software, and institutional routines, as a mutation of bio-power. Ethical deliberation does not disappear—it is outsourced, diluted, and made invisible. The ethical becomes logistical, an invisibility that also enables overlooked harms. Campello's account of forensic genetics in prisons highlights how seemingly routine genomic practices can amplify systemic violence. When DNA collection and use become mundane parts of criminal justice, they risk entrenching surveillance and injustice under the guise of neutrality.

Similarly, García-Deister critique the political overvaluation of genetic databases in Mexico. DNA is framed as a necropolitical promise: a tool of hope for families searching for the disappeared, even as its actual efficacy remains limited. Genomics becomes a symbolic technology that manages grief and expectation while masking institutional failure.

Finally, Tam and Raffaetà remind us that even seemingly benign, everyday genomic research can have geopolitical consequences. The mundane practices of nation-based studies help construct biosocial identities and national belonging, linking local participants to global imaginaries of science and citizenship. In this way, mundane genomics participates in both cultural and political projects.

In sum, as genomics becomes mundane, its power is not diminished—it is redistributed. Across domains, it reshapes not only how we understand health, identity, and responsibility, but also how we imagine the future, legislate the present, and relate to the past. The challenge is not simply to acknowledge this shift but to interrogate the ethical, social, and political arrangements it normalizes in its wake.

The Concept of the Mundane in Studying Genomics

Throughout this volume, we have seen how the expansion of genomics into everyday settings—whether in clinical care, reproduction, forensic practices, or ancestry testing—comes with remarkable consequences. Understanding genomics as "mundane" is a way to account for these consequences, but also a necessary conceptual tool to make sense of and analyse them. It does more than describe genomics' normalization; it enables us to examine how genomic practices take shape, gain meaning, and produce consequences within the flow of ordinary life. The mundane, thus, is both a *process*—a mode through which genomics is embedded into routines, emotions, and infrastructures—and a *conceptual lens* that sharpens our ability to see the subtle, often-overlooked dynamics at play.

This concept reminds us how the dramatic promises of genetic innovation can obscure the nuanced ways whereby genomics actually operates: through small shifts, contested meanings, quiet resistances, and diffuse social labour. The mundane reminds us that consequences do not only stem from regulatory decisions or clinical breakthroughs but also from everyday interactions, hesitations, and adaptations. It redirects our gaze to the *how* of genomics—how it is taken up, reframed, negotiated, or even ignored.

Argudo and García-Deister, for instance, show how treating the mundane as *modest* helps resist the decontextualization of genomics. By attending to the overlooked contributors to genomic practice—such as nurses, clothing, tattoos, or informal networks—they expose the collaborative and material grounding of what might otherwise appear as purely scientific work. In this way, the mundane foregrounds the multiplicity of actors and artefacts that coproduce genomic meaning and labour.

This reframing also supports a more grounded critique. García Sancho, for example, rejects the notion that genomics is a panacea. Instead, he frames it as a *companion practice*, one element in broader, collective pursuits of justice. Here, the mundane offers a way to decentre genomic authority, treating it as one tool among many rather than an inevitable or dominant force.

Hilberg, Stelmach, Kleinhout-Vliek, and Martin take this further by using the mundane to *question heroics*. While gene therapy is often narrated through triumphalist narratives, attention to the mundane highlights the economic, institutional, and logistical considerations that shape its real-world implementation. This shift challenges overblown promises and invites more measured, socially responsive storytelling.

Heinsen similarly treats the mundane as a way of revealing genomics' *situatedness*. Emphasizing frictions, dependencies, and small-scale adaptations, it shows that genomics rarely dominates the settings in which it appears—whether in hospitals, homes, or forensic contexts. These minor and often ambivalent shifts are not signs of failure but signs of how genomics is *actually* lived: negotiated, reinterpreted, and woven into diverse life worlds.

The power of the mundane also lies in its ability to accommodate *complexity and contradiction*. Santoro and Merlo demonstrate this by analysing the metaphors people use to make sense of DNA: genes as lotteries, switches, or networks. These metaphors are frequently voiced by the same person, capturing an oscillation between determinism and openness. Rather than seeking conceptual purity, the mundane recognizes this ambiguity as constitutive of how people relate to genomic knowledge.

Eliza Brown's analysis of PGT-A (Preimplantation Genetic Testing for Aneuploidies) reveals how the technology "leaks" beyond its intended genetic function. In her account, the mundane thus captures how genomic tools serve broader logics of care, control, and commercial reassurance, often without explicit recognition. Anaïs Martin provides a similar insight in the realm of direct-to-consumer (DTC) testing, showing how DNA matches do not speak for themselves; they require *interpretation* through family photos, storytelling, and emotional labour. Kinship, in this view, is not dictated by genetic similarity but built through mundane acts of meaning-making. Genomics here is dialogic, not determinative—its authority emerging through the social imaginaries and practices that sustain it.

In sum, the mundane offers more than a description of genomics' everyday presence—it offers a critical approach. It helps scholars uncover the ordinary infrastructures, affective investments, interpretive labour, and moral negotiations that shape the current genomic landscape. As genomics becomes more ubiquitous, the need to understand its mundane dimensions becomes not a secondary concern but a central task. Through this lens, we see genomics not only as a scientific frontier, but as a social process—incremental, relational, and deeply embedded in the texture of everyday life.

Taken together, the contributions in this volume show that to understand genomics today, we must look not only to laboratories, policy arenas, or moments of discovery, but to the slow, uneven, and often improvised ways it is woven into daily life. The concept of the mundane enables this shift in perspective. It brings into focus the subtle negotiations, affective attachments, material arrangements, and social meanings that animate genomic practices outside of spectacular moments. Far from being peripheral, the mundane is where genomics gathers force—through leaks, metaphors, habits, and hesitations. It reveals how genomic knowledge is never simply imposed but is always translated, refracted, and made liveable within particular contexts. As genomics continues to expand, its power will not lie solely in scientific innovation or regulatory authority, but in its ability to inhabit ordinary worlds. This volume is our modest contribution to showing how attending to the mundane is not only analytically necessary—it is ethically and politically vital. It equips us to ask different questions, hear different voices, and imagine genomic futures grounded in the complexity of how people actually live with, and through, their genes.

Index[1]

[1] Note: Page numbers followed by 'n' refer to notes.

V. Argudo-Portal et al. (eds.), *Mundane Genomics*,
https://doi.org/10.1007/978-981-95-4550-6

The manufacturer's authorised representative in the EU is Springer
Nature Customer Service Centre GmbH, Europaplatz 3, 69115 Heidelberg,
Germany. If you have any concerns regarding our products, please
contact ProductSafety@springernature.com

Printed and bound by CPI Group (UK) Ltd, Croydon, CR0 4YY
21/09/2026
02219945-0015